# Covariant Physics

# Covariant Physics

## From Classical Mechanics to General Relativity and Beyond

### Moataz H. Emam
State University of New York College at Cortland

OXFORD

UNIVERSITY PRESS

# OXFORD
## UNIVERSITY PRESS

Great Clarendon Street, Oxford, OX2 6DP,
United Kingdom

Oxford University Press is a department of the University of Oxford.
It furthers the University's objective of excellence in research, scholarship,
and education by publishing worldwide. Oxford is a registered trade mark of
Oxford University Press in the UK and in certain other countries

© Moataz H. Emam 2021

The moral rights of the author have been asserted

First Edition published in 2021

Published in the United States of America by Oxford University Press
198 Madison Avenue, New York, NY 10016, United States of America

British Library Cataloguing in Publication Data

Data available

Library of Congress Control Number: 2020951689

ISBN 978–0–19–886489–9 (hbk.)
ISBN 978–0–19–886500–1 (pbk.)

DOI: 10.1093/oso/9780198864899.001.0001

Printed and bound by
CPI Group (UK) Ltd, Croydon, CR0 4YY

Links to third party websites are provided by Oxford in good faith and
for information only. Oxford disclaims any responsibility for the materials
contained in any third party website referenced in this work.

*To Miriam, Maya, and Mourad*

# Preface

This is *not* "yet another book on the general theory of relativity." While the theory did inspire, and indeed plays a major role in, our discussion, it can be thought of as only one part of a greater whole. Initially developed as a theory of gravity, it was slowly realized that general relativity had a deeper meaning than just that. It teaches us, among other things, a completely new understanding of space and time, together renamed **spacetime**: a single fabric with its own particular set of symmetries and properties. While we learn that the familiar Euclidean geometry[1] that we have all learned (and loved?) in school no longer applies, we also discover that its most fundamental symmetries are not lost, but rather "upgraded" to new and deeper ones; leading to a "spacetime" that's a lot more dynamic and flexible than the one we learned about from Euclid, Galileo,[2] or Newton.[3] A byproduct of this dynamical flexibility, prominent in general relativity, is the phenomenon that we call "gravity;" but it is *only* a byproduct. It is actually possible, at least in principle, to go through an entire course on general relativity without mentioning the word "gravity" once, instead just focusing on its formal structure and the symmetries it inspires! Clearly such a choice would not be a very productive one, as it means that we would be missing out on what is undoubtedly the most important application of the theory; nevertheless, it *is* possible. The theory itself is the culmination of the efforts of many physicists and mathematicians, most notably Albert Einstein,[4] Hermann Minkowski,[5] Hendrik Lorentz,[6] Henri Poincaré,[7] David Hilbert,[8] and others. Einstein, of course, is responsible for the more revolutionary ideas in the theory. Today it is considered his greatest contribution to physics, as well as possibly the most amazing singular achievement by a human being in history.[9]

---

[1] Euclid of Alexandria was a Greek mathematician who lived around the third century BCE. His book *Elements* is considered the foundational text of modern geometry in two- and three- dimensional flat space. His axioms, theorems, and lemmas constitute the familiar geometry taught today to high school students. The term "Euclidean," coined in his honor, signifies flat space, as opposed to "non-Euclidean," or curved spaces.

[2] Galileo Galilei: Italian astronomer, physicist, engineer, philosopher, and mathematician (1564–1642).

[3] Sir Isaac Newton: English mathematician, physicist, astronomer, theologian, and author, requiring perhaps no introduction (1642–1726).

[4] German physicist (1879–1955).

[5] German mathematician (1864–1909).

[6] Dutch physicist (1853–1928).

[7] French mathematician and theoretical physicist (1854–1912).

[8] German mathematician (1862–1943).

[9] Charles Darwin's theory of natural selection might be a close contender.

One of the most fundamental characteristics of this new understanding of space and time that has arisen from, but, one can argue, has become even more important than, the theory of relativity, is what mathematicians sometimes call **diffeomorphism invariance**, which, as far as we are concerned, can be thought of as simply the "invariance of the laws of physics between coordinate systems," also known as **coordinate covariance**, or just **covariance**. This idea, which will be our main focus for most of this book, lies at the deepest foundations of physics and is certainly true in the special theory of relativity (**special covariance**), the general theory (**general covariance**), and even ordinary non-relativistic physics, i.e. the usual Euclidean/Galilean/Newtonian world view (**classical covariance**). What it means is this: The use of any type of coordinate system does *not* affect the physical outcome of any given problem. A specific physics formula applied to some problem should give *exactly* the same physical answers whether one has used Cartesian coordinates or any other type of coordinate system (and they are legion). This should be intuitively true even to the beginner, because coordinate systems are a human choice and nature should not, and indeed does not, care about our choices or conventions. On the other hand, the way the formula *looks* in Cartesian coordinates is significantly different from the way it looks in other coordinates. So an important question to ask is this: How can one write the laws of physics in a single form that is true in *any* coordinate system? In other words, in a form that is coordinate *covariant*, yet is easy enough to use once a choice of coordinates has been made? In this book we will approach our study of space and time from the perspective of covariance; as such, it is perhaps important to understand what we mean by this, even this early in the reading. Let's then clarify further: Consider Newton's second law as an example. It is usually first encountered in introductory courses in the following form:

$$\mathbf{F} = m\mathbf{a}, \tag{0.1}$$

where $\mathbf{F}$ is the vector representing the net sum of all external forces acting on a mass $m$ and resulting in an acceleration $\mathbf{a}$.[10] Written in this form, eqn (0.1) is correct in any coordinate system, i.e. in a sense it is *already* covariant. But to actually calculate numerical results one still needs to make a specific choice of coordinates which will necessarily split the equation into its constituent components and have a form that is highly dependent on the coordinate system used. For example, in two-dimensional Cartesian coordinates $(x, y)$, Newton's law reduces to the following set of equations:

$$
\begin{aligned}
F_x &= ma_x \\
F_y &= ma_y,
\end{aligned}
\tag{0.2}
$$

where $(F_x, F_y)$ and $(a_x, a_y)$ are the Cartesian components of the force and acceleration respectively. With the knowledge that acceleration is the second derivative of the position we can write

---

[10]We will mostly denote vectors by bold face type font, except when there is a possibility of confusion. In such cases we will use an arrow over the symbol to denote a vector; for example, $\vec{F}$.

$$F_x = m\frac{d^2x}{dt^2}$$
$$F_y = m\frac{d^2y}{dt^2}, \tag{0.3}$$

or in a more compact form

$$F_x = m\ddot{x}$$
$$F_y = m\ddot{y}, \tag{0.4}$$

where an over dot is the standard notation for a derivative with respect to time ($\dot{x} = \frac{dx}{dt}$, $\ddot{x} = \frac{d^2x}{dt^2}$, and so on). This should be familiar even to the reader who has only had an introductory physics course, but now consider writing the equivalent equations in polar coordinates $(\rho, \varphi)$. We find that they are *not*:

$$F_\rho = m\ddot{\rho}$$
$$F_\varphi = m\ddot{\varphi}. \quad \text{Wrong!} \tag{0.5}$$

In other words, one does not simply replace $x \to \rho$ or $y \to \varphi$ and expect to get the right answer. (In fact, it is clear that the second equation in (0.5) does not even have the correct units.) The correct relationships are actually

$$F_\rho = m\ddot{\rho} - m\rho\dot{\varphi}^2$$
$$F_\varphi = m\rho\ddot{\varphi} + 2m\dot{\rho}\dot{\varphi}. \tag{0.6}$$

The two sets of eqns (0.4) and (0.6) are not identical in form, as (0.4) is with the (wrong) eqns (0.5). Specifically, there are "extra" terms and multiplicative factors in (0.6), and it is not immediately obvious where these came from or whether there is a pattern that is common to them. It is, however, true that applying either one of these choices to the same problem will describe the same physical behavior and give the same physical results. But where did the extra quantities come from? Mathematically, they arise from the relationships that relate polar coordinates to Cartesian coordinates, i.e. $x = \rho\cos\varphi$ and $y = \rho\sin\varphi$, as can be easily derived from Fig. [1.3b]. All one needs to do is to use calculus to find how $\dot{x}$ and $\dot{\rho}$ (for example) are related, and so on. In principle one needs to rederive the Newtonian equations of motion each and every time one makes a specific choice of a coordinate system. This is indeed the way it is usually done in physics classes. However, what if the coordinate system is *not* defined ahead of time? What if it depends on the curvature of space and time that is not *a priori* known? What if the coordinate system *itself* arises as the solution of the problem under study? Can one rewrite (0.1) or any other law of physics in such a way that it remains coordinate invariant and gives the correct set of equations for *any* coordinate system, whether it was known from the start or still to be derived? This is the problem of coordinate invariance in physics, or covariance as it is more commonly referred to.

Learning how to do all of this is a useful thing, whether one is going to study relativity or not, because it provides the student with a recipe for dealing with any choice of coordinate system in any physics problem, whether such a choice is one of the more common ones, like polar coordinates in two dimensions or cylindrical and spherical coordinates in three, or less common, such as parabolic coordinates or elliptic coordinates. In short, it provides the physicist (or engineer) with a powerful mathematical tool applicable to a variety of situations. It also acts as an easy (*-ish*) introduction to the subject of **tensor** mathematics, rarely discussed in an undergraduate setting, yet encountered in almost every branch of physics. While these "tensors" are particularly useful in a relativistic setting, they can also be found in non-relativistic mechanics and in electrodynamics, and knowledge of their properties and rules would be particularly useful for the student wishing to delve deeper into these topics, or at least hopes for a better understanding than is normally delivered to undergraduates.[11] We will touch upon all of this, and by the time we are done we will have a general understanding of (almost) all of fundamental (non-quantum) physics in a covariant setting and arbitrary spacetime backgrounds. We conclude the book with three chapters focusing on topics rarely, if ever, found in similar treatments. These are: classical fields, differential forms, and modified theories of gravity. These chapters are particularly geared towards the students interested in reading modern research papers. They should provide just a bit of extra preparation as well as possibly whet the reader's appetite for more.

## A note to instructors:

This book is written from a minimalist's perspective. By no stretch of the imagination am I claiming that the discussions here are exhaustive or complete. In fact, I imagine that most mathematicians would be particularly appalled by the lack of rigor throughout, and most physicists will point out that many interesting applications and discussions commonly found in similar monographs are missing. This is all quite intentional. The intended audience of this book are those undergraduate students who may have only had introductory mechanics and electromagnetism at the level of Serway [1], for instance, as well as elementary calculus up to and including multivariable calculus. While the mathematical background gets a bit more intense in some places, it is my hope that students with only that much preparation can manage the majority of the book. As the students progress in their studies and reach more advanced courses in mechanics and electrodynamics, one hopes that this book, or the memory thereof, would provide a background tying all the various disciplines together in one continuous thread. The theme of covariance permeates all of physics, and yet is rarely discussed in an undergraduate setting. I have tried to make the book as self-contained as possible, such that a student reading it on their own would manage most of it with as little help as possible.

---

[11] It truly hurts when in teaching an undergraduate class I find myself forced to say something like "This is called the so-and-so tensor, but unfortunately we have no time in this class to explain what that *really* means."

One also hopes that this book can provide the minimum mathematical training needed by undergraduate students wishing to do research in certain areas of theoretical physics. In fact, the idea of the book, as well as the level of detail and sequence of topics, has arisen from my years of supervising undergraduate research. With graduation dates preset, one would like one's students to start their research experience as early as possible in their college years, and get to finish within the time they have available. In my undergraduate theoretical physics research program I have usually designed my students' learning experience in such a way that they would be able to get to the research part as fast as possible, amassing the needed mathematical training in the shortest possible period of time. In many cases, I have instructed them to focus on certain topics while skipping others, because of my preconceived knowledge of the nature of the research problem that I will eventually pose to them. For example, a student whose research will be calculating geodesic orbits of certain spacetime backgrounds need not waste their initial study sessions on understanding the details of how these spacetimes were derived in the first place; rather a cursory knowledge would be acceptable. If undergraduate research is the instructor's intention in assigning this book to their students, then it is advisable to plan their reading ahead of time. Of course, as they work on their projects they may go back and study in more detail those sections that they have initially only skimmed over. I have tried to make this book exhaustive enough for these purposes, although depending on what the research mentor has in mind, other books and/or notes can be supplemented.

## On the exercises in the book:

I have tried to include as many exercises for practice and/or homework assignments as I could without overwhelming the reader. Many of the exercises in the book are chosen not only because they provide practice on the topics we discuss, but also because they include some additional material of their own. For example, some of the exercises require the reader to teach themselves computational techniques not discussed in the text or read on a topic that we didn't include in detail. Some may even be long enough to count as a final project or something similar, e.g. Exercise 6.49, or Exercises 7.26 and 7.27 combined. These increase in frequency, as well as intensity, as we get to the final chapters. A word of warning, however: Please be aware that some of the exercises list *many* different ways of practice; so many, in fact, that the student need not do all. For example, an exercise may say "do such and such for the following cases" and list *10* different cases; e.g. Exercise 2.22 which in turn is referred to multiple times in subsequent exercises. The student really need *not* do all of the listed cases; only enough to feel they have reached a certain degree of comfort with the techniques. Likewise, if this book is used in a formal course, the instructor should choose some of the cases for homework assignments, or even introduce their own. Many exercises depend on previous ones, however. For example, Exercise 2.23 needs the results of Exercise 2.22, so if the reader chooses to do, say three, of the cases in 2.22, then they would only be able to do the same three cases in 2.23, and so on.

Some of the exercises assigned already have solutions available in various places around the internet. This is unavoidable. However, it is understood that it is in your best interest to work out the exercises yourself, *then* check against the available solutions. It is also true that there are freely available software packages that would solve many of the exercises in this book for you. This is the case for most mathematical topics in this day and age. I actually *encourage* the reader to find and download these packages, or better yet learn to design and write your own. You can do so from scratch using any programming language or from within one of the commercially available symbolic manipulation software packages. In other words, you should treat the digital computer as a tool for learning, *not* as a way of avoiding work. The choice of which exercises to do by hand and which to do using the computer is mostly left to the reader or their instructors, even if not specifically stated. For example, once the instructor is satisfied that the students are able to calculate a specific quantity that usually takes hours to find by hand, say the components of a particularly large tensor, then they might just allow the students to use software packages for the remainder of the course.

# Acknowledgements

It is very difficult, if not downright impossible, to give credit to every single person that I would like to. For I believe that any direct help I have had in writing this book is no more important than the indirect positive effect many people have had on my life in general; without which I would not have become the person who is able to write such a book in the first place. As such, I beg forgiveness to anyone that I won't mention and assure them that I appreciate them no less. To begin, I must acknowledge my parents' love, encouragement, and support. These are the foundations upon which we all grow and advance. Rest in peace mom and dad; I hope you are as proud of me as I am proud to be your son. Secondly, I acknowledge the support I have received from my family and close friends; you are too many to mention, but you were all supportive and understanding of my efforts in writing this book. I probably would have ended up nowhere without you. Thirdly I must acknowledge my students over at least the past fifteen years, in both the United States and Egypt. I have used the notes on which this book is based to teach them the material, and they have always honored me with valuable feedback and comments. I would also like to thank the colleagues, teachers, and friends who have had an immediate impact on this book, whether via discussions we have had, support and encouragement (sometimes such a long time ago), or direct feedback. These are, in no particular order:

- Dr. Mark Peterson of Mt. Holyoke College, MA: You allowed me to offer my very first class on the general theory of relativity. The notes I developed for that course were the earliest version of this book. Our discussions about teaching methods have had a lasting effect on me as a teacher and as an author. Your encouragement and faith in my abilities have stayed with me since then.
- Dr. Ahmed Alaa Eldin Abouelsaood of Cairo University, Egypt: Now where would I have ended without your advice and encouragement? You took a chance on a young man who came to you with nothing other than a passion for physics, and your gamble paid off (I hope). Not only did you allow me to absorb from you as much knowledge as I possibly could have at the time, you were also the one who provided the initial guidance which, quite literally, started the rest of my life.
- Dr. David Kastor of the University of Massachusetts at Amherst, MA: As my PhD advisor you did much more than supervise me. You allowed me to pick your brains in many topics in theoretical physics and patiently listened to me as I described how I understood things, even if my ideas sounded simple or mundane. These thoughts, which you encouraged, have slowly developed and matured, and became major contributions to this work. More recently, thanks for taking the time to read the manuscript and coming up with useful comments as well as suggestions for new sections. I am particularly grateful to you for suggesting the subtitle of the book.
- Dr. Eric Edlund of the State University of New York College at Cortland, NY: Thanks for the time you dedicated to reading the manuscript and providing excellent suggestions and comments.

- Mr. Tariq Saeed of the University of Mansoura, Egypt: Thanks for providing valuable input from a student perspective. Your suggestions on sections §3.3, §4.7, and others were particularly useful. I also truly appreciated your taking the time to check many of the calculations in this book.
- Eng Muhammad El-Nashash'ee, and Eng Akram Masoud: Thank you for taking the time to read the manuscript and providing valuable feedback and suggestions.
- My cousin and dear friend Brigadier General Eng Omar Farouk: Many thanks for the big push you gave me at a time I desperately needed one.
- Last, but certainly not least, my dear friend Mr. Ahmed Abuali: I will always be grateful for your valuable input, help, as well as many hours of discussion.

# Contents

With the exception of Fig. [5.12] and [8.7], all the figures are original by the author. Software used: "Wolfram Mathematica® 11.3," and "Inkscape® 0.92."

# 1

# Coordinate Systems and Vectors

*Geometry has two great treasures: one is the Theorem of Pythagoras; the other, the division of a line into extreme and mean ratio. The first we may compare to a measure of gold; the second we may name a precious jewel.*

Johannes Kepler

## Introduction

Coordinate covariance is defined as the invariance of the laws of physics under change of coordinate systems, whether they differ from each other by translation, rotation, or both. In this chapter we begin by introducing the various two- and three-dimensional coordinate systems usually used in problems of physics and engineering. Some of these should be quite familiar to the reader, while others may be completely new. We specifically highlight the concept of "line element," also known as the "metric," as a defining principle of coordinate systems as well as a unifying theme amongst them. This should allow for a smoother transition into physics later. In principle, all that an experienced physicist needs in order to figure out what type of covariance is at play in a given situation is knowledge of the metric. Hence introducing the concept early on makes sense and allows us time to get used to it before things get more interesting. Next we will turn to defining the basics of vector mathematics and discussing their formulation in a new language simply known as the "index" or "component" notation.[1] This notation by itself connects vectors to the coordinate systems they are defined in and highlights their covariance. Furthermore, the notation provides the majority of the mathematical material that we need to collect in our search for a fully covariant formulation of physics, relativistic or not. In fact, I claim that the first two chapters *by themselves* constitute at least 80 percent of the mathematics needed to study *both* theories of relativity. As the reader will see, none of this mathematics is too far beyond ordinary algebra and a bit of multivariable calculus.

---

[1] Also sometimes referred to as the "tensor notation;" however, this is somewhat misleading, so we will avoid using it too much.

*Covariant Physics: From Classical Mechanics to General Relativity and Beyond.* Moataz H. Emam,
Oxford University Press (2021). © Moataz H. Emam.
DOI: 10.1093/oso/9780198864899.003.0001

## 1.1 Coordinate Systems

For the purposes of many science and engineering applications, the most commonly used system of coordinates is the **Cartesian**,[2] also known as **box** or **rectangular** coordinates. By their very nature, Cartesian coordinates emphasize the intuitive concepts of width, depth, and height. It is usually the coordinate system of choice for **Euclidean** geometry; i.e. the ordinary geometry we all studied in school. It is, however, only *one* choice among many, as we will see. Euclidean geometry describes flat space, where parallel lines stay parallel and the angles of a triangle sum up to exactly 180°. If one happens to be interested in curved or **non-Euclidean** geometries, like cartographers doing geometry on the surface of the spherical globe, then not only do Cartesian coordinates stop making sense, but they actually become impossible to define, except approximately on a very small scale. To clarify this, consider the surface of planet Earth as an example, assumed a perfect sphere; one can draw straight lines in a grid on the floor in a room or in the street without trouble, but if you want to draw longer "straight" lines between cities, countries, or even continents, you find that you cannot do so and are forced to work with circles, like those of latitude or longitude. In a sense, then, most curved surfaces of interest are approximately flat on small scales; one says they are *locally* flat, as opposed to *globally* so.

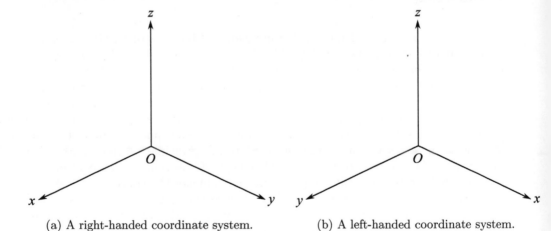

(a) A right-handed coordinate system.          (b) A left-handed coordinate system.

Fig. 1.1: Cartesian coordinates.

Traditionally, we use the symbols $x$, $y$, and $z$ to denote the three Cartesian axes in three-dimensional space, ordered in the so-called "right-handed" system, as shown in Fig. [1.1a]. The right-handedness convention is particularly useful in physics, as it is the basis of various mathematical operations such as vector cross products. For comparison a left-handed system is shown, but we will not be using that. Generally,

---

[2]Invented in the seventeenth century by the French philosopher and mathematician René Descartes (1596-1650), whose name in Latinized form is Cartesius.

coordinate systems are defined via the geometric shapes they seem to trace. So in the case of Cartesian coordinates, surfaces of constant $x$, $y$, and $z$ are infinite flat planes as shown in Fig. [1.2], hence the alternative title "box coordinates." Now the conventional choice of the letters $x$, $y$, and $z$ to denote the Cartesian axes is very familiar to most people; however, a more useful notation, and one that we might as well start using right away, employs *indices* to denote the different axes. So $x$ becomes $x^1$ (i.e. the *first* axis), $y$, being the second axis, becomes $x^2$, and $z \rightarrow x^3$, where the reader is warned *not* to confuse indices with powers, so $x^2$ is not $x$ squared, but is simply pronounced "ex two" and similarly for "ex one" and "ex three." If we wish to denote exponentiation, then we can simply use parentheses; for example, the square of "ex three" is $\left(x^3\right)^2 = x^3 \times x^3 = z^2$. If there is a possibility of confusion, we can temporarily go back to the old $(x, y, z)$ system, depending on the context. This "numbering" system, also known as the **index notation** or the **component notation**, is a much more powerful choice than the conventional one for a variety of reasons, as we will see.

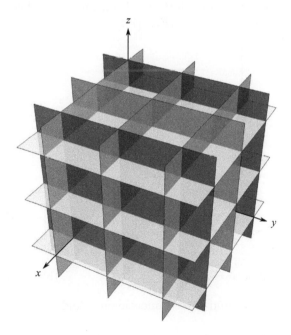

Fig. 1.2: The Cartesian grid: planes of constant $x$, $y$, and $z$.

The next most commonly used system of coordinates in three dimensions is **cylindrical coordinates** (which become **polar coordinates** in the $x$–$y$ plane), also known as **cylindrical polar coordinates**. They are denoted by $(\rho, \varphi, z)$ and are related to the Cartesian coordinates via

$$x^1 = \rho \cos \varphi$$
$$x^2 = \rho \sin \varphi$$
$$x^3 = z, \tag{1.1}$$

and their inverse

$$\rho^2 = \left(x^1\right)^2 + \left(x^2\right)^2$$
$$\tan \varphi = \frac{x^2}{x^1}$$
$$z = x^3, \tag{1.2}$$

as is easy to confirm by inspecting Fig. [1.3]. Cylindrical coordinates are a particularly useful choice in physics when the problem under study has cylindrical symmetry, i.e. studies a situation with cylindrical shapes or a reasonable approximation thereof. The name comes from the fact that surfaces of constant $\rho$ describe infinite cylinders spanning the entire $z$ direction. Surfaces of constant $z$ are of course still infinite planes, and so are surfaces of constant $\varphi$ (Fig. [1.4]).

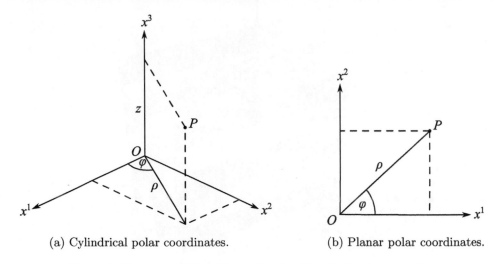

(a) Cylindrical polar coordinates.  (b) Planar polar coordinates.

Fig. 1.3: Two- and three-dimensional cylindrical polar coordinates.

An important difference between the cylindrical (as well as two-dimensional polar) and Cartesian coordinate systems is that the cylindrical system does *not* cover all of three-dimensional space. What this means is that while any point in space can be described by a specific set of numbers in Cartesian coordinates $(x^1, x^2, x^3)$, in the cylindrical coordinate system there is a single point, out of the infinity of points in space, that *cannot* be described by a unique set of the three numbers $(\rho, \varphi, z)$. This is the point located exactly at the origin of coordinates, which we usually denote by $O$. While the coordinates of $O$ in the Cartesian system are simply $(0,0,0)$, in the cylindrical system only $\rho = 0$ and $z = 0$ can be defined but $\varphi$ remains ambiguous, because *any* value can be used for it. In other words, there is no unique set of three numbers describing the location of $O$; the set of numbers $(\rho = 0, \varphi = 0°, z = 0)$ does imply the origin, but so do the sets $(\rho = 0, \varphi = 10°, z = 0)$, $(\rho = 0, \varphi = 20°, z = 0)$, $(\rho = 0, \varphi = 312.56°, z = 0)$, and so on. There is an infinite number of possibilities, all describing a single point! This phenomenon is a *defect* in the coordinate system itself and cannot be remedied; the system simply fails at that specific point. One then says that while Cartesian coordinates *cover*, *map*, or *chart* all three-dimensional space, cylindrical polar coordinates do not; exactly one point defies definition. Coordinate systems in this regard can be thought of as a grid covering three-dimensional space; the Cartesian grid or map is infinite, smooth, and complete, while the polar one has a "hole" in it located at $O$. If one then wishes to do a problem that studies the point at the origin using such a "defective" coordinate system, one has no choice but to shift the coordinate system itself such that the point in question no longer coincides with the origin. There is simply no other way. This issue also further emphasizes that spatial points and coordinate maps are *not* the same thing; an obvious issue, perhaps, but one that is sometimes confused in the minds of beginners.

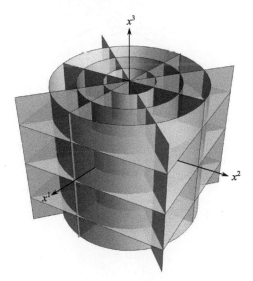

Fig. 1.4: The cylindrical coordinates grid: cylinders of constant $\rho$ and planes of constant $z$ and $\varphi$.

The remaining most commonly used system of coordinates in three-dimensional space is the **spherical coordinate system**, particularly useful in physics when problems exhibit spherical symmetry. As shown in Fig. [1.5], we define the numbers $(r, \theta, \varphi)$. They are related to the Cartesian coordinates as follows:

$$x^1 = r \sin\theta \cos\varphi$$
$$x^2 = r \sin\theta \sin\varphi$$
$$x^3 = r \cos\theta \tag{1.3}$$

$$r^2 = \left(x^1\right)^2 + \left(x^2\right)^2 + \left(x^3\right)^2$$

$$\tan \varphi = \frac{x^2}{x^1}$$

$$\cos \theta = \frac{x^3}{r}. \tag{1.4}$$

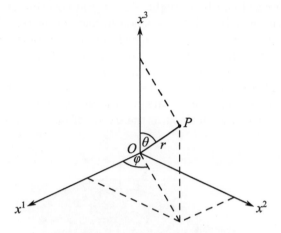

Fig. 1.5: Spherical coordinates.

It is also possible to define transformation relations between cylindrical coordinates and spherical coordinates. These are:

$$\rho = r \sin \theta$$

$$z = r \cos \theta$$

$$r^2 = \rho^2 + z^2$$

$$\tan \theta = \frac{\rho}{z}. \tag{1.5}$$

Note that the spherical coordinate system also suffers from the same defect we noted for cylindrical coordinates. Once again, the point at the origin cannot be specified by three unique numbers: only $r = 0$ can be used, while *both* $\theta$ and $\varphi$ are ambiguous. As before, the only way to deal with this in calculations is to avoid addressing the origin altogether or just shift the coordinate system. Now the defining property of spherical coordinates is that surfaces of constant $r$ describe spheres, concentric at the origin, while surfaces of constant $\theta$ and $\varphi$ define cones and flat surfaces respectively, as shown in Fig [1.6].

As noted, coordinate systems are defined based on geometric shapes: flat planes for Cartesian, cylinders for cylindrical (or circles for polar), and spheres for spherical. We note here that these are *not* the only options; other shapes can also be used to

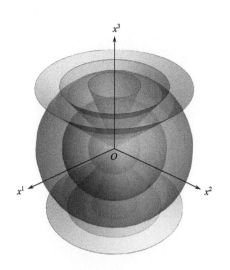
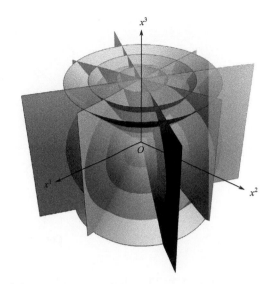

(a) Spheres of constant $r$ and cones of constant $\theta$ in spherical coordinates.

(b) The complete spherical coordinates grid.

Fig. 1.6: Spherical coordinates.

define coordinate systems. For example, consider the so-called **elliptic cylindrical coordinates** $(u, v, z)$, where surfaces of constant $u$ define concentric elliptic cylinders and surfaces of constant $v$ are hyperbolas, as shown in Fig. [1.7]. They are related to the Cartesian coordinates via

$$x^1 = a \cosh u \cos v$$
$$x^2 = a \sinh u \sin v$$
$$x^3 = z, \tag{1.6}$$

where $a$ is an arbitrary constant.

There are coordinate systems based on the parabolic shape, such as **parabolic cylindrical coordinates**, as shown in Fig. [1.8]:

$$x^1 = \xi \eta$$
$$x^2 = \frac{1}{2} \left( \eta^2 - \xi^2 \right)$$
$$x^3 = z. \tag{1.7}$$

A final example is the **parabolic coordinates**, as shown in Fig. [1.9]:

$$x^1 = uv \cos \theta$$
$$x^2 = uv \sin \theta$$
$$x^3 = \frac{1}{2} \left( u^2 - v^2 \right). \tag{1.8}$$

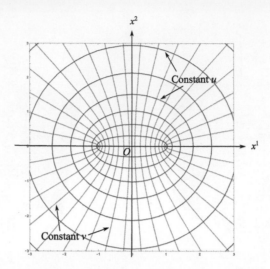

Fig. 1.7: The elliptic cylindrical coordinates grid in the $x$–$y$ plane: curves of constant $u$ (ellipses) and $v$ (hyperbolas) in the $z = 0$ plane. The extension to three dimensions is straightforward.

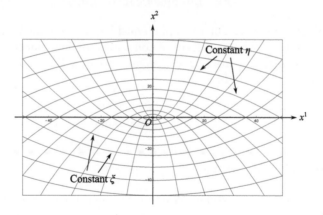

Fig. 1.8: The parabolic cylindrical coordinates grid in the $x$–$y$ plane: parabolic curves of constant $\xi$ (upright) and $\eta$ (inverted) in the $z = 0$ plane. The extension to three dimensions is straightforward.

All of the systems of coordinates we discussed, and many others that exist, are collectively known as **orthogonal coordinate systems**, i.e. perpendicular. In other words, their respective grid lines are perpendicular to each other at every point. There do exist non-orthogonal, or **skew**, coordinate systems, a simple example of which is given in the exercises.

**Exercise 1.1** Using a computer, see if you can reproduce some of the coordinate grids in this section, such as Fig. [1.4], [1.7], or [1.9]. Once you have had enough practice, plot the

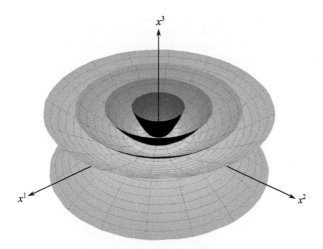

Fig. 1.9: The parabolic coordinate grid: The paraboloidal surfaces of constant $v$ (up-right) and $u$ (inverted) are shown. They can be thought of as the surfaces of revolution of the parabolas of constant $\xi$ and $\eta$ respectively in the parabolic cylindrical coordinates.

complete coordinates grids for the following systems. For the numerical values of the given constants, you may wish to experiment with a few and see which ones give you the best plots.

1. Paraboloidal coordinates $(\mu, \nu, \lambda)$ defined by

$$x^2 = \frac{4}{a-b}\,(\mu - a)\,(a - \nu)\,(a - \lambda)$$
$$y^2 = \frac{4}{a-b}\,(\mu - b)\,(b - \nu)\,(\lambda - b)$$
$$z = \mu + \nu + \lambda - a - b, \qquad\qquad (1.9)$$

where the constants $a$ and $b$ set the following limits on the coordinates:

$$\mu > a > \lambda > b > \nu > 0.$$

2. Oblate spheroidal coordinates $(\mu, \nu, \varphi)$ defined by

$$x = a \cosh \mu \cos \nu \cos \varphi$$
$$y = a \cosh \mu \cos \nu \sin \varphi$$
$$z = a \sinh \mu \sin \nu, \qquad\qquad (1.10)$$

where $a$ is an arbitrary constant.

3. Toroidal coordinates $(\sigma, \tau, \varphi)$ defined by

$$x = \frac{a \sinh \tau}{\cosh \tau - \cos \sigma}\,\cos \varphi$$
$$y = \frac{a \sinh \tau}{\cosh \tau - \cos \sigma}\,\sin \varphi$$
$$z = \frac{a \sin \sigma}{\cosh \tau - \cos \sigma}, \qquad \text{where } a \text{ is an arbitrary constant.} \qquad (1.11)$$

**Exercise 1.2** As an example of non-orthogonal, or skew, coordinates, consider a Cartesian coordinate system where the $z$ axis has been rotated in the $y$–$z$ plane by an angle $\alpha$, as shown in Fig. [1.10]. Derive the transformation equations relating the original orthogonal $x$, $y$, and $z$ coordinates to the skew coordinates $x'$, $y'$, and $z'$.

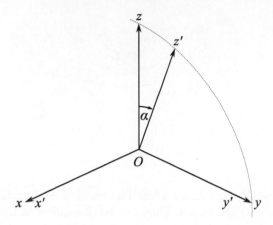

Fig. 1.10: The simplest possible example of a non-orthogonal, or skew, coordinate system.

## 1.2   Measurements and the Metric

A fundamental question in geometry is the question of spatial measurement—specifically: How does one measure a distance in space and relate it to the coordinate system in use? In Cartesian coordinates this is fully described by exploiting the familiar **Pythagorean theorem.**[3] Consider, to start, the two-dimensional plane with Cartesian coordinates shown in Fig. [1.11]. We define $\Delta l$ as the distance between two specific points with coordinates $(x^1, x^2)$ and $(x^{1'}, x^{2'})$. The question is: How is $\Delta l$ related to the changes in coordinates $\Delta x^1 = x^{1'} - x^1$ and $\Delta x^2 = x^{2'} - x^2$ between the points in question? This is clearly a problem for Pythagoras:

$$\Delta l^2 = \left(\Delta x^1\right)^2 + \left(\Delta x^2\right)^2. \tag{1.12}$$

Typically, we would be more interested in *infinitesimal* changes of coordinates (if we wanted to do calculus), so making the distance $\Delta l$ arbitrarily small; i.e. $\Delta l \to dl$ gives

$$dl^2 = \left(dx^1\right)^2 + \left(dx^2\right)^2. \tag{1.13}$$

---

[3]Pythagoras: Greek philosopher and mathematician (571–495 BCE).

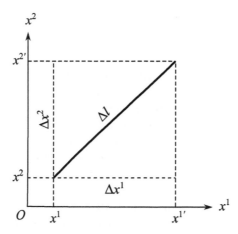

Fig. 1.11: The Pythagorean theorem.

The explicit use of indices in (1.13) is clearly becoming cumbersome. But this is exactly where the index notation becomes truly powerful. To see this, let us begin by using the summation symbol $\Sigma$ in (1.13):

$$dl^2 = \sum_{i=1}^{2} \left(dx^i\right)^2. \tag{1.14}$$

We can easily generalize to three dimensions by simply adding an extra term:

$$dl^2 = \left(dx^1\right)^2 + \left(dx^2\right)^2 + \left(dx^3\right)^2 \tag{1.15}$$

$$= \sum_{i=1}^{3} \left(dx^i\right)^2. \tag{1.16}$$

Interestingly, if, for whatever reason, one imagines a higher-dimensional space, in, say, arbitrary $N$ dimensions, then (1.15) is easily generalized by just adding more terms $x^4$, $x^5$, all the way up to $x^N$,

$$dl^2 = \left(dx^1\right)^2 + \left(dx^2\right)^2 + \left(dx^3\right)^2 + \left(dx^4\right)^2 + \left(dx^5\right)^2 + \cdots + \left(dx^N\right)^2, \tag{1.17}$$

or just changing the summation's upper limit to whatever the value of $N$ is:

$$dl^2 = \sum_{i=1}^{N} \left(dx^i\right)^2. \tag{1.18}$$

So the same formula (1.18) can be used to represent the Pythagorean theorem in *any* number of dimensions. All one needs to do is set the value of $N$. This simplicity is one reason why the index notation is a preferred choice. Now, expressions such as $\left(dx^i\right)^2$ may be a bit confusing. A common method used to avoid such confusion is to write (1.18) as follows:

$$dl^2 = \sum_{i=1}^{N} dx^i dx^i, \tag{1.19}$$

which should help a little. However, let us take this quest for simplification even further: Summations such as (1.19) arise all the time—sometimes double, triple, or more summations (as we will see). They are used to compactly write what would otherwise be very long equations (sometimes *pages* long). This is very common in both non-relativistic and relativistic physics. As he was pondering his theory of relativity, Albert Einstein himself came up with another simplification to expressions such as (1.19). This is now known as the **Einstein summation convention**, a particularly simple idea that goes a long way in simplifying the form of the equations. This is done by removing the summation symbol entirely:

$$dl^2 = dx^i dx^i. \tag{1.20}$$

The summation is hidden but still exists; eqn (1.20) is identical in meaning to (1.19).[4] But when one encounters such an expression, how does one know whether or not there is a summation if it is not explicitly written? The rule is this: If there are two *identical* indices per term, then a summation over them exists (*exactly* two; no more, no less). Either the limits of the summation are known beforehand or one can simply express them like this:

$$dl^2 = dx^i dx^i, \qquad i = 1, 2, \dots N, \tag{1.21}$$

where $N$ can be 2 (1.13), 3 (1.15), or just arbitrary. This language is very commonly used in theoretical physics and has more rules to it that we will explore later. On first encounter, an equation such as (1.21) may cause a great deal of confusion to the reader, so it may be advisable to go back and read from the beginning of this section just to remind oneself that (1.21) is nothing more than a shorthand for the Pythagorean theorem.

The quantity $dl$ is a distance between two (infinitesimally close) points, so as far as pure geometry is concerned it is always **positive definite**, as a distance should be. Generally, $dl$ is known as **the line element**. It is also sometimes referred to as the **metric** (from Latin *metricus*, and Greek *metrikos* or *metron*, meaning "measure"). In differential geometry and multivariable calculus $dl$ represents the basic element for measuring distances on curves. A familiar equation that arises from it is the formula for distance between two points on a given curve in the $x$–$y$ plane which one encounters in calculus:

$$l = \int_{x=a}^{x=b} dx \sqrt{1 + \left(\frac{dy}{dx}\right)^2}, \tag{1.22}$$

---

[4]I imagine Einstein simply got tired of writing summation symbols all the time.

where in this context $y(x)$ is the function describing the curve whose length is being measured. For our purposes, however, the quantities $l$ or $dl$ are unessential; it is $dl^2$ that is more important, for reasons that will become clear later. Hence, despite the possible notational confusion this might cause, in physics textbooks the terms "line element" and "metric" are more commonly used in reference to $dl^2$, *not dl*.

Now, what about other coordinate systems? What does the Pythagorean theorem look like in, say, two-dimensional polar coordinates? This is easily found; simply consider Fig. [1.12]. Infinitesimally, the shaded area approaches a right triangle. Applying the Pythagorean theorem to it gives

$$dl^2 = d\rho^2 + \rho^2 d\varphi^2. \tag{1.23}$$

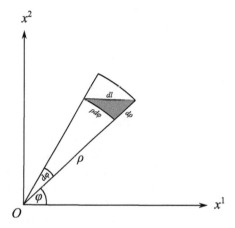

Fig. 1.12: The Pythagorean theorem applied to $dl$ in planar polar coordinates.

Now, we could have found eqn (1.23) analytically if we had considered how Cartesian coordinates are related to polar coordinates, i.e. by using the transformations (1.1). Simply put, take the differentials of both sides of (1.1), then plug the result into (1.13). Explicitly,

$$\begin{aligned} dx^1 &= d\rho \cos\varphi - \rho d\varphi \sin\varphi \\ dx^2 &= d\rho \sin\varphi + \rho d\varphi \cos\varphi, \end{aligned} \tag{1.24}$$

leading to

$$\begin{aligned} dl^2 &= \left(dx^1\right)^2 + \left(dx^2\right)^2 \\ &= d\rho^2 \cos^2\varphi + \rho^2 d\varphi^2 \sin^2\varphi - 2\rho\, d\rho\, d\varphi \sin\varphi \cos\varphi \\ &\quad + d\rho^2 \cos^2\varphi + \rho^2 d\varphi^2 \sin^2\varphi + 2\rho\, d\rho\, d\varphi \sin\varphi \cos\varphi \\ &= \left(d\rho^2 + \rho^2 d\varphi^2\right)\left(\sin^2\varphi + \cos^2\varphi\right) \\ &= d\rho^2 + \rho^2 d\varphi^2, \end{aligned} \tag{1.25}$$

where we have used Euler's[5] extremely useful identity $\sin^2 \varphi + \cos^2 \varphi = 1$, itself another form of Pythagoras' theorem. Cylindrical coordinates are a straightforward extension; just add $dz^2$:

$$dl^2 = d\rho^2 + \rho^2 d\varphi^2 + dz^2. \tag{1.26}$$

A similar calculation, albeit a bit longer, leads to the equivalent spherical coordinates expression by either plugging (1.3) into (1.15) or (1.5) into (1.26) to find the line element:

$$dl^2 = dr^2 + r^2 d\theta^2 + r^2 \sin^2 \theta d\varphi^2. \tag{1.27}$$

---

**Exercise 1.3** Find the metric $dl^2$ using the analytical method for:

1. Spherical coordinates using (1.3).
2. Spherical coordinates using (1.5).
3. Elliptic cylindrical coordinates (1.6).
4. Parabolic cylindrical coordinates (1.7).
5. The skew coordinates system of Exercise 1.2.

---

The analytical method just described is extremely useful because it provides a systematic way of transforming or generating metrics from one system of coordinates to another without having to draw sketches and agonizing over the trigonometry. We can easily generalize the method to any arbitrary system of coordinates $(u, v, w)$ whose relationship to the Cartesian (or any other coordinate system for that matter) is known; i.e. one can write the Cartesian coordinates as functions in, or transformations to, the new coordinates, $x^i(u, v, w)$, such as (1.1) or (1.3). To find the metric in the new coordinate system all one needs to do is take the appropriate derivatives and plug into (1.15). For this purpose let's define the so-called **scale factors** $h_u$, $h_v$, and $h_w$:

$$h_u^2 = \left(\frac{\partial x^1}{\partial u}\right)^2 + \left(\frac{\partial x^2}{\partial u}\right)^2 + \left(\frac{\partial x^3}{\partial u}\right)^2$$

$$h_v^2 = \left(\frac{\partial x^1}{\partial v}\right)^2 + \left(\frac{\partial x^2}{\partial v}\right)^2 + \left(\frac{\partial x^3}{\partial v}\right)^2$$

$$h_w^2 = \left(\frac{\partial x^1}{\partial w}\right)^2 + \left(\frac{\partial x^2}{\partial w}\right)^2 + \left(\frac{\partial x^3}{\partial w}\right)^2. \tag{1.28}$$

If we define a parameter $k$ that can be either $u$, $v$, or $w$, we can more compactly write

$$h_k^2 = \left(\frac{\partial x^i}{\partial k}\right)\left(\frac{\partial x^i}{\partial k}\right), \tag{1.29}$$

---

[5]Leonhard Euler: Swiss mathematician, physicist, astronomer, geographer, logician, and engineer (1707–1783).

with the summation convention on $i$ assumed. These scale factors lead directly to the metric/line element in the $(u, v, w)$ coordinate system:

$$dl^2 = h_u^2 du^2 + h_v^2 dv^2 + h_w^2 dw^2. \tag{1.30}$$

For example, for spherical coordinates (1.3)

$$
\begin{aligned}
h_r^2 &= \sin^2\theta\cos^2\varphi + \sin^2\theta\sin^2\varphi + \cos^2\theta = 1 \\
h_\theta^2 &= r^2\cos^2\theta\cos^2\varphi + r^2\cos^2\theta\sin^2\varphi + r^2\sin^2\theta = r^2 \\
h_\varphi^2 &= r^2\sin^2\theta\sin^2\varphi + r^2\sin^2\theta\cos^2\varphi = r^2\sin^2\theta,
\end{aligned}
\tag{1.31}
$$

which leads to (1.27) as required. Note that the scale factors method, as described, works for orthogonal coordinates *only*.

**Exercise 1.4** Find the metric $dl^2$ using the scale factors method for:
1. Parabolic coordinates (1.8).
2. Paraboloidal coordinates (1.9).
3. Oblate spheroidal coordinates (1.10).
4. Toroidal coordinates (1.11).

**Exercise 1.5** Consider the following metric:

$$dl^2 = du^2 + dv^2 + dw^2 - \left(\frac{3}{13}du + \frac{4}{13}dv + \frac{12}{13}dw\right)^2. \tag{1.32}$$

It appears that this is a three-dimensional metric in $u$, $v$, and $w$. This is not true! This metric is in fact two-dimensional in disguise. Prove this by finding a transformation between the "coordinates" $u$, $v$, and $w$ and the Cartesian coordinates $x$ and $y$, similar to (1.1), for example, that reduces (1.32) to

$$dl^2 = dx^2 + dy^2. \tag{1.33}$$

## 1.3   Vectors in Cartesian Coordinates

In introductory physics courses, vectors are defined as physical quantities that have both magnitude and direction. They are abstractedly represented by directional arrows whose lengths signify their magnitudes. This description, while correct, suffers from the defect that the visual image that goes along with it is only useful in two or three dimensions. In higher-dimensional physics, which we will consider soon, one can still define vectors in the same way, but the visual is lost. This is not because it doesn't exist, but because our brains are not built to visualize more than three dimensions. It is also true that the concept of vectors generalizes into a variety of more exotic mathematical objects that go by the names "tensors," "differential forms," and "spinors," all of which are quite useful in theoretical physics.[6] These objects, by their very nature, resist simple visualization except in the most restricted of special cases or analogies. Given all of this, we need to redefine the language used to describe vectors to make the transition to higher dimensions, as well as to the related objects, as smooth as possible, while at the same time preserving the coordinate-invariant nature of vector mathematics. We can still imagine, as well as sketch, vectors as "arrows" in space, but the emphasis shifts toward the calculational over the graphical.

To start, let's define vectors and their components in Cartesian coordinates in the usual way:

$$\mathbf{V} = V_x\hat{\mathbf{x}} + V_y\hat{\mathbf{y}} + V_z\hat{\mathbf{z}}, \tag{1.34}$$

where $\hat{\mathbf{x}}$, $\hat{\mathbf{y}}$, and $\hat{\mathbf{z}}$ are called the **Cartesian basis vectors**, usually defined as unit vectors pointing in the directions of the $x$, $y$, and $z$ axes respectively.[7] Addition and subtraction of vectors is defined pictorially as usual by completing the triangle formed by two vectors to find the result, as shown in Fig. [1.13].

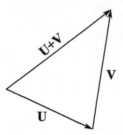

Fig. 1.13: Addition of vectors.

So, given two vectors $\mathbf{V}$ and $\mathbf{U} = U_x\hat{\mathbf{x}} + U_y\hat{\mathbf{y}} + U_z\hat{\mathbf{z}}$, their sum is

$$\mathbf{V} + \mathbf{U} = \left(V_x + U_x\right)\hat{\mathbf{x}} + \left(V_y + U_y\right)\hat{\mathbf{y}} + \left(V_z + U_z\right)\hat{\mathbf{z}}. \tag{1.35}$$

Furthermore, the so-called **dot product** of two vectors $\mathbf{V}$ and $\mathbf{U}$, also known as the **scalar** or **inner** product, is defined as follows:

$$\mathbf{V} \cdot \mathbf{U} = V_xU_x + V_yU_y + V_zU_z, \tag{1.36}$$

where the result no longer carries the vector status but is rather a **scalar**; a quantity with magnitude only. Another equivalent definition of the dot product is

$$\mathbf{V} \cdot \mathbf{U} = |\mathbf{V}|\,|\mathbf{U}|\cos\theta, \tag{1.37}$$

where $\theta$ is the angle between $\mathbf{V}$ and $\mathbf{U}$, as shown in Fig. [1.14].

---

[6]Two of which, tensors and differential forms, we will study in this book.
[7]In introductory physics textbooks the notation $\hat{\mathbf{i}}$, $\hat{\mathbf{j}}$, and $\hat{\mathbf{k}}$ is more commonly used.

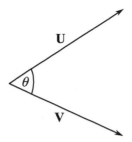

Fig. 1.14: The scalar or dot product of two vectors.

Now the magnitude of a vector, also known as its **norm**, is defined as the square root of its dot product with itself:

$$V = |\mathbf{V}| = \sqrt{\mathbf{V} \cdot \mathbf{V}} = \sqrt{V_x^2 + V_y^2 + V_z^2}. \tag{1.38}$$

Note that we define everything in the most general sense, so the vectors considered can be functions in space as well as time — in other words, they can be **vector fields**, i.e. $\mathbf{V}(t, \mathbf{r})$ with components $V_x(t, x, y, z)$, $V_y(t, x, y, z)$, and $V_z(t, x, y, z)$.

The main language needed to achieve our objectives in this chapter is exactly the index notation briefly introduced in §1.2. As pointed out earlier, it is also sometimes referred to as the tensor notation, since it becomes truly powerful when dealing with tensors, which we will introduce later. For now let's see how Cartesian vectors can be recast in indices. Consider the definition (1.34). Rename the basis vectors as follows:

$$\hat{\mathbf{x}} \to \hat{\mathbf{e}}_1, \qquad \hat{\mathbf{y}} \to \hat{\mathbf{e}}_2, \qquad \hat{\mathbf{z}} \to \hat{\mathbf{e}}_3. \tag{1.39}$$

Also rename the components of the vector

$$V_x \to V^1, \qquad V_y \to V^2, \qquad V_z \to V^3, \tag{1.40}$$

where once again it is extremely important *not* to confuse the superscripts with exponents. The numbers 1, 2, and 3 simply keep track of which component of the vector $\mathbf{V}$ we are talking about. We also note that we gave the basis vectors subscripts rather than superscripts. At this point this is just a matter of convention, but it will acquire a deeper meaning later.

It is now clearly possible to write the vector $\mathbf{V}$ in the following compact form:

$$\mathbf{V} = V^1 \hat{\mathbf{e}}_1 + V^2 \hat{\mathbf{e}}_2 + V^3 \hat{\mathbf{e}}_3 = \sum_{i=1}^{3} V^i \hat{\mathbf{e}}_i. \tag{1.41}$$

Alternatively, by adopting the Einstein summation convention (i.e. dropping the summation symbol) we get the elegant shorthand

$$\mathbf{V} = V^i \hat{\mathbf{e}}_i. \qquad (1.42)$$

Recall that the summation is still there, it is just hidden; once again, the rule is: *If an index appears twice in one term, it is summed over*. Here is another example: the vector addition formula (1.35) can now be written as follows:

$$\mathbf{V} + \mathbf{U} = V^i \hat{\mathbf{e}}_i + U^i \hat{\mathbf{e}}_i = \left( V^i + U^i \right) \hat{\mathbf{e}}_i. \qquad (1.43)$$

Note that which letter to use as an index in a summation is *completely* arbitrary. For example, $V^j \hat{\mathbf{e}}_j$ or $V^a \hat{\mathbf{e}}_a$ (or even $U^\heartsuit \hat{\mathbf{e}}_\heartsuit$ if one wishes) still means *exactly* the same thing. In fact, it is perfectly acceptable, as well as useful sometimes, to change the letters of the indices in mid calculation. Consider that we can write

$$\mathbf{V} + \mathbf{U} = V^i \hat{\mathbf{e}}_i + U^j \hat{\mathbf{e}}_j = \left( V^i + U^i \right) \hat{\mathbf{e}}_i \qquad (1.44)$$

and it is still equivalent to (1.43), even though we changed the index $j$ into $i$ in the last step. Because of this freedom, summation indices are sometimes referred to as **dummy indices**. There is another type of index that we will introduce soon.

The process of performing summations using the Einstein convention is known as **index contraction**, so (1.42) may be described as "$V^i$ contracted with $\hat{\mathbf{e}}_i$," which is exactly the same thing as saying multiply each individual $V^i$ with its respective $\hat{\mathbf{e}}_i$ and sum them up. For ordinary two- or three-dimensional physics, the indices will count over either $1, 2$ or $1, 2, 3$ depending on whether we are looking at two-dimensional vectors in the plane or the full three-dimensional structure. For example, the position vector

$$\mathbf{r} = x\hat{\mathbf{x}} + y\hat{\mathbf{y}} + z\hat{\mathbf{z}} \qquad (1.45)$$

can now be rewritten as

$$\mathbf{r} = x^i \hat{\mathbf{e}}_i, \qquad (1.46)$$

where $x = x^1$, $y = x^2$, and $z = x^3$ as before. We now use this language to reconstruct everything we have ever learned about vectors. Note that we will be working purely in Cartesian coordinates for a while. Attention to vectors in curvilinear coordinates will be given eventually.

Suppose we want to find the dot product of the two vectors

$$\mathbf{V} = V^i \hat{\mathbf{e}}_i, \quad \mathbf{U} = U^i \hat{\mathbf{e}}_i \qquad (1.47)$$

by putting them together as usual:

$$\mathbf{V} \cdot \mathbf{U} = \left( V^i \hat{\mathbf{e}}_i \right) \cdot \left( U^i \hat{\mathbf{e}}_i \right). \qquad \text{Wrong!} \qquad (1.48)$$

We remember that we are not allowed to have more than two repeating indices in any one term of an equation, since this should be a double sum: one summation is for vector **V** and the other for **U**. As it stands now, expression (1.48) implies a single sum, which is generally incorrect. To solve this problem we use the dummy property of the summation indices and give one of them a different name, signifying a different summation:

$$\mathbf{V} \cdot \mathbf{U} = \left( V^i \hat{\mathbf{e}}_i \right) \cdot \left( U^j \hat{\mathbf{e}}_j \right), \qquad (1.49)$$

where the index $j$ also runs from 1 to 3. In other words, this is equivalent to

$$\mathbf{V} \cdot \mathbf{U} = \sum_{i=1}^{3} \sum_{j=1}^{3} \left[ \left( V^i \hat{\mathbf{e}}_i \right) \cdot \left( U^j \hat{\mathbf{e}}_j \right) \right]. \qquad (1.50)$$

Now, let's continue by reordering:

$$\mathbf{V} \cdot \mathbf{U} = V^i U^j \left( \hat{\mathbf{e}}_i \cdot \hat{\mathbf{e}}_j \right). \qquad (1.51)$$

If an equation such as (1.51), or any other subsequent equation for that matter, is not immediately obvious, then the reader is advised to write it out in detail to see how the summation convention works. In this case (1.51) is simply shorthand for

$$\begin{aligned}
\mathbf{V} \cdot \mathbf{U} = {} & V^1 U^1 \left( \hat{\mathbf{e}}_1 \cdot \hat{\mathbf{e}}_1 \right) + V^1 U^2 \left( \hat{\mathbf{e}}_1 \cdot \hat{\mathbf{e}}_2 \right) + V^1 U^3 \left( \hat{\mathbf{e}}_1 \cdot \hat{\mathbf{e}}_3 \right) \\
& + V^2 U^1 \left( \hat{\mathbf{e}}_2 \cdot \hat{\mathbf{e}}_1 \right) + V^2 U^2 \left( \hat{\mathbf{e}}_2 \cdot \hat{\mathbf{e}}_2 \right) + V^2 U^3 \left( \hat{\mathbf{e}}_2 \cdot \hat{\mathbf{e}}_3 \right) \\
& + V^3 U^1 \left( \hat{\mathbf{e}}_3 \cdot \hat{\mathbf{e}}_1 \right) + V^3 U^2 \left( \hat{\mathbf{e}}_3 \cdot \hat{\mathbf{e}}_2 \right) + V^3 U^3 \left( \hat{\mathbf{e}}_3 \cdot \hat{\mathbf{e}}_3 \right).
\end{aligned} \qquad (1.52)$$

Now we have known since introductory physics that the expression $(\hat{\mathbf{e}}_i \cdot \hat{\mathbf{e}}_j)$ can only be either 1 or zero, because these are orthogonal (i.e. perpendicular) unit vectors. So, if $i = j$, then $(\hat{\mathbf{e}}_i \cdot \hat{\mathbf{e}}_j) = 1$, while for $i \neq j$ we have $(\hat{\mathbf{e}}_i \cdot \hat{\mathbf{e}}_j) = 0$ (previously known as $\hat{\mathbf{x}} \cdot \hat{\mathbf{x}} = \hat{\mathbf{y}} \cdot \hat{\mathbf{y}} = \hat{\mathbf{z}} \cdot \hat{\mathbf{z}} = 1$ and $\hat{\mathbf{x}} \cdot \hat{\mathbf{y}} = \hat{\mathbf{y}} \cdot \hat{\mathbf{z}} = \hat{\mathbf{z}} \cdot \hat{\mathbf{x}} = 0$). So applying this to (1.52) leads to

$$\mathbf{V} \cdot \mathbf{U} = V^1 U^1 + V^2 U^2 + V^3 U^3, \qquad (1.53)$$

as required. The orthogonality property of the basis vectors allows us to symbolically incorporate this information in (1.51) by introducing the following new symbol:

$$\delta_{ij} = \hat{\mathbf{e}}_i \cdot \hat{\mathbf{e}}_j \quad \text{such that} \quad \delta_{ij} \begin{cases} = 1 \text{ for } i = j \\ = 0 \text{ for } i \neq j \end{cases}. \qquad (1.54)$$

The symbol $\delta_{ij}$ is known as the **Kronecker delta**,[8] and can be understood at this point as a "book keeping" method of deciding which expressions are vanishing and which aren't. We can then rewrite (1.51):

$$\mathbf{V} \cdot \mathbf{U} = V^i U^j \delta_{ij}. \qquad (1.55)$$

---

[8]Leopold Kronecker: Prussian/German mathematician (1823–1891).

This is a double sum over $i$ and $j$, which, if we perform faithfully, gives

$$\mathbf{V} \cdot \mathbf{U} = V^i U^j \delta_{ij} = \sum_{i=1}^{3} \sum_{j=1}^{3} V^i U^j \delta_{ij}$$

$$= V^1 U^1 \delta_{11} + V^1 U^2 \delta_{12} + V^1 U^3 \delta_{13} + V^2 U^1 \delta_{21} + V^2 U^2 \delta_{22} + V^2 U^3 \delta_{23}$$
$$+ V^3 U^1 \delta_{31} + V^3 U^2 \delta_{32} + V^3 U^3 \delta_{33}$$
$$= V^1 U^1 + V^2 U^2 + V^3 U^3, \quad \text{since only } \delta_{11}, \delta_{22}, \text{ and } \delta_{33} \text{ are non-vanishing.}$$
$$(1.56)$$

The use of the Kronecker delta is extremely useful, as expressions tend toward more complicated index manipulations. In reference to eqn (1.20), for example, one can rewrite the Cartesian metric as follows:

$$dl^2 = \delta_{ij} dx^i dx^j. \tag{1.57}$$

The reader can verify that it means exactly the same thing as before.

---

**Exercise 1.6** Expand (1.57) and show that it leads to (1.15).

---

Now, as briefly discussed earlier, the fact that we are using both upper and lower indices may be slightly confusing at this point. As far as summations are concerned, which is the reason the index language was invented in the first place, it apparently doesn't matter whether we use upper or lower indices. So the following expressions are essentially all equivalent:

$$V^i U^j \delta_{ij} = V_i U_j \delta^{ij} = V^i U_j \delta^i_j = V_i U_j \delta_{ij}. \tag{1.58}$$

The Kronecker delta does not change meaning, either; for instance, $\delta^{22} = \delta_{22} = \delta^2_2 = 1$ and $\delta^{23} = \delta_{23} = \delta^2_3 = 0$, and so on. So if they are all the same, why are we making a distinction between quantities with upper indices and quantities with lower indices? The reason is that quantities with upper or lower indices are equivalent to each other *only* in Cartesian coordinates! When we make the transition to non-Cartesian coordinates we will find that vectors with upper indices and vectors with lower indices are no longer exactly the same. So we might as well get used to the concept of upper and lower indices from the start. At this point we can, however, define a relationship between vector components with upper indices and those with lower indices as follows. Let

$$U^1 \delta_{11} = U_1$$
$$U^2 \delta_{22} = U_2$$
$$U^3 \delta_{33} = U_3, \tag{1.59}$$

or symbolically

$$U^j \delta_{ij} = U_i \quad \text{and} \quad V^i \delta_{ij} = V_j \tag{1.60}$$

for the components of any vectors **U** and **V**. Also,

$$U_j \delta^{ij} = U^i \quad \text{or} \quad V_i \delta^{ij} = V^j. \tag{1.61}$$

The process defined by (1.60) and (1.61) is usually known as the **lowering** or the **raising** of indices respectively. If we begin with an upper index component $U^i$, we can contract it with a Kronecker delta to define a lower index $U_j$. We may even take this further by defining the **hybrid Kronecker delta** as follows

$$\delta^i_j = \delta^{ik} \delta_{kj}, \tag{1.62}$$

which can be used to change the names of the indices *without* raising or lowering

$$U^j \delta^i_j = U^i, \quad U_i \delta^i_j = U_j. \tag{1.63}$$

Notice the appearance of a single non-repeating index in (1.60), (1.61), and (1.63). This is called a **free index**. *The rule for free indices is that if there is one of them in a term of the equation, then there should be one by the same name in each other term*. They are obviously not used in summations, but rather to denote that an equation such as (1.60) is in fact three equations, one for each value of the free index; in other words, (1.60) means the same thing as all three in (1.59). Note that eqn (1.62) has *two* free indices and only one dummy index, which implies that it is in fact equivalent to nine separate equations:

$$\begin{aligned}
\delta^1_1 &= \delta^{1k} \delta_{k1} \\
\delta^2_1 &= \delta^{2k} \delta_{k1} \\
\delta^3_2 &= \delta^{3k} \delta_{k2}, \quad \text{and so on.}
\end{aligned} \tag{1.64}$$

Now in the literature a vector whose components carry upper indices is known as a **contravariant** vector, while a vector whose components carry lower indices is known as a **covariant** vector. Another terminology is in calling a lower-index vector the **dual vector**, while the upper-index vector is simply the **vector**. There is no need to further the reader's confusion, though, so we will just call them "upper and lower indices." Once again, as far as Cartesian vectors are concerned, the components are the same for both types of vectors anyway, so $U^1 = U_1$, $U^2 = U_2$, and $U^3 = U_3$. If it makes it clearer, here is how one reads a formula such as (1.60):

$$U^j \delta_{ij} = U^1 \delta_{i1} + U^2 \delta_{i2} + U^3 \delta_{i3} = U_i. \tag{1.65}$$

Only one of the three terms in the last equation will be non-zero: the one where the free index $i$ has the same value as the other index on the Kronecker delta, yielding $U_i$. Explicitly, (1.65) is equivalent to

$$U^j \delta_{1j} = U^1 \delta_{11} + U^2 \delta_{12} + U^3 \delta_{13} = U^1 \delta_{11} = U_1$$
$$U^j \delta_{2j} = U^1 \delta_{21} + U^2 \delta_{22} + U^3 \delta_{23} = U^2 \delta_{22} = U_2$$
$$U^j \delta_{3j} = U^1 \delta_{31} + U^2 \delta_{32} + U^3 \delta_{33} = U^3 \delta_{33} = U_3. \tag{1.66}$$

A similar process happens with (1.63):

$$U^j \delta^i_j = U^1 \delta^i_1 + U^2 \delta^i_2 + U^3 \delta^i_3 = U^i. \tag{1.67}$$

So, we conclude this rather lengthy definition of the dot product by noting that, based on the preceding,

$$\mathbf{V} \cdot \mathbf{U} = V^i U^j \delta_{ij} = V_i U_j \delta^{ij} = V^i U_j \delta^j_i = V^i U_i = V_i U^i \tag{1.68}$$

are all equivalent, where the last two are derived by using either one of the equations in (1.60) and (1.61) and the dummy indices can carry any labels as usual.

The language and rules defined here are extremely important for the rest of the book. Yes, it is a lot to absorb all at once, but the reader is *strongly* advised not to move forward without deeply understanding everything mentioned in this section so far. In the hope of making all of this slightly more intuitive, let's take it one step further. The rather funny, and potentially confusing, upper and lower index notation is closely related to something you may already be familiar with.[9] In terms of matrices, a vector is usually defined as a column matrix:

$$\mathbf{V} = \begin{pmatrix} V_x \\ V_y \\ V_z \end{pmatrix}. \tag{1.69}$$

In order to, say, dot this vector with another,

$$\mathbf{U} = \begin{pmatrix} U_x \\ U_y \\ U_z \end{pmatrix}, \tag{1.70}$$

one must follow the rules of matrix multiplication which insist that the transpose of one of these vectors is found first, so:

$$\mathbf{V} \cdot \mathbf{U} = \begin{pmatrix} V_x & V_y & V_z \end{pmatrix} \begin{pmatrix} U_x \\ U_y \\ U_z \end{pmatrix}$$
$$= \begin{pmatrix} U_x & U_y & U_z \end{pmatrix} \begin{pmatrix} V_x \\ V_y \\ V_z \end{pmatrix} = V_x U_x + V_y U_y + V_z U_z \tag{1.71}$$

[9] If the words "matrix" and "matrix multiplication" do not mean anything to you, then feel free to *temporarily* skip the following to the end of the section, although eventually some knowledge of matrices will be needed. Good tutorials in the language of matrices may be easily found online.

What is the difference between a vector as a row matrix and the same vector as a column matrix? In this context they are exactly the same! We are simply following the rules of matrix multiplication by demanding two different forms for the same vector. In the same spirit, requiring the rules of index contraction to be between upper and lower indices even though the components are exactly the same (in Cartesian coordinates only) is simply a matter of following the rules. So one can think of an upper-index vector as simply a column matrix, while a lower-index vector is its transpose; a row matrix, or the other way around, it doesn't really matter. So:

$$V_i U^i \leftrightarrow \begin{pmatrix} V_x & V_y & V_z \end{pmatrix} \begin{pmatrix} U_x \\ U_y \\ U_z \end{pmatrix} \tag{1.72}$$

and

$$V^i U_i \leftrightarrow \begin{pmatrix} U_x & U_y & U_z \end{pmatrix} \begin{pmatrix} V_x \\ V_y \\ V_z \end{pmatrix}. \tag{1.73}$$

Finally, the Kronecker delta itself may be thought of as representing the components of a matrix—specifically the **identity matrix**—where the indices label the rows and the columns as follows:

$$[\delta] = \begin{matrix} j \\ \downarrow \end{matrix} \begin{pmatrix} i \rightarrow \\ 1 & 0 & 0 \\ 0 & 1 & 0 \\ 0 & 0 & 1 \end{pmatrix} \quad \text{or} \quad \begin{matrix} i \\ \downarrow \end{matrix} \begin{pmatrix} j \rightarrow \\ 1 & 0 & 0 \\ 0 & 1 & 0 \\ 0 & 0 & 1 \end{pmatrix}. \tag{1.74}$$

Clearly, the Kronecker delta has the property of being symmetric, or in other words,

$$\delta_{ij} = \delta_{ji}. \tag{1.75}$$

Note that if we perform the following (double) summation:

$$\delta^{ij} \delta_{ij} = \delta^i_i = \delta^1_1 + \delta^2_2 + \delta^3_3 = 3, \tag{1.76}$$

the result is exactly the same as the trace of the identity matrix. In fact, it is true that the trace of *any* matrix $A_{ij}$ may be found by contracting all of its indices with the Kronecker delta,

$$\delta^{ij} A_{ij} = \delta_{ij} A^{ij} = A^i_i \overset{\text{or}}{=} A^j_j = \text{Tr}\,[A] \tag{1.77}$$

where $A_{ij}$ are the components of the matrix

$$[A] = \begin{pmatrix} A_{11} & A_{12} & A_{13} \\ A_{21} & A_{22} & A_{23} \\ A_{31} & A_{32} & A_{33} \end{pmatrix}. \tag{1.78}$$

The analogy between the index notation of vectors and the matrix notation is tempting, and one may consider switching completely to matrices and abandoning the index language altogether. But in the long run this would actually be counterproductive, since the index language is a lot more powerful as far as our objectives are concerned. So we will only use matrices (and later determinants also) to explain certain properties of the index notation, simplify some calculations where appropriate, and possibly provide a "visual" to index-based calculations.[10]

**Exercise 1.7** Rewrite the following expressions in the index notation using *all* possible forms as in (1.58):

1. For four arbitrary vectors: $(\mathbf{A} \cdot \mathbf{B})(\mathbf{C} \cdot \mathbf{D})$.
2. $(\mathbf{A} - \mathbf{B}) \cdot \hat{\mathbf{n}} = 0$, where $\mathbf{A}$ and $\mathbf{B}$ are arbitrary vectors and $\hat{\mathbf{n}}$ is a unit vector. In the index notation what is the condition that $\hat{\mathbf{n}}$ has unit magnitude?
3. $\cos\theta = \frac{\mathbf{A} \cdot \mathbf{B}}{AB}$, where $A$ and $B$ are the magnitudes of the vectors $\mathbf{A}$ and $\mathbf{B}$, and $\theta$ is the angle between them.
4. Coulomb's law of electrostatics, $\mathbf{E} = k\frac{\mathbf{r}}{r^3}$, where $\mathbf{E}$ is the electric field vector, $k$ is Coulomb's constant, $\mathbf{r}$ is the position vector (1.46) of some arbitrary point, and $r$ is its magnitude.
5. $dW = \mathbf{F} \cdot d\mathbf{r}$, where $\mathbf{F}$ is some arbitrary force and $d\mathbf{r} = dx\hat{\mathbf{x}} + dy\hat{\mathbf{y}} + dz\hat{\mathbf{z}}$ is the displacement vector.
6. Kinetic energy $= \frac{1}{2}mv^2$, where $v^2$ is the velocity squared of some point mass $m$.

**Exercise 1.8** Organize the following expansions in the index notation using contractions over dummy indices and free indices where appropriate:

1. $A_3^1 B_1^2 + A_3^2 B_2^2 + A_3^3 B_3^2$.
2. $A_{11}^1 + A_{12}^2 + A_{13}^3$.
3. The three expressions:

$$A^{11}B_1 + A^{12}B_2 + A^{13}B_3 = C^1,$$
$$A^{21}B_1 + A^{22}B_2 + A^{23}B_3 = C^2,$$
$$A^{31}B_1 + A^{32}B_2 + A^{33}B_3 = C^3.$$

**Exercise 1.9** Expand the following contractions:

1. $A_i^j B_j C^i$.
2. $A_i B_k^j C^k D_j^i$.

**Exercise 1.10** Consider the following index expressions. There are some intentional mistakes as well as missing labels (denoted by question marks). Following the rules of the index notation, fix the mistakes and insert the correct labels. In the cases where there may be more than one way to fix the issue, explain what the mistake is and choose one possible way of fixing it. With the single exception of the Kronecker delta, all matrices such as $A$, $B$, or $R$ are arbitrary.

1. $\mathbf{B} \cdot \mathbf{C} = B^i C_j$.
2. $A^{ij} = B^{ik} C_k^?$.
3. $N_i = R_i^k D_k^p C_?$.

---

[10]To this author's knowledge there is only one textbook that uses matrices exclusively to explore the theories of relativity: [2].

4. $L_k^i D^k = R_m^n N_n^? P^?$.
5. $\delta^{ij} A_j B_i^k = A^? B_i^k = C^?$.
6. $N_i = R_i^k \delta_k^p C_? = R_?^? C^?$.

**Exercise 1.11** For an arbitrary vector $B^i$, show that $A_{ij} B^i B^j = 0$ if the components of the matrix $A$ are $A_{ij} = i - j$.

## 1.4 Derivatives in the Index Notation

In many physical situations, we need to calculate the derivative of a vector with respect to the coordinates. For that purpose, the **del** or **nabla** operator is defined. In Cartesian coordinates, this is

$$\vec{\nabla} = \hat{\mathbf{x}} \frac{\partial}{\partial x} + \hat{\mathbf{y}} \frac{\partial}{\partial y} + \hat{\mathbf{z}} \frac{\partial}{\partial z}. \tag{1.79}$$

The **gradient** of a function $f(x, y, z)$ is defined by

$$\vec{\nabla} f = \hat{\mathbf{x}} \frac{\partial f}{\partial x} + \hat{\mathbf{y}} \frac{\partial f}{\partial y} + \hat{\mathbf{z}} \frac{\partial f}{\partial z} \tag{1.80}$$

and the **divergence** of a vector **V** is

$$\vec{\nabla} \cdot \mathbf{V} = \frac{\partial V^x}{\partial x} + \frac{\partial V^y}{\partial y} + \frac{\partial V^z}{\partial z}$$

$$\text{or} = \frac{\partial V^1}{\partial x^1} + \frac{\partial V^2}{\partial x^2} + \frac{\partial V^3}{\partial x^3}. \tag{1.81}$$

Note that the divergence,[11] being essentially a dot product, easily lends itself to the index notation, so

$$\vec{\nabla} \cdot \mathbf{V} = \frac{\partial V^i}{\partial x^i}. \tag{1.82}$$

We will further simplify the notation by defining:

$$\partial_i = \frac{\partial}{\partial x^i}. \tag{1.83}$$

Generally, vectors are initially defined as upper-index objects, while the derivative $\partial_i$ is defined as a vector with a lower index from the start. This is because $x^i$ is in the

---

[11]If none of this is familiar to you, perhaps it is time to review your knowledge of multivariable calculus.

denominator on the right-hand side of (1.83) and the free index $i$ has to be balanced on both sides. One can, however, raise the index on $\partial_i$ using (1.61),

$$\partial^i = \delta^{ij}\partial_j, \tag{1.84}$$

where in this case the order of the symbols is significant. One should not write $\partial_i \delta^{ij}$ since it implies taking the derivative of $\delta^{ij}$, which is not our intention. We can now finally write the divergence of a vector and the gradient of a scalar as follows:

$$\vec{\nabla} \cdot \mathbf{V} = \partial_i V^i \quad \text{and} \quad \vec{\nabla}f = \delta^{ij}\hat{\mathbf{e}}_i\left(\partial_j f\right) = \hat{\mathbf{e}}_i\left(\partial^i f\right). \tag{1.85}$$

The so-called **directional derivative** is defined via

$$df = d\mathbf{r} \cdot \vec{\nabla}f = dx^i \frac{\partial f}{\partial x^i} = dx^i\left(\partial_i f\right), \tag{1.86}$$

where $d\mathbf{r}$ is the displacement vector along a curve defined by the position vector $\mathbf{r}\left(s\right) = x^i \hat{\mathbf{e}}_i$ and $s$ is some parameter on the curve (as in Fig. [3.1]). In physics, position vectors and displacements are usually parameterized by time, i.e. $s \to t$, but mathematically speaking $s$ can be any parameter. The directional derivative, which is really just the chain rule applied on $f$, gives the change of the function $f$ along the direction of the curve, as defined by $d\mathbf{r}$.

Another useful differential operator in physics is the **Laplacian** $\nabla^2$, which can be thought of as the dot product of del with itself;[12] i.e.

$$\nabla^2 = \vec{\nabla} \cdot \vec{\nabla} = \frac{\partial^2}{\partial x^2} + \frac{\partial^2}{\partial y^2} + \frac{\partial^2}{\partial z^2}. \tag{1.87}$$

In the index notation this becomes any of the following equivalent expressions:

$$\nabla^2 = \delta^{ij}\partial_i\partial_j = \delta_{ij}\partial^i\partial^j = \partial_i\partial^i = \partial^i\partial_i. \tag{1.88}$$

**Exercise 1.12** Express the following in the index notation using all possible forms. $\mathbf{A}$ and $\mathbf{B}$ are vectors, while $f$ is a scalar function:

1. $\mathbf{A} \cdot \left(\vec{\nabla}f\right)$.

2. $\vec{\nabla} \cdot \vec{\nabla}f$.

3. $\mathbf{A}\left(\vec{\nabla} \cdot \mathbf{B}\right)$.

4. The wave equation $\nabla^2 f - \frac{1}{v^2}\frac{\partial^2 f}{\partial t^2} = 0$. You may use the additional notation $\partial_t = \partial/\partial t$. The quantity $v$ is a constant signifying the speed of the wave.

---

[12]The Laplacian is named after Pierre-Simon, marquis de Laplace, a French mathematician and physicist (1749–1827).

**Exercise 1.13**

1. Given the expression $a_{ij} A^i A^j$, where the coefficients $a_{ij}$ are constants and $A^i \left( x^1, x^2, x^3 \right) = A^i \left( x^k \right)$ are the components of a vector field, show that $\partial_k \left( a_{ij} A^i A^j \right) = 2 a_{ij} A^i \left( \partial_k A^j \right)$.

2. What are the *numerical* values of $\partial_i x_j$ and $\partial_i x^j$ for all the possible combinations of indices? How are these expressions related to the Kronecker delta?

3. Using your result from part 2 show that $\vec{\nabla} \cdot \mathbf{r} = 3$, where $\mathbf{r}$ is the Cartesian position vector (1.46).

4. Prove that $\vec{\nabla} \left( \ln |\mathbf{r}| \right) = \frac{\mathbf{r}}{r^2}$ using the index notation.

5. Prove that $\nabla^2 \left( \ln |\mathbf{r}| \right) = \frac{1}{r^2}$ using the index notation.

Although you proved them in Cartesian coordinates, the identities in parts 3, 4, and 5 concerning the position vector are true for *any* coordinate system. This is because the expressions $\vec{\nabla} \cdot \mathbf{r}$, $\vec{\nabla} \left( \ln |\mathbf{r}| \right)$ and $\nabla^2 \left( \ln |\mathbf{r}| \right)$ are covariant (coordinate invariant). If they are true in one coordinate system they are true in *all* coordinate systems; the expressions in the index notation, however, such as $\partial_i x^j$, are only true in Cartesian coordinates. Indeed, one of our major objectives is to figure out how to write covariant index expressions.

To conclude this section, I would like to make a note of a mysterious notation that arises in some books and research papers. I have hesitated in including it in our discussion here because I have always felt that it is unnecessarily confusing and would personally never use it. However, it is such a common language in the literature that it might be worthwhile just making a note of it. This is the claim that the derivative $\partial_i$ is a "basis vector" in the direction $i$. In other words, it plays the role of $\hat{\mathbf{e}}_i$. What exactly does that mean? How can a differential operator be a basis vector? This strange language arises from the directional derivative (1.86). This equation has the form $\mathbf{V} = V^i \hat{\mathbf{e}}_i$ *if* we "think" of $df$ as a vector with components $dx^i$ such that the basis vectors would just be $\partial_i f$. From that perspective $\partial_i f$, or more economically just $\partial_i$, can be thought of as the basis vectors defining a particular direction in space. It is common enough to read statements such as "the basis vector $\partial_i$" in some sources. This is used in Cartesian as well as curvilinear coordinates, so for example $\partial_r$ or $\partial_\theta$ may be used to indicate basis vectors in spherical coordinates. Perhaps the reader may agree with me that this is a somewhat "forced" language and can be highly confusing. But if you do encounter it in your studies, then now you know what it means.

## 1.5 Cross Products

We separate the discussion on cross products into its own section for a couple of very good reasons. Firstly, the cross product as defined in introductory courses is only correct in three-dimensional space. Once we make the transition to higher dimensions, the cross product can no longer be defined in the way we are used to. Secondly, the cross product in the index notation requires a bit of extra patience, so readers are advised to go through this section only after they are completely comfortable with the index language.

In component form, the cross product is defined by the determinant:[13]

$$\mathbf{C} = \mathbf{A} \times \mathbf{B} = \begin{vmatrix} \hat{\mathbf{e}}_1 & \hat{\mathbf{e}}_2 & \hat{\mathbf{e}}_3 \\ A^1 & A^2 & A^3 \\ B^1 & B^2 & B^3 \end{vmatrix}. \tag{1.89}$$

Or, if the reader is not comfortable with determinants, we can alternatively write:

$$\begin{aligned} \mathbf{C} &= \mathbf{A} \times \mathbf{B} \\ &= \left(A^2 B^3 - A^3 B^2\right) \hat{\mathbf{e}}_1 + \left(A^3 B^1 - B^3 A^1\right) \hat{\mathbf{e}}_2 + \left(A^1 B^2 - A^2 B^1\right) \hat{\mathbf{e}}_3, \end{aligned} \tag{1.90}$$

such that the components of vector $\mathbf{C}$ are

$$\begin{aligned} C^1 &= A^2 B^3 - A^3 B^2 \\ C^2 &= A^3 B^1 - B^3 A^1 \\ C^3 &= A^1 B^2 - A^2 B^1. \end{aligned} \tag{1.91}$$

The right-hand sides of (1.91) have the property of **antisymmetry**. In the first equation, for instance, the $A^3 B^2$ carries a minus sign, whereas $A^2 B^3$ is positive. This means that an exchange of indices $2 \to 3$ and $3 \to 2$ requires a change of sign. Explicitly,

$$A^2 B^3 - A^3 B^2 = - \left(A^3 B^2 - A^2 B^3\right). \tag{1.92}$$

An object with indices that changes sign upon index exchange is called **totally antisymmetric**, as opposed to a totally symmetric object. We have already seen an example of a totally symmetric object: the Kronecker delta. Dot products are inherently symmetric: $\mathbf{A} \cdot \mathbf{B} = \mathbf{B} \cdot \mathbf{A}$, hence the use of a Kronecker delta in their definition makes sense. Cross products, on the other hand are, by their very nature, antisymmetric: $\mathbf{A} \times \mathbf{B} = -\mathbf{B} \times \mathbf{A}$. We conclude, then, that a Kronecker delta *cannot* be used to define the cross product. In addition, we need an object that has more than two indices because each equation of (1.91) requires two indices plus an extra free index to denote which of the three equations (1.91) we mean. This is because cross products are vectors, while dot products are scalars, i.e. have no components and need no free indices. The expression $(\mathbf{A} \cdot \mathbf{B})^i$ has no meaning, while one can easily write $C^i = (\mathbf{A} \times \mathbf{B})^i$ to mean the $i^{\text{th}}$ component of vector $\mathbf{C}$. For this purpose we define the so-called **Levi-Civita totally antisymmetric symbol** as follows:[14]

$$\varepsilon_{ijk} = \hat{\mathbf{e}}_i \cdot (\hat{\mathbf{e}}_j \times \hat{\mathbf{e}}_k) \tag{1.93}$$

---

[13] Although this definition is very common in physics textbooks, mathematicians don't like it much. In linear algebra, the strict definition of a determinant is over numbers, *not* vectors. So technically eqn (1.89) doesn't make sense from a purely mathematical perspective; however, the physicist's attitude is, "Hey, it works."

[14] Tullio Levi-Civita: Italian mathematician (1873–1941).

or explicitly:

$$\varepsilon_{ijk} = \begin{cases} \text{zero} & \text{for } i = j, \text{ or } j = k, \text{ or } i = k \\ +1 & \text{for even permutations} \\ -1 & \text{for odd permutations} \end{cases} \tag{1.94}$$

and  $\varepsilon_{123} = +1$.

This requires a bit of explanation. First note that the symbol $\varepsilon_{ijk}$ carries three indices; how can one visualize such an object in terms of matrices? Well, this is where the matrix analogy breaks down. For a matrix, such as $\delta_{ij}$ as defined in (1.74), each index runs over $1, 2, 3$; hence we have $3 \times 3 = 9$ components. The $\varepsilon_{ijk}$, on the other hand has $3 \times 3 \times 3 = 27$ components, and hence *cannot* be a square matrix. To visualize it, a sort of "cubic matrix" may be invoked; an object with three rows, three columns, and three stacks of rows and columns, like a Rubik's Cube®. But this is not a very useful analogy, so we will not pursue it any further.

The components of $\varepsilon_{ijk}$ are as follows: Define the first component $\varepsilon_{123}$ as $+1$; then the components with even permutations of the indices are also $+1$, or, explicitly,

$$\varepsilon_{123} = \varepsilon_{312} = \varepsilon_{231} = +1. \tag{1.95}$$

The components with odd permutations are

$$\varepsilon_{132} = \varepsilon_{321} = \varepsilon_{213} = -1, \tag{1.96}$$

and finally the "diagonal" components, the ones with at least *any* two indices the same, are vanishing; for example,

$$\varepsilon_{112} = \varepsilon_{222} = \varepsilon_{233} = 0 \ \dots \ \text{etc.} \tag{1.97}$$

This means that even if we keep the indices symbolic, one gets the rule that any exchange of two neighboring symbols picks up a minus sign; for example,

$$\varepsilon_{ijk} = -\varepsilon_{ikj} = -\varepsilon_{jik} = \varepsilon_{jki}, \tag{1.98}$$

and so on. Now let's see how this helps with cross products. Equation (1.90) is then

$$\mathbf{C} = \delta^{ni} \varepsilon_{ijk} A^j B^k \hat{\mathbf{e}}_n, \tag{1.99}$$

or equivalently eqns (1.91) become

$$C^n = \delta^{ni} \varepsilon_{ijk} A^j B^k, \tag{1.100}$$

or, in two steps,

$$C_i = \varepsilon_{ijk} A^j B^k, \qquad C^n = \delta^{ni} C_i. \tag{1.101}$$

The reader is *strongly* encouraged to verify that (1.100) does indeed give (1.91) for the choices $n = 1$, $n = 2$, and $n = 3$. In (1.100), $(i, j, k)$ are all dummy indices following the rules of the summation (three sums here), while $n$ is a free index that must appear once in each term of the equation. In the first equation of (1.101), the $j, k$ indices are dummy and $i$ is free. In the second equation, the index $i$ is dummy while $n$ is free. In eqn (1.100) the Kronecker delta does not contribute to the definition of the cross product; its only purpose is to make sure that all vector components on either side of the equation carry upper indices. This is emphasized in (1.101) by defining the cross product in the first equation, then using the second equation to raise the index on the result.

A particularly important characteristic of $\varepsilon_{ijk}$ is that the number of indices coincides with the number of dimensions we are working in; namely three. So we have three indices and each one of them counts over $1, 2, 3$. This was not true of the Kronecker delta, where it had two indices each still running over $1, 2, 3$. So, if it happens that one needs to work in a two-dimensional plane, where the indices run over 1 and 2 only, then the Kronecker delta is still the same object with two indices, but the Levi-Civita symbol would have to be redefined as the two-index object $\varepsilon_{ij}$, such that $\varepsilon_{12} = 1$, $\varepsilon_{21} = -1$, and $\varepsilon_{11} = \varepsilon_{22} = 0$, which in this case *can* be represented by a matrix:

$$[\varepsilon_{ij}] = \begin{pmatrix} 0 & +1 \\ -1 & 0 \end{pmatrix}. \tag{1.102}$$

A two-dimensional $\varepsilon$ may have its uses every once in a while, but the ability to define a two-dimensional cross product is *not* among them. If one tries to do so, one gets

$$\begin{aligned} f &= \varepsilon_{ij} A^i B^j \qquad i, j = 1, 2 \\ &= A^1 B^2 - A^2 B^1, \end{aligned} \tag{1.103}$$

where $f$ is a scalar *not* a vector, since it carries no indices. If (1.100) represents three equations (the components of the vector **C**), then eqn (1.103) represents a single equation (the value of the scalar $f$). Consider further the case of four dimensions; the straightforward generalization

$$C_{ij} = \varepsilon_{ijkl} A^k B^l \qquad i, j, k, l = 1, \dots, 4 \tag{1.104}$$

also does *not* result in a new vector, but rather gives a two-index matrix, $C_{ij}$. In other words, (1.104) represents 16 equations, one for each component of $C_{ij}$. It is a property specific to three dimensions, and *only* three dimensions, that a "cross product" of two vectors yields another vector! While cross products play an important role in three-dimensional physics, they have little meaning in other dimensions. Having said that, let me re-emphasize: Yes we can define operations such as (1.103) and (1.104), and yes they may have certain uses every once in a while, but they do *not* represent cross products. The operations can exist, but their *interpretation* varies.

The Levi-Civita symbol follows the same rules as any other object with indices. For instance, its indices can be raised and lowered by appropriate contractions with the Kronecker delta (in Cartesian coordinates only). So we can write expressions like $\varepsilon^i_{\ jk} = \delta^{il}\varepsilon_{ljk}$, $\varepsilon_i^{\ j}_{\ k} = \delta^{jl}\varepsilon_{ilk}$, $\varepsilon^{ij}_{\ \ k} = \delta^{il}\delta^{jm}\varepsilon_{lmk}$, and so on. A relation between the Levi-Civita symbol and the Kronecker delta exists in determinant form as follows:

$$\varepsilon_{ijk}\varepsilon^{mnl} = \begin{vmatrix} \delta^m_i & \delta^n_i & \delta^l_i \\ \delta^m_j & \delta^n_j & \delta^l_j \\ \delta^m_k & \delta^n_k & \delta^l_k \end{vmatrix}. \tag{1.105}$$

Close inspection of (1.105) reveals the pattern: The upper indices correspond to the columns, while the lower indices correspond to the rows. To expand this,

$$\varepsilon_{ijk}\varepsilon^{mnl} = \delta^m_i\left(\delta^n_j\delta^l_k - \delta^l_j\delta^n_k\right) - \delta^n_i\left(\delta^m_j\delta^l_k - \delta^l_j\delta^m_k\right) + \delta^l_i\left(\delta^m_j\delta^n_k - \delta^n_j\delta^m_k\right). \tag{1.106}$$

Using (1.106), one can deduce the following useful identities:

$$\varepsilon_{ijk}\varepsilon^{mnk} = \left(\delta^m_i\delta^n_j - \delta^n_i\delta^m_j\right)$$
$$\varepsilon_{ijk}\varepsilon^{mjk} = 2\delta^m_i$$
$$\varepsilon_{ijk}\varepsilon^{ijk} = 6. \tag{1.107}$$

**Exercise 1.14**

1. Verify that (1.99), (1.100), and (1.101) give the correct expressions for the cross product.
2. Verify (1.106) component by component. In other words, show that the left-hand side and the right-hand side give the same numerical values for any combination of the free indices.
3. Starting with (1.106), proceed to derive the identities (1.107) by setting the appropriate indices equal to each other on both sides.
4. Guess the formula equivalent to (1.106) for the case of the two-dimensional Levi-Civita symbol defined by (1.102). Check your guess component by component. Derive the identities similar to (1.107). Are you starting to see a pattern?

An important theorem that follows from the difference in properties between symmetric and antisymmetric objects goes as follows: The full contraction of a symmetric object with an antisymmetric object *always* vanishes. The proof goes like this: Suppose we contract a symmetric matrix $A^{ij}$ with a fully antisymmetric matrix $B_{ij}$, i.e. we want to calculate the quantity $A^{ij}B_{ij}$. Now the very properties of symmetry and antisymmetry allow us to write

$$A^{ij} = A^{ji}, \qquad B_{ij} = -B_{ji}, \tag{1.108}$$

and hence

$$A^{ij}B_{ij} = -A^{ji}B_{ji}, \tag{1.109}$$

but both indices are dummy and may be renamed, so we rename $i \to j$ and $j \to i$ on the right-hand side to find that

$$A^{ij} B_{ij} = -A^{ij} B_{ij}. \tag{1.110}$$

In other words, the quantity $A^{ij} B_{ij}$ is equal to negative itself, but this can only be true if and only if the quantity itself is zero $A^{ij} B_{ij} = 0$; hence the contraction of symmetric and antisymmetric objects vanishes identically. One can then show that it is always true that $\varepsilon_{ijk} \delta^{jk} = 0$, since $\varepsilon$ is antisymmetric in any pair of indices. If one writes such a calculation explicitly, for example as

$$\varepsilon_{ijk} \delta^{jk} = \sum_{j=1}^{3} \sum_{k=1}^{3} \varepsilon_{i11} \delta^{11} + \varepsilon_{i12} \delta^{12} + \ldots = 0, \tag{1.111}$$

it is easy to see that no terms can possibly have simultaneously non-vanishing $\varepsilon$ and $\delta$. If $j = k$ then $\varepsilon$ vanishes, if $j \neq k$ it is $\delta$ that vanishes, and so on.

Generally speaking, any matrix $A^{ij}$ can be decomposed into a sum of two matrices— one totally symmetric and the other totally antisymmetric—thus

$$A^{ij} = A^{(ij)} + A^{[ij]}, \tag{1.112}$$

where the parentheses ( ) and [ ] are standard notation for totally symmetric and totally antisymmetric indices respectively; i.e. $A^{(ij)} = A^{(ji)}$, while $A^{[ij]} = -A^{[ji]}$. The inverse relations can be derived as follows:

$$A^{(ij)} = \frac{1}{2} \left( A^{ij} + A^{ji} \right)$$
$$A^{[ij]} = \frac{1}{2} \left( A^{ij} - A^{ji} \right). \tag{1.113}$$

**Exercise 1.15** Derive (1.113) from (1.112).

**Exercise 1.16** Decompose the matrix

$$[A] = \begin{bmatrix} 1 & 1 & 2 \\ -3 & 2 & -3 \\ 4 & 3 & 3 \end{bmatrix} \tag{1.114}$$

into the sum of two matrices, one totally symmetric and the other totally antisymmetric.

**Exercise 1.17** In the following expressions, the components of a matrix $B$ are given in terms of $A$. Invert these relations; in other words, find the components of $A$ in terms of $B$. *Hint*: Use the identities (1.107).

1. $\varepsilon_{ijk} A^k = B_{ij}$.
2. $\varepsilon_{ijk} A^{jk} = B_i$, if $A^{jk}$ is totally antisymmetric.

3. $\varepsilon_{ijk} A^{jk} = B_i$, if $A^{jk}$ is totally symmetric.
4. $\varepsilon_{ijk} A^{jk} = B_i$, if $A^{jk}$ is *neither* totally symmetric nor totally antisymmetric.
5. $\delta^{ik} \varepsilon_{ijn} A^n = B_j^k$.

Finally, and for completeness, we use the language developed in this section and the previous one to define the curl of a vector. Recall that in Cartesian coordinates this is:

$$\mathbf{U} = \vec{\nabla} \times \mathbf{V} = \left( \frac{\partial V_3}{\partial x^2} - \frac{\partial V_2}{\partial x^3} \right) \hat{\mathbf{e}}_1 + \left( \frac{\partial V_1}{\partial x^3} - \frac{\partial V_3}{\partial x^1} \right) \hat{\mathbf{e}}_2 + \left( \frac{\partial V_2}{\partial x^1} - \frac{\partial V_1}{\partial x^2} \right) \hat{\mathbf{e}}_3. \quad (1.115)$$

The reader can verify that this can be written most compactly as

$$\mathbf{U} = \vec{\nabla} \times \mathbf{V} = \varepsilon^{ij}{}_k \partial_j V^k \hat{\mathbf{e}}_i, \quad (1.116)$$

or in components as

$$U^i = \left( \vec{\nabla} \times \mathbf{V} \right)^i = \varepsilon^{ij}{}_k \partial_j V^k. \quad (1.117)$$

Note how we used the Kronecker delta to raise and lower indices appropriately. One can easily see that the following expressions are all equivalent:

$$U^i = \varepsilon^i{}_{jk} \partial^j V^k = \varepsilon^i{}_j{}^k \partial^j V_k = \varepsilon^{ijk} \partial_j V_k$$
$$= -\varepsilon^i{}_{kj} \partial^j V^k = -\varepsilon^{ik}{}_j \partial^j V_k = -\varepsilon^{jik} \partial_j V_k. \quad (1.118)$$

**Exercise 1.18** Write Maxwell's eqns (4.168), (4.169), (4.170), and (4.171) in the index notation. You may use the additional definition $\partial_t = \partial/\partial t$.

**Exercise 1.19** Prove the following identities using the properties of $\varepsilon$ and $\delta$. Unless otherwise specified, all vectors and scalars are arbitrary.

1. $\vec{\nabla} \times \mathbf{r} = 0$, where $\mathbf{r}$ is the position vector (1.46).
2. $\mathbf{A} \cdot (\mathbf{B} \times \mathbf{C}) = \mathbf{C} \cdot (\mathbf{A} \times \mathbf{B}) = \mathbf{B} \cdot (\mathbf{C} \times \mathbf{A})$.
3. $\mathbf{A} \times (\mathbf{B} \times \mathbf{C}) = \mathbf{B} (\mathbf{A} \cdot \mathbf{C}) - \mathbf{C} (\mathbf{A} \cdot \mathbf{B})$. This identity is popularly known, for obvious reasons, as the "BAC–CAB" rule.
4. $(\mathbf{A} \times \mathbf{B}) \cdot (\mathbf{C} \times \mathbf{D}) = (\mathbf{A} \cdot \mathbf{C}) (\mathbf{B} \cdot \mathbf{D}) - (\mathbf{A} \cdot \mathbf{D}) (\mathbf{B} \cdot \mathbf{C})$.
5. $\vec{\nabla} \times \vec{\nabla} f = 0$.
6. $\vec{\nabla} \cdot \left( \vec{\nabla} \times \mathbf{A} \right) = 0$.
7. $\vec{\nabla} \cdot (\mathbf{A} \times \mathbf{B}) = \mathbf{B} \cdot \left( \vec{\nabla} \times \mathbf{A} \right) - \mathbf{A} \cdot \left( \vec{\nabla} \times \mathbf{B} \right)$.

## 1.6   Vectors in Curvilinear Coordinates

We now turn our attention, at long last, to studying vector components using the index notation in coordinate systems other than Cartesian. We begin by considering how basis vectors arise in such systems. The strict definition of basis vectors is that they are vectors that point in the direction of *changing* coordinates. As such, they are necessarily orthogonal in orthogonal coordinate systems. In Cartesian coordinates, our old friends $(\hat{\mathbf{x}}, \hat{\mathbf{y}}, \hat{\mathbf{z}})$, also known as $(\hat{\mathbf{e}}_1, \hat{\mathbf{e}}_2, \hat{\mathbf{e}}_3)$, are pointing in the directions of changing $x^1$, $x^2$, and $x^3$ coordinates respectively. Notice that the definition does *not* necessarily require the basis vectors to be unit vectors! We could have just as easily defined $(\hat{\mathbf{e}}_1, \hat{\mathbf{e}}_2, \hat{\mathbf{e}}_3)$ to be non-unity, or even non-constant—changing in magnitude as they get further and further from the origin. Had we done that, we may have had to change a few later formulae and vector components would have had different values, but the new basis vectors would still be useful and should still have led to the same physical results. Unit basis vectors are sometimes called the **physical basis**, while non-unit basis vectors can go by the name of the **natural basis**. In most of the physics encountered in an undergraduate curriculum the physical basis is used, due to its simplicity and ease of visualization. Usually all one needs to do is draw a sketch of the coordinate system desired and plot the unit vectors right on it, then use ordinary trigonometry to figure out their explicit form.

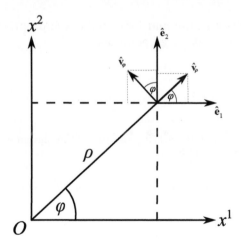

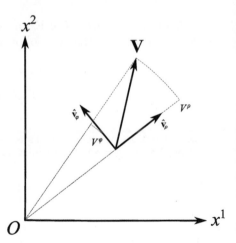

(a) Relating the polar basis vectors to the more familiar Cartesian basis.

(b) The components of a vector **V** in polar basis.

Fig. 1.15: Planar polar coordinates.

Consider the example of planar polar coordinates shown in Fig. [1.15a]. The physical basis vectors $\hat{\mathbf{v}}_\rho$ and $\hat{\mathbf{v}}_\varphi$ are related to the Cartesian ones as follows:

$$\hat{\mathbf{v}}_\rho = \cos\varphi\hat{\mathbf{x}} + \sin\varphi\hat{\mathbf{y}}$$
$$\hat{\mathbf{v}}_\varphi = -\sin\varphi\hat{\mathbf{x}} + \cos\varphi\hat{\mathbf{y}}, \tag{1.119}$$

as can be verified by checking the trigonometry. The reader can easily confirm that these are indeed physical basis vectors in the sense that they have unit magnitudes: $\hat{\mathbf{v}}_\rho \cdot \hat{\mathbf{v}}_\rho = \hat{\mathbf{v}}_\varphi \cdot \hat{\mathbf{v}}_\varphi = 1$. They are also orthogonal: $\hat{\mathbf{v}}_\rho \cdot \hat{\mathbf{v}}_\varphi = 0$. Hence a vector $\mathbf{V}$ can now be written in polar components as follows (Fig. [1.15b])

$$\mathbf{V} = V^\rho \hat{\mathbf{v}}_\rho + V^\varphi \hat{\mathbf{v}}_\varphi. \tag{1.120}$$

**Exercise 1.20** Explicitly verify that the polar physical basis vectors (1.119) satisfy $\hat{\mathbf{v}}_\rho \cdot \hat{\mathbf{v}}_\rho = \hat{\mathbf{v}}_\varphi \cdot \hat{\mathbf{v}}_\varphi = 1$ and $\hat{\mathbf{v}}_\rho \cdot \hat{\mathbf{v}}_\varphi = 0$.

On the other hand, one can define the natural basis vectors $\hat{\mathbf{g}}_\rho$ and $\hat{\mathbf{g}}_\varphi$ that are still orthogonal $\hat{\mathbf{g}}_\rho \cdot \hat{\mathbf{g}}_\varphi = 0$ but do not necessarily have unit magnitudes: $\hat{\mathbf{g}}_\rho \cdot \hat{\mathbf{g}}_\rho \neq 1$ and $\hat{\mathbf{g}}_\varphi \cdot \hat{\mathbf{g}}_\varphi \neq 1$. Visually, they would look the same as $\hat{\mathbf{v}}_\rho$ and $\hat{\mathbf{v}}_\varphi$ do in Fig. [1.15] and a vector $\mathbf{V}$ can still be written as

$$\mathbf{V} = V^{\rho'} \hat{\mathbf{g}}_\rho + V^{\varphi'} \hat{\mathbf{g}}_\varphi, \tag{1.121}$$

except that its components $V^{\rho'}$ and $V^{\varphi'}$ are not necessarily the same as $V^\rho$ and $V^\varphi$. All of this reminds us that while a vector is a unique quantity, its components are not; their values highly depend not only on the choice of coordinate system but also on the choice of basis. Note that the natural and physical basis vectors must be related by

$$\frac{\hat{\mathbf{g}}_i}{|\hat{\mathbf{g}}_i|} = \hat{\mathbf{v}}_i, \qquad \text{(no sum)} \tag{1.122}$$

since they share the same directions. In this last formula, there is no summation over repeated indices; the formula simply means $\frac{\hat{\mathbf{g}}_1}{|\hat{\mathbf{g}}_1|} = \hat{\mathbf{v}}_1$ (and similarly for $i = 2$ and $i = 3$). The magnitude of $\hat{\mathbf{g}}_i$ is found, of course, via $|\hat{\mathbf{g}}_i| = \sqrt{\hat{\mathbf{g}}_i \cdot \hat{\mathbf{g}}_i}$ (no sum).

We now ask: is there a way to analytically derive the basis vectors in terms of the Cartesian basis without having to draw diagrams and do trigonometry? This is especially important when we start discussing coordinate systems that are more complex than polar coordinates. Our intention in this book is to relate everything to the metric, so let us ask how the basis vectors of *any* given coordinate system are related to the metric in that system. In Cartesian coordinates, this can be easily seen as follows: Consider the Cartesian position vector (1.46). Next, if one assumes that the object toward which $\mathbf{r}$ is pointing is displaced over an infinitesimal distance, one can define the so-called **displacement vector** as the differential change of the position vector, so

$$d\mathbf{r} = d\left(x^i \hat{\mathbf{e}}_i\right) = x^i d\hat{\mathbf{e}}_i + dx^i \hat{\mathbf{e}}_i = dx^i \hat{\mathbf{e}}_i, \tag{1.123}$$

where we have used the fact that the Cartesian basis vectors are constant, i.e. $d\hat{\mathbf{e}}_i = 0$. Next, take the dot product of the displacement vector with itself:

$$d\mathbf{r} \cdot d\mathbf{r} = |d\mathbf{r}|^2 = \left(dx^i \hat{\mathbf{e}}_i\right) \cdot \left(dx^j \hat{\mathbf{e}}_j\right) = \left(\hat{\mathbf{e}}_i \cdot \hat{\mathbf{e}}_j\right) dx^i dx^j = \delta_{ij} dx^i dx^j = dl^2, \tag{1.124}$$

which leads to *exactly* the Cartesian metric. This is clearly no surprise, since a change in position, i.e. a displacement, is exactly what $dl$ is. In other words, $dl$ is the *magnitude* of the displacement vector: $dl = |d\mathbf{r}|$. This relates the basis vectors directly to metric.

Now, can we generate a similar procedure in any *other* coordinate system? In other words, can we define the basis vectors (whether natural or physical) such that the dot product of the displacement vector in *that* system with itself gives the correct metric? So for an arbitrary system of coordinates $(u^1, u^2, u^3)$ with physical basis vectors $\hat{\mathbf{v}}_i$ and natural basis vectors $\hat{\mathbf{g}}_i$, can we write

$$(\hat{\mathbf{v}}_i \cdot \hat{\mathbf{v}}_j)\, du^i du^j = (\hat{\mathbf{g}}_i \cdot \hat{\mathbf{g}}_j)\, du'^i du'^j = dl^2, \tag{1.125}$$

as in (1.124)? Starting with the knowledge of how the new coordinates are related to the Cartesian coordinates $x^1 \left(u^1, u^2, u^3\right)$, $x^2 \left(u^1, u^2, u^3\right)$, and $x^3 \left(u^1, u^2, u^3\right)$, or just $x^i \left(u^j\right)$, let us substitute in (1.46). The position vector is now

$$\begin{aligned}
\mathbf{r} &= x^1 \left(u^1, u^2, u^3\right) \hat{\mathbf{e}}_1 + x^2 \left(u^1, u^2, u^3\right) \hat{\mathbf{e}}_2 + x^3 \left(u^1, u^2, u^3\right) \hat{\mathbf{e}}_3 \\
&= x^i \left(u^j\right) \hat{\mathbf{e}}_i.
\end{aligned} \tag{1.126}$$

Plugging this into (1.123) and using the chain rule,

$$\begin{aligned}
d\mathbf{r} &= dx^i \hat{\mathbf{e}}_i \\
&= \left(\frac{\partial x^i}{\partial u^j}\right) du^j \hat{\mathbf{e}}_i \\
&= \left(\frac{\partial}{\partial u^j} x^i \hat{\mathbf{e}}_i\right) du^j \\
&= \left(\frac{\partial \mathbf{r}}{\partial u^j}\right) du^j.
\end{aligned} \tag{1.127}$$

Hence the dot product of the displacement vector with itself is

$$|d\mathbf{r}|^2 = \left(\frac{\partial \mathbf{r}}{\partial u^i}\right) \cdot \left(\frac{\partial \mathbf{r}}{\partial u^j}\right) du^i du^j. \tag{1.128}$$

Comparing with (1.125), it is clear that $\left(\frac{\partial \mathbf{r}}{\partial u^i}\right) \cdot \left(\frac{\partial \mathbf{r}}{\partial u^j}\right)$ in the new coordinate system plays the role of $(\hat{\mathbf{e}}_i \cdot \hat{\mathbf{e}}_j)$ in Cartesian coordinates, so we define the natural basis vectors in the $\left(u^1, u^2, u^3\right)$ coordinate system by

$$\hat{\mathbf{g}}_i = \frac{\partial \mathbf{r}}{\partial u^i}. \tag{1.129}$$

As in (1.54) we further define

$$g_{ij} = \hat{\mathbf{g}}_i \cdot \hat{\mathbf{g}}_j, \tag{1.130}$$

such that (1.128) becomes

$$dl^2 = g_{ij}\, du^i du^j, \tag{1.131}$$

as intended. The reader can easily verify that in orthogonal coordinates the components $g_{ij}$, where $i = j$, are just the scale factors squared, as defined by (1.30), so

$g_{11} = h_1^2$, $g_{22} = h_2^2$, and $g_{33} = h_3^2$, while $g_{12} = g_{23} = \cdots = 0$. Clearly this new matrix $g_{ij}$ plays the *same* role in a given curvilinear coordinate system that the Kronecker delta plays in Cartesian coordinates. It is also a diagonal square matrix, i.e.

$$[g] = \begin{pmatrix} h_1^2 & 0 & 0 \\ 0 & h_2^2 & 0 \\ 0 & 0 & h_3^2 \end{pmatrix}. \tag{1.132}$$

For Cartesian coordinates, $[g]$ simply reduces to $[\delta]$ as the scale factors become ones. At this point let us note that it is possible to choose a coordinate system with "off-diagonal" components, i.e. non-vanishing $g_{12}$, $g_{32}$, etc. Contrary to popular belief among beginners in this subject, this is not necessarily a sign of the curvature of space. It is a sign, however, of the non-orthogonality of the coordinate system. In other words, it is possible to choose a coordinate system where the basis vectors are not perpendicular to each other such that $g_{ij} = \hat{\mathbf{g}}_i \cdot \hat{\mathbf{g}}_j \neq 0$ for $i \neq j$. As briefly mentioned earlier, this is known as a skew coordinate system.

Finally, if one wants to find a physical system of basis vectors, one can just apply (1.122). And so there we have it: an analytical method of calculating both natural and physical basis vectors simply from the knowledge of how coordinate systems are related to each other. In the process, we defined the quantity $g_{ij}$, which will eventually play a major role in our quest for covariance. In relativistic physics, it is traditional to work with the natural basis, and this is what we will do (unless explicitly mentioned, all vectors shall be written in the natural basis). This, as we will see later, may lead to some minor differences between formulae used in this book and others used in other textbooks, particularly books on mechanics and electromagnetism. One then should be careful when applying formulae across textbooks.

As an example, let's find the natural basis vectors for cylindrical polar coordinates and verify that we get the correct metric components. Substituting (1.1) in (1.126) gives

$$\mathbf{r} = (\rho \cos \varphi)\, \hat{\mathbf{e}}_1 + (\rho \sin \varphi)\, \hat{\mathbf{e}}_2 + z\hat{\mathbf{e}}_3 \tag{1.133}$$

and applying (1.129) gives

$$\hat{\mathbf{g}}_\rho = \frac{\partial \mathbf{r}}{\partial \rho} = \cos \varphi \hat{\mathbf{e}}_1 + \sin \varphi \hat{\mathbf{e}}_2$$

$$\hat{\mathbf{g}}_\varphi = \frac{\partial \mathbf{r}}{\partial \varphi} = -(\rho \sin \varphi)\, \hat{\mathbf{e}}_1 + (\rho \cos \varphi)\, \hat{\mathbf{e}}_2$$

$$\hat{\mathbf{g}}_z = \frac{\partial \mathbf{r}}{\partial z} = \hat{\mathbf{e}}_3. \tag{1.134}$$

Finally, (1.130) gives $g_{11} = g_{\rho\rho} = \hat{\mathbf{g}}_\rho \cdot \hat{\mathbf{g}}_\rho = 1$, $g_{22} = g_{\varphi\varphi} = \hat{\mathbf{g}}_\varphi \cdot \hat{\mathbf{g}}_\varphi = \rho^2$, and $g_{33} = g_{zz} = \hat{\mathbf{g}}_z \cdot \hat{\mathbf{g}}_z = 1$, which immediately lead to the correct metric (1.26) via (1.131).

**Exercise 1.21** Derive the natural basis vectors $\hat{\mathbf{g}}_i$ for the following coordinate systems. Then use (1.122) to find their physical basis vectors $\hat{\mathbf{v}}_i$.

1. Spherical coordinates (1.3).
2. Elliptic cylindrical coordinates (1.6).
3. Parabolic cylindrical coordinates (1.7).
4. Parabolic coordinates (1.8).
5. Paraboloidal coordinates (1.9).
6. Oblate spheroidal coordinates (1.10).
7. Toroidal coordinates (1.11).
8. The skew coordinates of Exercise 1.2.

Now that we have a recipe for basis vectors in any coordinate system, let us continue with our original objective of defining vectors in a given curvilinear coordinate system. As shown in (1.42), we can write

$$\mathbf{V} = V^i \hat{\mathbf{g}}_i, \quad \mathbf{U} = U^j \hat{\mathbf{g}}_j, \quad i = 1, 2, 3. \tag{1.135}$$

Note that the numbers $1, 2, 3$ no longer denote the Cartesian coordinates $(x^1, x^2, x^3)$. They now represent the set of curvilinear coordinates defined by the $\hat{\mathbf{g}}$'s; for example, in spherical coordinates, $1 \to r$, $2 \to \varphi$, and $3 \to \theta$. The dot product between $\mathbf{V}$ and $\mathbf{U}$ is a straightforward generalization of (1.51),

$$\mathbf{V} \cdot \mathbf{U} = V^i U^j \left( \hat{\mathbf{g}}_i \cdot \hat{\mathbf{g}}_j \right) = V^i U^j g_{ij}, \tag{1.136}$$

where we have used (1.130). There is a remaining notational issue: In many textbooks, the arbitrary coordinate $u^i$ in (1.131), which may be a distance or an angle, is usually replaced with $x^i$ or $z^i$. Although this is a potentially confusing notation, it is quite standard in the literature, so from this point onward we will adopt this language even when non-Cartesian coordinates are used. For example, the following expression is the most general form of the metric:

$$dl^2 = g_{ij} dx^i dx^j, \tag{1.137}$$

where $x^i$ and $x^j$ may be Cartesian, cylindrical, spherical, or any other system of coordinates. The components $g_{ij}$ are what decides which coordinate system we are in. So if $g_{ij} = \delta_{ij}$, then the $x$'s are the usual Cartesian coordinates, and so on. Since our intention is to write everything in a coordinate-invariant way, all subsequent formulae, unless specifically mentioned, will have $x^i$ to mean *any* system of coordinates.

Generally speaking, the rules of the index notation apply to $g_{ij}$ in the same way they have applied to $\delta_{ij}$. For example, the rules of raising and lowering of indices (1.60) and (1.61) are now upgraded to

$$U^j g_{ij} = U_i \quad \text{or} \quad V^i g_{ij} = V_j$$
$$U_j g^{ij} = U^i \quad \text{or} \quad V_i g^{ij} = V^j \qquad (1.138)$$

and the dot product may take any of these alternative, but totally equivalent, forms:

$$\mathbf{V} \cdot \mathbf{U} = V^i U^j g_{ij} = V_i U_j g^{ij} = V^i U_j \delta^j_i = V^i U_i = V_i U^i. \qquad (1.139)$$

(Keep reading to find out why $\delta^j_i$ is still used, rather than something like $g^j_i$.) Notice that we can no longer assume that $U^i \equiv U_i$, as was the case in Cartesian coordinates, as the metric is no longer a unit matrix. Hence the distinction between upper- and lower-index vectors and other quantities becomes essential. The quantity known as the "vector" which we are all familiar with, is the one defined here with an upper index (sometimes referred to as "contravariant"). That other vector with a lower index (covariant or dual) is a new entity that we will have to get used to. Upper- and lower-index vectors are defined in this way in order to give the correct answer as far as dot products of vectors are concerned. In fact, as the reader may have already concluded, it *is* the dot product, in more ways than one, that *defines* what a vector is—not really the other way around.

As a simple example, consider a vector $\mathbf{V} = 5\hat{\mathbf{g}}_\rho + \pi\hat{\mathbf{g}}_\varphi + 2\hat{\mathbf{g}}_z$ in cylindrical polar coordinates. Since it is clear that the upper index components are $V^\rho = 5$, $V^\varphi = \pi$, and $V^z = 2$, the lower index components are $V_\rho = V^\rho g_{\rho\rho} = 5$, $V_\varphi = V^\varphi g_{\varphi\varphi} = \pi\rho^2$, and $V_z = V^z g_{zz} = 2$. This leads to the norm $V^i V_i = V^\rho V_\rho + V^\varphi V_\varphi + V^z V_z = 25 + \pi^2\rho^2 + 4 = 29 + \pi^2\rho^2$. Hence at a specific distance from the origin, say $\rho = 2$, this vector's norm is $29 + 4\pi^2 = 68.478$.

**Exercise 1.22** Given the vector $\mathbf{V} = 2\hat{\mathbf{g}}_r - \hat{\mathbf{g}}_\varphi + 3\pi\hat{\mathbf{g}}_\theta$ in spherical coordinates, find its lower index components and calculate the vector's norm at the point $(4, \pi/2, \pi/3)$.

One thing is left to discuss: What is the interpretation of the quantity $g^{ij}$? Let's figure that out. Take one of the eqns in (1.138), say $U^j g_{ij} = U_i$. Contract both sides with $g^{ik}$:

$$U^j g_{ij} g^{ik} = U_i g^{ik}$$
$$= U^k. \qquad (1.140)$$

But we also know from (1.63) that $U^j \delta^k_j = U^k$. The inevitable conclusion is

$$g^{ik} g_{kj} = \delta^i_j, \qquad (1.141)$$

known simply as the **orthogonality condition** because of its similarity to the behavior of orthogonal vectors (1.54). Notice that (1.141), when interpreted as a matrix multiplication, means that $g^{ik}$ multiplied with $g_{kj}$ gives the identity matrix! In other words, $g^{ij}$ is simply the **inverse matrix** of $g_{ij}$. Since $g_{ij}$ as defined by (1.132) is a diagonal matrix for orthogonal coordinates, its inverse $g^{ij}$ in matrix form has the components

$$[g]^{-1} = \begin{pmatrix} \frac{1}{h_1^2} & 0 & 0 \\ 0 & \frac{1}{h_2^2} & 0 \\ 0 & 0 & \frac{1}{h_3^2} \end{pmatrix}, \tag{1.142}$$

such that the matrix form of (1.141) is

$$[g]^{-1}[g] = \begin{pmatrix} \frac{1}{h_1^2} & 0 & 0 \\ 0 & \frac{1}{h_2^2} & 0 \\ 0 & 0 & \frac{1}{h_3^2} \end{pmatrix} \begin{pmatrix} h_1^2 & 0 & 0 \\ 0 & h_2^2 & 0 \\ 0 & 0 & h_3^2 \end{pmatrix} = \begin{pmatrix} 1 & 0 & 0 \\ 0 & 1 & 0 \\ 0 & 0 & 1 \end{pmatrix}. \tag{1.143}$$

Since the quantities $g_{ij}$ are the coefficients of a given metric, it is quite common in the literature to call $g_{ij}$ the **metric tensor** and $g^{ij}$ the **inverse metric tensor**, where the term "tensor" describes a new mathematical entity that we will define in the next chapter. Technically the $g_{ij}$'s are components of the metric tensor, just as the $V^i$'s are components of a given vector. The reader may have encountered tensors in their studies of classical mechanics or electricity and magnetism, but a strict definition of what makes a specific collection of numbers a "tensor" will have to wait until we have answered another important question: What makes a specific set of numbers a "vector?"

**Exercise 1.23** Using your results for Exercise 1.21, find the components of the metric tensor $g_{ij}$ and its inverse $g^{ij}$ for each case and explicitly verify orthogonality.

**Exercise 1.24** Rewrite the following expressions as matrix multiplications. Check you got it right by assuming numerical values for the components. The given matrices may or may not, strictly speaking, be "tensors," as we will define them in the next chapter, and as such may or may not have any physical meanings. But this is acceptable for the time being, since our interest in this exercise and the next is to just practice the rules of indices.

1. $D_{ij} = A_{ik}B^{kn}C_{jn}$.
2. $D_{ij} = A_{ik}B_j{}^n C_n^k$.

**Exercise 1.25** For the following, find the components of $A$ in terms of the given arbitrary quantities, where $g$ is the metric or inverse metric.

1. $g^{ij}A^k{}_i = B^{kj}$.
2. $A^{ij}g_{jk} = B_{kn}g^{ni}$.
3. $g^{ki}g^{nj}A_{ij} = B^{kn}$.
4. $g_{im}A_n{}^k g^{nm} = B_{im}C^{km}$.

# 2

# Tensors

*Ultra-modern physicists [are tempted to believe] that Nature in all her infinite variety needs nothing but mathematical clothing [and are] strangely reluctant to contemplate Nature unclad. Clothing she must have. At the least she must wear a matrix, with here and there a tensor to hold the queer garment together.*

Sydney Evershed

## Introduction

The mathematical concept of "vector" describes processes that have cause and effect occurring along the same straight line. But there exist physical phenomena where this is not true; the "cause" in these cases may result in several effects acting along *different* directions. To understand what this means consider the following example. It is well known that the behavior of simple elastic materials such as springs or metal rods follows **Hooke's law**,[1]

$$\mathbf{F} = -k\mathbf{x}, \tag{2.1}$$

where $\mathbf{F}$ is the force applied along the direction of the spring's axis, causing a contraction or expansion $\mathbf{x}$, and $k$ is a proportionality constant known as the stiffness or the spring's constant. Now the equation is balanced in both magnitude *and* direction. What this means is that the direction of the vector $\mathbf{F}$ on the left-hand side is along the same straight line as that of the vector $\mathbf{x}$ on the right-hand side; the spring contracts or expands along the same straight line as the force, as shown in Fig. [2.1a]. This is true only in the simplest of elastic materials. In contrast, consider a rectangular-shaped eraser made of some elastic material held between the thumb and forefinger. Now squeeze your fingers together, as shown in Fig. [2.1b]. The force you are exerting acts along the straight line connecting your fingers; let's call this the "line of action." However, the response of the eraser is multidirectional; while the eraser does shrink between your fingers along the line of action just as a spring would, it also expands a bit in a plane *perpendicular* to the line of action. In other words, the effect acts along several directions simultaneously. Clearly a phenomenon such as this cannot be explained by (2.1); in fact, it cannot be explained by any formula involving vectors

---

[1]Robert Hooke: an English scientist and a contemporary of Newton's (1635–1703).

*Covariant Physics: From Classical Mechanics to General Relativity and Beyond*. Moataz H. Emam, Oxford University Press (2021). © Moataz H. Emam.
DOI: 10.1093/oso/9780198864899.003.0002

alone, since there is no way of writing a vector equation that has different directions on either side. An entirely new mathematical concept, allowing for a different type of balance between the left- and right-hand sides, is needed. This is the concept of the **tensor**: a mathematical representation of physical phenomena that allows for causes and effects to take on different directions. Mathematically a tensor is defined in terms of how it changes from one coordinate system to another. In order to understand this, we have to ask the same question about vectors first, and then we will see that vectors turn out to be simply a special case of tensors.

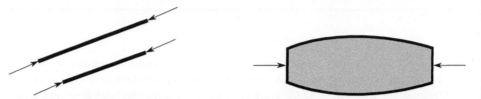

(a) Hooke's law: A rod under compression contracts along the force's line of action.

(b) The multidirectional response of more complex elastic materials.

Fig. 2.1: The failure of vectors to represent multidirectional responses.

## 2.1  What's a Vector, Really?

It might sound strange that after all the work we did in the first chapter we are still asking a question such as this. But it does pay to look at the basics every once in a while. Since tensors are generalizations of vectors, one needs to first understand in more detail what constitutes a vector. Let us begin, even more elementarily, by defining what a scalar is. A scalar is a quantity that is invariant under both translations and rotations. What does this mean? Suppose you have a position-dependent function, $M(\mathbf{r})$, representing, say, the distribution of mass of some rigid object and defined at each spatial point $\mathbf{r}$ in some given coordinate system. Now translating the coordinate system parallel to itself and/or rotating it with a specific angle does not affect the *value* of $M(\mathbf{r})$ in each point in space, even when the coordinates of said points themselves change, as of course they will. Since a rotation and/or translation of coordinates is exactly what happens when one goes from one coordinate system to another, we conclude that we can define a scalar as a quantity completely invariant under any change of coordinate system. In other words, scalars are coordinate covariant (or just covariant for short). A vector, on the other hand, could be affected by translations, rotations, or a combination thereof. Although its magnitude, being a scalar, stays constant, its direction does not. An observer $O'$ looking at an angle with respect to another observer $O$ will see a vector pointing in a *different* direction than that observed by $O$. In other words, my right is not the same as your right. This intuitive concept can be used to rigorously define what a vector is; in other words, we will use it to decide what makes a specific set of numbers a vector.

Let us study this notion in more detail. Focusing on rotations, consider the two **frames of reference** (another word for coordinate system) shown in Fig. [2.2] rotated with respect to each other by an arbitrary angle $\theta$. A given point in space $P$, has coordinates $x^1$ and $x^2$, as designated by observer $O$, but has coordinates $x^{1'}$ and $x^{2'}$, as seen by observer $O'$.

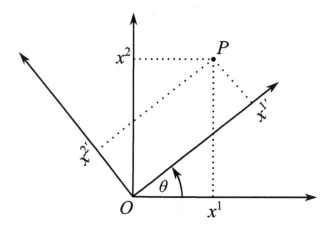

Fig. 2.2: Rotated frames of reference.

It is an exercise in trigonometry to show that the two frames of reference are related by[2]

$$x^{1'} = x^1 \cos\theta + x^2 \sin\theta$$
$$x^{2'} = -x^1 \sin\theta + x^2 \cos\theta, \tag{2.2}$$

which may be expressed in matrix notation by:

$$\begin{pmatrix} x^{1'} \\ x^{2'} \end{pmatrix} = \begin{pmatrix} \cos\theta & \sin\theta \\ -\sin\theta & \cos\theta \end{pmatrix} \begin{pmatrix} x^1 \\ x^2 \end{pmatrix}. \tag{2.3}$$

The converse is also true:

$$x^1 = x^{1'} \cos\theta - x^{2'} \sin\theta$$
$$x^2 = x^{1'} \sin\theta + x^{2'} \cos\theta \tag{2.4}$$

or

$$\begin{pmatrix} x^1 \\ x^2 \end{pmatrix} = \begin{pmatrix} \cos\theta & -\sin\theta \\ \sin\theta & \cos\theta \end{pmatrix} \begin{pmatrix} x^{1'} \\ x^{2'} \end{pmatrix}, \tag{2.5}$$

where we note that the rotation matrix in (2.5) is both the transpose *and* the inverse of the matrix in (2.3).

---

[2]Primes over the indices do *not* denote differentiation; they denote a different coordinate system.

---

**Exercise 2.1**

1. Derive eqns (2.2) by exploring the trigonometry of Fig. [2.2].
2. Explicitly demonstrate that the rotation matrix in (2.5) is the inverse of the matrix in (2.3).

---

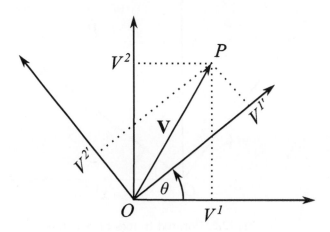

Fig. 2.3: Rotated vectors.

Now consider a vector **V**. In components the vector may be expressed in either of the two reference frames, as follows:

$$\mathbf{V} = V^i \hat{\mathbf{g}}_i = V^{i'} \hat{\mathbf{g}}_{i'} \qquad i = 1, 2. \tag{2.6}$$

The components of the vector can be written in terms of each other in the same way as in (2.2),

$$\begin{aligned} V^{1'} &= V^1 \cos\theta + V^2 \sin\theta \\ V^{2'} &= -V^1 \sin\theta + V^2 \cos\theta, \end{aligned} \tag{2.7}$$

as shown in Fig. [2.3]. To write (2.7) in the index notation, let us define a new quantity; call it the **transformation tensor** $\lambda^{i'}_j$.[3] It is a matrix whose components in this specific example are $\lambda^{1'}_1 = \cos\theta$, $\lambda^{1'}_2 = \sin\theta$, $\lambda^{2'}_1 = -\sin\theta$, and $\lambda^{2'}_2 = \cos\theta$; or, in matrix notation:

$$[\lambda] = \begin{pmatrix} \cos\theta & \sin\theta \\ -\sin\theta & \cos\theta \end{pmatrix}, \tag{2.8}$$

representing the rotational matrix needed to rotate any two-dimensional vector **V** through a specific angle $\theta$. In other words:

---

[3]The term "tensor" will be defined shortly; at this point you can think of $\lambda^{i'}_j$ as just a matrix.

$$\begin{pmatrix} V^{1'} \\ V^{2'} \end{pmatrix} = \begin{pmatrix} \cos\theta & \sin\theta \\ -\sin\theta & \cos\theta \end{pmatrix} \begin{pmatrix} V^1 \\ V^2 \end{pmatrix}. \tag{2.9}$$

In three dimensions, rotations can be around any of the three axes, so in rotating a vector by an angle $\theta$ about the $x^1$ axis one simply generalizes (2.8) to be

$$[\lambda]_{x^1} = \begin{pmatrix} 1 & 0 & 0 \\ 0 & \cos\theta & -\sin\theta \\ 0 & \sin\theta & \cos\theta \end{pmatrix}. \tag{2.10}$$

Rotations about the $x^2$ and $x^3$ axes are respectively

$$[\lambda]_{x^2} = \begin{pmatrix} \cos\theta & 0 & -\sin\theta \\ 0 & 1 & 0 \\ \sin\theta & 0 & \cos\theta \end{pmatrix} \tag{2.11}$$

$$[\lambda]_{x^3} = \begin{pmatrix} \cos\theta & -\sin\theta & 0 \\ \sin\theta & \cos\theta & 0 \\ 0 & 0 & 1 \end{pmatrix}. \tag{2.12}$$

More complex rotations require multiple use of any combination of those three $\lambda$'s. For example, if one wishes to rotate a given vector by an angle $\varphi$ around the $x^3$ axis and *then* rotate it by an angle $\theta$ around the new axis $x^1$, then all one needs to do is construct a "compound" $\lambda$ by multiplying the appropriate $\lambda$'s like this:

$$[\lambda] = \begin{pmatrix} 1 & 0 & 0 \\ 0 & \cos\theta & -\sin\theta \\ 0 & \sin\theta & \cos\theta \end{pmatrix}_{x^1} \begin{pmatrix} \cos\varphi & -\sin\varphi & 0 \\ \sin\varphi & \cos\varphi & 0 \\ 0 & 0 & 1 \end{pmatrix}_{x^3}$$
$$= \begin{pmatrix} \cos\varphi & -\sin\varphi & 0 \\ \cos\theta\sin\varphi & \cos\theta\cos\varphi & -\sin\theta \\ \sin\theta\sin\varphi & \sin\theta\cos\varphi & \cos\theta \end{pmatrix}, \tag{2.13}$$

where it is important to keep the multiplication order: The first rotation matrix is placed *last* in the order of multiplication. In other words, rotational matrices are *not* commutative: $[\lambda]_A [\lambda]_B \neq [\lambda]_B [\lambda]_A$.

**Exercise 2.2** Given the vector $\mathbf{V} = (5, 3, 2)$:

1. Calculate its components after rotation in the sequence: First through an angle of 30° around the $x^3$ axis, then through an angle of 50° around the $x^2$ axis.
2. Find its components after a rotation the other way around; i.e. first through an angle of 50° around the $x^2$ axis, *then* through an angle of 30° around the $x^3$ axis.
3. Obviously the two matrices you found are not the same, but to further drive the point sketch the vector before and after both rotations and explicitly show that it ends up pointing in different directions.

4. As we will see in more detail, although rotations change the direction of a vector, they do not change its magnitude or norm. Explicitly verify this by finding the magnitude of the vector in this exercise before *and* after both rotations.

Hence eqns (2.7) and their three-dimensional counterparts can be written as

$$V^{i'} = \lambda_j^{i'} V^j. \tag{2.14}$$

Notice that one may say that we have either rotated the coordinates by an angle $\theta$ and used (2.14) to find the components of a vector in the new coordinates, or fixed the coordinates and rotated the vector *itself* by an angle $\theta$, then used (2.14) to find the new components of the vector. Both are obviously equivalent statements. Another important thing to note is that all curvilinear coordinate systems can relate to each other by some sort of rotation or series of rotations. So a relation such as (2.14) may be used to find the components $V^{i'}$ of a vector in one coordinate system from its components $V^i$ in another coordinate system if we knew the tensor $\lambda_j^{i'}$. So one may easily move from, say, spherical coordinates to elliptic coordinates analytically, as opposed to drawing a picture and agonizing over the trigonometry. All one needs to do is to figure out the components $\lambda_j^{i'}$ of the transformation tensor. Although we will mostly focus on rotations, it is easily shown that translating a vector $V^i$ can be done by just adding numbers to its components thus:

$$V^i = V^i + a^i, \tag{2.15}$$

where $a^i$ are the components of a constant vector. In general, rotating as well as translating a vector can then be written as

$$V^{i'} = \lambda_j^{i'} V^j + a^{i'}. \tag{2.16}$$

Transformation tensors in general (whether rotational or otherwise, as we will see) have specific properties, well studied in mathematics. For example, if one rotates a vector using $\lambda_j^{i'}$ and then rotates it back to its original direction, one must obviously get the same components; in other words, applying (2.14) twice should yield

$$V^{i'} = \lambda_j^{i'} V^j \qquad V^j = \lambda_{k'}^{j} V^{k'}$$
$$V^{i'} = \lambda_j^{i'} \lambda_{k'}^{j} V^{k'} = V^{i'}, \tag{2.17}$$

which immediately implies the following very important property of such quantities:

$$\lambda_j^{i'} \lambda_{k'}^{j} = \delta_{k'}^{i'}. \tag{2.18}$$

This is the **orthogonality condition** of the transformation tensor $\lambda$. Notice that in terms of matrix components $\lambda_j^{i'} \lambda_{k'}^{j}$ is the matrix multiplication of $\lambda_j^{i'}$ with its transpose, which also happens to be its inverse.[4]

---

[4]The reader versed in matrix manipulation may recall that a matrix whose transpose is the same as its inverse is called an **orthogonal matrix** for the very same reasons described here.

How does one then find the components $\lambda^{i'}_j$? In other words, given two different coordinate systems, $O$ and $O'$, and with the knowledge of how they relate to each other, such as (1.1) and similar equations, how can one analytically calculate the transformation tensor? This can be done by using the following recipe: If we know how the coordinates $(x^1, x^2, x^3)$ are related to another system of coordinates $\left(x^{1'}, x^{2'}, x^{3'}\right)$ where neither may be Cartesian, then it can be shown that the recipe to find the components of the transformation tensor is as follows:

$$\lambda^{i'}_j = \frac{\partial x^{i'}}{\partial x^j} = \partial_j x^{i'},$$  (2.19)

with the inverse

$$\lambda^{i}_{j'} = \frac{\partial x^i}{\partial x^{j'}} = \partial_{j'} x^i.$$  (2.20)

The definitions (2.19) and (2.20) are quite general; they apply to any number of dimensions as well as any system of coordinates. As an aside, note that those derivatives are always with respect to a *different* coordinate system; in fact, you can easily check that if one takes a derivative of coordinates with respect to themselves in the *same* coordinate system, one gets

$$\partial_j x^i = \delta^i_j,$$  (2.21)

since a derivative of a given $x^i$ with itself ($i = j$) is unity, while with respect to another ($i \neq j$) it is zero.[5] It is important to understand the difference between (2.19) and (2.21).

We are finally in a position to define vectors: A vector in three dimensions is a set of three functions that satisfy both of the following criteria:

- Their norm (dot product with itself) is invariant under transformation from one system of coordinates to another, so

$$V^{i'} V_{i'} = V^i V_i.$$  (2.22)

Since the dot product is a scalar then this property is automatically satisfied.
- Their components transform following the rule

$$V^{i'} = \lambda^{i'}_j V^j = \left(\frac{\partial x^{i'}}{\partial x^j}\right) V^j = \left(\partial_j x^{i'}\right) V^j.$$  (2.23)

As such, we say that any set of three numbers or functions do *not* constitute a vector unless both of these rules are fully satisfied. Then and only then can one call these three quantities the "components of a vector." This is the basic definition of what a vector is, and this is the exact same definition that we will use to define that which we call a "tensor."

[5]If you have done the first two parts of Exercise 1.13 then you already knew this.

**Exercise 2.3** Show that using (2.23) together with the orthogonality condition (2.18) leads to (2.22). Alternatively, one can argue that (2.22) *requires* (2.18).

As an example, let's consider finding the transformation tensors from Cartesian coordinates to cylindrical polar coordinates and back. If we let the Cartesian coordinates be the primed coordinates, then (1.1) leads to $\lambda_1^{1'} = \frac{\partial x}{\partial \rho} = \cos\varphi$, $\lambda_2^{1'} = \frac{\partial x}{\partial \varphi} = -\rho\sin\varphi$, $\lambda_1^{2'} = \frac{\partial y}{\partial \rho} = \sin\varphi$, $\lambda_2^{2'} = \frac{\partial y}{\partial \varphi} = \rho\cos\varphi$, and $\lambda_3^{3'} = 1$:

$$\left[\lambda_j^{i'}\right] = \begin{pmatrix} \cos\varphi & \sin\varphi & 0 \\ -\rho\sin\varphi & \rho\cos\varphi & 0 \\ 0 & 0 & 1 \end{pmatrix}. \tag{2.24}$$

Finding $\lambda_{j'}^i$ via (1.2),

$$\left[\lambda_{j'}^i\right] = \begin{pmatrix} \cos\varphi & -\frac{1}{\rho}\sin\varphi & 0 \\ \sin\varphi & \frac{1}{\rho}\cos\varphi & 0 \\ 0 & 0 & 1 \end{pmatrix}. \tag{2.25}$$

Direct matrix multiplication between (2.24) and (2.25) shows that orthogonality is satisfied:

$$\left[\lambda_j^{i'}\right]\left[\lambda_{i'}^i\right] = \begin{pmatrix} \cos\varphi & \sin\varphi & 0 \\ -\rho\sin\varphi & \rho\cos\varphi & 0 \\ 0 & 0 & 1 \end{pmatrix}\begin{pmatrix} \cos\varphi & -\frac{1}{\rho}\sin\varphi & 0 \\ \sin\varphi & \frac{1}{\rho}\cos\varphi & 0 \\ 0 & 0 & 1 \end{pmatrix} \tag{2.26}$$

$$= \begin{pmatrix} 1 & 0 & 0 \\ 0 & 1 & 0 \\ 0 & 0 & 1 \end{pmatrix}. \tag{2.27}$$

In this case it was straightforward, albeit a bit nasty, to find $\lambda_{j'}^i$ by direct differentiation. Sometimes it is difficult to do so. In the cylindrical/Cartesian case another way of finding one of the two transformation tensors is by using (2.18) directly:

$$\begin{pmatrix} \cos\varphi & \sin\varphi & 0 \\ -\rho\sin\varphi & \rho\cos\varphi & 0 \\ 0 & 0 & 1 \end{pmatrix}\begin{pmatrix} \lambda_{1'}^1 & \lambda_{1'}^2 & 0 \\ \lambda_{2'}^1 & \lambda_{2'}^2 & 0 \\ 0 & 0 & 1 \end{pmatrix} = \begin{pmatrix} 1 & 0 & 0 \\ 0 & 1 & 0 \\ 0 & 0 & 1 \end{pmatrix}. \tag{2.28}$$

This leads to four equations in the four unknown components of $\lambda_{j'}^i$. In this case, which is essentially two-dimensional, the four equations are not too difficult to solve and lead directly to (2.25). In other cases, as we will explore in Exercise 2.6, finding $\lambda_{j'}^i$ is somewhat more challenging.

**Exercise 2.4** Find and solve the four simultaneous equations that arise from (2.28) and confirm that the result is exactly (2.25).

**Exercise 2.5** Assuming that the Cartesian coordinates are the primed coordinates of (2.23), find the transformation tensor $\lambda_j^{i'}$ for each of the following cases:

1. Spherical coordinates (1.3).
2. Elliptic cylindrical coordinates (1.6).
3. Parabolic cylindrical coordinates (1.7).
4. Parabolic coordinates (1.8).
5. Paraboloidal coordinates (1.9).
6. Oblate spheroidal coordinates (1.10).
7. Toroidal coordinates (1.11).

**Exercise 2.6** Explore finding the *inverse* transformation tensor $\lambda_{j'}^{i}$ for the coordinates systems in Exercise 2.5. This may not be as easy as it sounds, as was discussed briefly in the text. Feel free to teach yourself how to do this using a computer, particularly by using one of the commercially available symbolic manipulation software packages.

1. Find $\lambda_{j'}^{i}$ by direct differentiation of the inverse transformations to the given coordinate systems. For spherical coordinates these are just (1.4); for the others they must be computed first.
2. Use (2.18) in its matrix form to find $\lambda_{j'}^{i}$. This, in most cases, would require reviewing the procedure of finding inverse matrices in any textbook on linear algebra, such as [3].

**Exercise 2.7** Using the following vectors defined in the given coordinates systems, compute their components in Cartesian coordinates using the $\lambda$'s you found in Exercise 2.5.

1. $\mathbf{A} = (5, \pi/3, \pi/4)$ in spherical coordinates (1.3).
2. $\mathbf{B} = (-1, 2, -2)$ in elliptic cylindrical coordinates (1.6).
3. $\mathbf{C} = (2\xi, -\eta, 3z)$ in parabolic cylindrical coordinates (1.7).
4. $\mathbf{D} = (-u, 4v, 2\theta)$ in parabolic coordinates (1.8).
5. $\mathbf{E} = \left(3\varphi\mu^2, 2\nu, -\nu + 3\mu\right)$ in oblate spheroidal coordinates (1.10).
6. $\mathbf{F} = \left(2\sigma\tau, -\sigma^3, 3\tau\right)$ in toroidal coordinates (1.11).

## 2.2 Defining Tensors

Simply put, a tensor is a multi-directional object with components that follow the same transformation rules as vectors, i.e. eqns (2.22) and (2.23). Viewed as such, vectors are simply a special case of tensors. But in general the components of tensors can have more than one index; for example, $g_{ij}$ is a two-index tensor, while $V^i$ is a one-index tensor. To include tensors with more than one index, eqns (2.22) and (2.23) will have to be appropriately modified, but the basic principles are the same: Tensors represent physical quantities whose "magnitude" does not change upon coordinate transformation but its components transform according to the nature of the coordinate systems in question. Generally, a tensor is an object whose components have an arbitrary number of upper and lower indices. We classify them in terms of **rank**; so a tensor $\mathbf{T}$ of rank $(p, q)$ has components carrying $p$ upper indices and $q$ lower indices thus: $T^{i_1 \cdots i_p}_{\quad\quad j_1 \cdots j_q}$. In this language, vectors are tensors of either rank $(1, 0)$ or $(0, 1)$, depending on whether their index is upper or lower respectively. Other examples of tensors we have already seen are the metric $g_{ij}$ (a tensor of rank $[0, 2]$), its inverse $g^{ij}$ (rank $[2, 0]$), and $\delta^i_j$ (of rank $[1, 1]$). In fact, even scalar functions, with no indices, are tensors of rank $(0, 0)$. We have not yet met any higher-rank tensors. For example, the Levi-Civita symbol $\varepsilon_{ijk}$ is in fact *not* a tensor, since it can be shown that it does not satisfy the appropriate component transformation rules; in other words, it is not covariant. Not every object with indices is necessarily a tensor. In the particular case of $\varepsilon_{ijk}$, this can, in fact, be *made* into a tensor that performs the same job, as we will see later.

To avoid a common misconception, it is perhaps important to emphasize that the quantities with indices—for example $T^{i_1 \cdots i_p}_{\quad\quad j_1 \cdots j_q}$—are *not* themselves *the* tensor but rather its *components*. This is just as the $V^i$'s are not the vector but rather the components of one. The true vector is really $\mathbf{V} = V^i \hat{\mathbf{g}}_i$, or $\mathbf{V} = V_i \hat{\mathbf{g}}^i$, where $\hat{\mathbf{g}}^i = g^{ik} \hat{\mathbf{g}}_i$ in the usual way. A true tensor, then, may be written out in some coordinate basis as follows:

$$\mathbf{T} = T^{i_1 \cdots i_p}_{\quad\quad j_1 \cdots j_q} \hat{\mathbf{g}}_{i_1} \cdots \hat{\mathbf{g}}_{i_p} \hat{\mathbf{g}}^{j_1} \cdots \hat{\mathbf{g}}^{j_q}. \tag{2.29}$$

The tensor $\mathbf{T}$ can now simply be called a tensor of rank $(p + q)$, since the notation $(p, q)$ applies to the location of the indices on the components and this can become ambiguous in the language of eqn (2.29). For example, a given tensor $\mathbf{A}$ of rank 4 can be written in the $(4, 0)$ form $\mathbf{A} = A^{ijkl} \hat{\mathbf{g}}_i \hat{\mathbf{g}}_j \hat{\mathbf{g}}_k \hat{\mathbf{g}}_l$, but it can also be expressed in the forms $\mathbf{A} = A^{ijk}_{\quad l} \hat{\mathbf{g}}_i \hat{\mathbf{g}}_j \hat{\mathbf{g}}_k \hat{\mathbf{g}}^l$, $\mathbf{A} = A^{ij}_{\quad kl} \hat{\mathbf{g}}_i \hat{\mathbf{g}}_j \hat{\mathbf{g}}^k \hat{\mathbf{g}}^l$, or $\mathbf{A} = A^i_{\quad jkl} \hat{\mathbf{g}}_i \hat{\mathbf{g}}^j \hat{\mathbf{g}}^k \hat{\mathbf{g}}^l$. They are all different representations of the *same* tensor. As such, the term "tensor of rank $(p, q)$" really refers to the components rather than the tensor itself. Because this "complete" form of what a tensor really is is rarely needed, we find that it is also rarely emphasized in the literature,[6] and one finds oneself referring to $T^{i_1 \cdots i_p}_{\quad\quad j_1 \cdots j_q}$ as *the* tensor. This, I am sure, has arisen to avoid saying "the components of tensor such and such" all the time, and I beg the reader's forgiveness since I will most definitely be using this erroneous language quite often.

So what are the transformation rules for tensors?[7] We already know the answer for rank $(1, 0)$ tensors, i.e. upper index, or contravariant, vectors. These are given by (2.23). For $(0, 1)$ tensors, i.e. lower-index or covariant vectors, the transformation rule is appropriately fixed as follows:

---

[6]At least not in the physics literature. Mathematicians have their own ways of talking about this.

[7]See? What I should really say is "What are the transformation rules for the *components* of tensors."

$$V_{i'} = \lambda_{i'}^j V_j = \left( \frac{\partial x^j}{\partial x^{i'}} \right) V_j = \left( \partial_{i'} x^j \right) V_j. \tag{2.30}$$

Notice how the primes were adjusted on both sides to match the indices they are on (the primes are on the indices *not* on the vectors). The analogous transformation formulae for higher-rank tensors are found by adding the appropriate transformation derivatives. For example, a tensor $A^{ij}$ of rank $(2,0)$ transforms as follows:

$$A^{i'j'} = \lambda_m^{i'} \lambda_n^{j'} A^{mn} = \left( \partial_m x^{i'} \right) \left( \partial_n x^{j'} \right) A^{mn}, \tag{2.31}$$

while tensors of rank $(0,2)$ and $(1,1)$ transform by

$$A_{i'j'} = \left( \partial_{i'} x^m \right) \left( \partial_{j'} x^n \right) A_{mn} \tag{2.32}$$

$$A_{j'}^{i'} = \left( \partial_{j'} x^m \right) \left( \partial_n x^{i'} \right) A_m^n. \tag{2.33}$$

Notice how the summation and free indices balance on both sides of these equations as they should. The inverse metric tensor transforms like (2.31), while the metric tensor transforms like (2.32). In other words,

$$g^{i'j'} = \lambda_m^{i'} \lambda_n^{j'} g^{mn},$$
$$g_{i'j'} = \lambda_{i'}^m \lambda_{j'}^n g_{mn}. \tag{2.34}$$

Equations (2.34) may be useful in finding the components of a new metric tensor in terms of the components of an old metric tensor. This can be used in place of the methods discussed in §1.2. They are also general enough to cover non-orthogonal coordinate systems. The method of the scale factors (1.28) is in fact a special case of (2.34) for orthogonal coordinates, as the reader can show in Exercise 2.8.

**Exercise 2.8** In the second equation of (2.34), let $g_{mn} = \delta_{mn}$, i.e. Cartesian coordinates. Expand the double sum over $m$ and $n$ and show that it gives the expected expressions for the scale factors (1.28).

**Exercise 2.9** Show that the metrics of the coordinate systems of Exercise 2.6 arise from the Cartesian metric by applying (2.34). Alternatively you can also show that the Cartesian metric arises from the metrics of (2.5).

A more complex example would be of a mixed-index tensor such as a tensor of rank $(2,3)$:

$$A^{i'j'}{}_{k'l'm'} = \lambda_o^{i'} \lambda_g^{j'} \lambda_{k'}^n \lambda_{l'}^r \lambda_{m'}^s A^{og}{}_{nrs}. \tag{2.35}$$

A few minutes of staring at these equations should be enough for the reader to observe the correct pattern of derivatives and indices.[8] In general, a tensor of rank $(p, q)$ transforms as follows:

$$A^{i'_1 \cdots i'_p}{}_{j'_1 \cdots j'_q} = \lambda^{i'_1}_{m_1} \cdots \lambda^{i'_p}_{m_p} \lambda^{n_1}_{j'_1} \cdots \lambda^{n_q}_{j'_q} A^{m_1 \cdots m_p}{}_{n_1 \cdots n_q}. \tag{2.36}$$

Finally, just as one defines the norm of a vector by $V^i V_i$, so does one define the "norm" of a tensor by contracting *all* indices as follows:

$$A^{ij} A_{ij}, \qquad B^i{}_{jk} B_i{}^{jk}, \qquad T^{i_1 \cdots i_p}{}_{j_1 \cdots j_q} T_{i_1 \cdots i_p}{}^{j_1 \cdots j_q}, \tag{2.37}$$

where the indices are raised and lowered by the metric $g_{ij}$ in any coordinate system. Since these "norms" are scalars (all indices contracted), they do not change value upon transformation from one coordinate system to another, as seen in (2.22). Covariant quantities other than the norms can be constructed from tensors. In fact *any* contraction of all indices leading to a scalar is covariant. For example, some such quantities (useful in relativity) can be built by "self-contractions"

$$A^i{}_i, \qquad C^{ij}{}_{ij}, \qquad T^{i_1 \cdots i_p}{}_{i_1 \cdots i_p}. \tag{2.38}$$

---

**Exercise 2.10** A tensor of rank $(1, 1)$ in spherical coordinates has the non-vanishing components: $A^r{}_\varphi = 2$ and $A^\theta{}_r = -1$. Find its norm and self-contraction at the point $(4, \pi/2, \pi/3)$.

---

Tensor manipulations follow the rules of the index notation, so most of the calculations we have done so far in this book are in fact tensor calculations. The important distinction is that tensors must transform following the above rules. If a quantity, never mind how many indices it carries, does *not* transform this way, it is not a tensor. Physically what this means is that the non-tensorial quantity in question is something that changes depending on the choice of coordinates. Since nature does not care about our choices of coordinates, tensors are particularly important quantities; they guarantee the covariance of the equations of physics.

As pointed out earlier, we are focusing mainly on rotations; however, translations can be easily included in the transformation definitions as follows: Consider a vector

---

[8]Equation (2.31) represents nine independent equations: one for each of the components $A^{11}$, $A^{12}$, etc. In each of these equations there is a double sum over the repeated indices $m$ and $n$. Since each index counts from 1 to 3, each of the nine equations contains nine terms, giving a total of 27 terms altogether. On the other hand, eqn (2.35) represents 120 independent equations, each of which contains *five* summations over the repeated indices $(o, g, n, r, s)$. *Each* of the 120 equations then contains $3^5 = 243$ terms for a total of $120 \times 243 = 29,160$ terms altogether! Perhaps the reader can now appreciate the power of the index notation and the Einstein summation convention; just imagine having to write all of these equations explicitly! And of course it gets worse in dimensions higher than 3.

$V^i$; translate it to a new coordinate system that differs from the old system by a constant vector $a^i$, hence

$$U^i = V^i + a^i. \tag{2.39}$$

The norm of the "new" vector is then

$$U^i U_i = \left(V^i + a^i\right)\left(V_i + a_i\right). \tag{2.40}$$

Now transform $U^i$ to a new coordinate system using (2.23) and find its norm $U^{i'} U_{i'}$; this leads directly back to $U^i U_i$ as follows:

$$
\begin{aligned}
U^{i'} U_{i'} &= \left(V^{i'} + a^{i'}\right)\left(V_{i'} + a_{i'}\right) = \left(\lambda_j^{i'} V^j + \lambda_j^{i'} a^j\right)\left(\lambda_{i'}^k V_k + \lambda_{i'}^k a_k\right) \\
&= \lambda_j^{i'} V^j \lambda_{i'}^k V_k + \lambda_j^{i'} a^j \lambda_{i'}^k V_k + \lambda_j^{i'} V^j \lambda_{i'}^k a_k + \lambda_j^{i'} a^j \lambda_{i'}^k a_k \\
&= V^i V_i + a^i V_i + V^i a_i + a^i a_i = \left(V^i + a^i\right)\left(V_i + a_i\right) = U^i U_i. \quad\tag{2.41}
\end{aligned}
$$

One can construct tensors from tensors by simple multiplications and/or contractions. An example we already know of is the generation of a scalar (rank 0 tensor) from two vectors (rank 1 tensors) by the ordinary dot product; i.e. contracting tensors $A^i$ and $B_i$ generates a third tensor $f$:

$$f = A^i B_i. \tag{2.42}$$

We can also generate a tensor from $A^i$ and $B_i$ by multiplication rather than contraction:[9]

$$R^i{}_j = A^i B_j. \tag{2.43}$$

Contracting a second-rank tensor with a vector gives

$$K^i = C^{ij} B_j. \tag{2.44}$$

It is even possible to generate a new tensor from contracting indices on a single tensor, such as

$$R_{jk} = P^i{}_{jik}. \tag{2.45}$$

Finally, here is a more complex example:

---

[9]The quantity $R^i{}_j$ in (2.43) is sometimes referred to as a "dyad:" a matrix formed by the product of two vectors which is neither the dot product nor the cross product.

$$N^k = A^{ijk}{}_{lm} H^l M^m{}_{ij}. \tag{2.46}$$

Now, are we absolutely sure that quantities constructed from other tensors are, in fact, themselves tensors? The criterion is to check whether or not the newly constructed quantities transform following the appropriate transformation rules. Consider the following example: Let $C^{ij}$ be a tensor (i.e. it transforms according to [2.31]) and let $B_j$ also be a tensor (i.e. transforms according to [2.30]). Now, constructing the quantity $K^i$ via $K^i = C^{ij} B_j$, we need to prove that $K^i$ is also a tensor. First we write $C^{ij}$ and $B_j$ in a different coordinate system than the one they were defined in, and then we construct a *new* quantity $K^{i'}$ as follows:

$$K^{i'} = C^{i'j'} B_{j'}. \tag{2.47}$$

Now, since we know that $C^{ij}$ and $B_j$ are tensors, we can substitute their transformation rules:

$$K^{i'} = \lambda^{i'}_k \lambda^{j'}_m C^{km} \lambda^n_{j'} B_n. \tag{2.48}$$

Using the orthogonality property $\lambda^{j'}_m \lambda^n_{j'} = \delta^n_j$, (2.48) becomes

$$K^{i'} = \lambda^{i'}_k C^{kn} B_n, \tag{2.49}$$

but the quantity $C^{kn} B_n$ is exactly $K^k$, so we end up with

$$K^{i'} = \lambda^{i'}_k K^k, \tag{2.50}$$

which is the correct transformation rule for a rank $(1,0)$ tensor; in other words, $K^i$ is indeed a tensor. Clearly this calculation can be generalized to prove that any of the previous examples are tensors if and only if they were constructed from the contractions or multiplications of tensors.

---

**Exercise 2.11** For each of the following cases, prove that $K^i{}_j$ transforms as a tensor if the quantities on the right-hand side transform like ones.

1. $K^i{}_j = B^i C_j$.
2. $K^i{}_j = B^{ik} C_{jk}$.

---

So what are these higher-rank tensors? What physical quantities do they represent? The metric is an example of tensors that have a specific purpose: to define the relationship between a distance in space $dl$ and its equivalent coordinate displacements $dx^i$.[10] As briefly discussed in the introduction, higher-rank tensors are used when a

---

[10]There is a minor technicality concerning the definition of the transformation tensor $\lambda$ itself. Although we call it a "tensor," and it does in fact transform like one, it is technically *not* a tensor. A basic requirement of tensors which we have not emphasized, although it may turn out to be obvious, is that they are supposed to be defined in a *single* coordinate system. The "tensor" $\lambda$, however, is a transformation *between* coordinate systems. Hence it is not quite a tensor in the strictest sense. However we will continue to call it so, since for all intents and purposes what transforms as a tensor is a tensor, despite minor technicalities.

given phenomenon has different directions representing cause and effect. They are very important in relativistic physics; in fact, one cannot get too far studying relativity without them. Also in ordinary non-relativistic physics one encounters such objects every once in a while. In mechanics the $3 \times 3$ matrix of moments of inertia is an example of a rank 2 tensor. A tensor that goes by the name of the "stress tensor" is used in a generalization of Hooke's law to solve the eraser example we discussed in the introduction. There is even a *fourth*-rank tensor used in connection with the generalized Hooke's law. We will look more closely at these examples in the next chapter.

**Exercise 2.12** Using the following tensors defined in the given coordinates systems, compute their components in Cartesian coordinates using the $\lambda$'s you found in Exercise 2.5.

1. The rank $(1,1)$ tensor whose non-vanishing components in spherical coordinates $(1.4)$ are $A^1{}_2 = 2$ and $A^2{}_1 = -1$.
2. The rank $(2,1)$ tensor whose non-vanishing components in elliptic cylindrical coordinates $(1.6)$ are $B^{11}{}_2 = 2v$ and $B^{21}{}_3 = -1u^2$.
3. The rank $(1,2)$ tensor whose non-vanishing components in parabolic cylindrical coordinates $(1.7)$ are $C^1{}_{11} = \xi^2$, $C^1{}_{22} = e^{-\xi}$, $C^1{}_{33} = \xi\eta$, $C^2{}_{12} = -\eta$, and $C^3{}_{33} = z$.

**Exercise 2.13** Insert the missing indices on the tensor $C$ in each of the following cases:

1. $A^i B^j{}_i = C^?_?$.
2. $A^i B^j{}_i D^n{}_{jk} = C^?_?$.
3. $A^{ij}{}_{jkl} B^n{}_i E^l{}_m D^k = C^?_?$.

**Exercise 2.14** Prove that the Levi-Civita symbol $(1.94)$ is not a tensor.

**Exercise 2.15** Consider a matrix $[A]$ whose components are defined by $A^{ij} = B^i + C^j$, where $B$ and $C$ are vectors. Is it a tensor? Clearly the very definition of this matrix violates the laws of indices, but is that enough to prove that it is not a tensor?

## 2.3 Derivatives and Parallel Transport

There is one very subtle piece of the definition of vector operations that is usually missing in introductory treatments. Consider vector subtraction (or addition). The way it is usually done is that one assumes that the vectors are placed tail end to tail end then a triangle is completed, as in Fig. [1.13]; i.e. they are assumed to start at the *same* point. However, the definition of a vector does not require it to be located at any particular point in space: it's only defined by its magnitude and direction, *not* by where it is located. This discrepancy is accounted for by noting that we can freely move vectors around as long as we preserve their lengths and directions. This process is known as **parallel transport**. One says that we can parallel transport a vector **U** in space, i.e. move it around parallel to itself, until we get its tail end to coincide with the tail end of vector **V**, *then* perform the required vector operation, in this case subtraction as in Fig. [2.4].

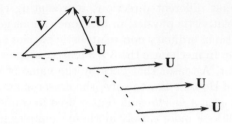

Fig. 2.4: Parallel transporting a vector along a curve in space to perform a vector subtraction.

Notice that it does not make any sense to do the operation without bringing the two vectors together first. The same is also true when performing a dot product, as it does not make sense to ask what is the angle $\theta$ between the vectors unless one is parallel transported first, as shown in Fig. [2.5].

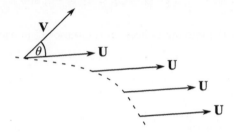

Fig. 2.5: Parallel transporting a vector to perform a dot product.

Parallel transporting a vector clearly preserves its magnitude as well as its direction; in other words, it remains the *same vector*. In Cartesian coordinates (and *only* in Cartesian coordinates) the components of the vector also remain unchanged. In other words, *nothing* changes in the vector upon parallel transport. Hence a detailed discussion of this process is simply either ignored or just skimmed over in introductory physics courses where Cartesian coordinates are almost exclusively used. However, it turns out that when parallel transporting a vector in non-Cartesian coordinate systems its components will actually change, despite the fact that its magnitude and direction don't. Consider as a simple example the vector **V** shown in Fig. [2.6]. Originally the vector is chosen to start at point $A$. It has a specific length and is pointing exclusively to the right. In Cartesian coordinates it has an $x$ component that is equal to its length and a vanishing $y$ component. When transported parallel to itself to point $B$, it continues to have the same Cartesian components. Now if one considers the *exact* same vector in polar coordinates (the polar grid is sketched for reference), it starts out at point $A$ having a purely radial component and no angular component but when transported to point $B$ the exact opposite becomes true: it now has a vanishing radial component but a non-zero angular component! It can also be made to have non-vanishing values for both components by parallel transporting it to a new point

$C$. It is the *same* vector, but its components have changed upon parallel transport!

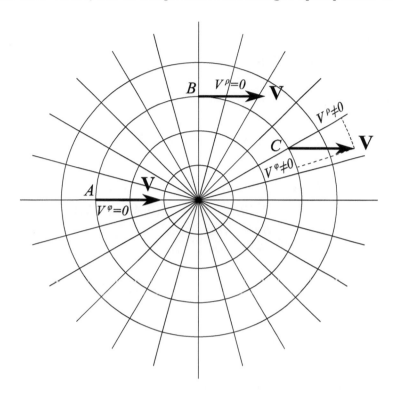

Fig. 2.6: The polar components of a vector change upon parallel transport.

Hence it becomes a problem of vector analysis to figure out the new components of the transported vector. We note that this issue is not only crucial for additions, subtractions, and products of vectors, but it is also *particularly* important when taking the derivative of a vector; this is because derivatives involve vector subtraction. Recall that the derivative of a vector $\mathbf{V}(s)$ is defined in the following standard way:

$$\frac{d\mathbf{V}}{ds} = \lim_{h \to 0} \frac{\mathbf{V}(s+h) - \mathbf{V}(s)}{h}, \tag{2.51}$$

where $s$ is some parameter; for example, a distance. Notice that the definition, the very *definition*, of vector differentiation involves the subtraction of vector $\mathbf{V}(s)$ from $\mathbf{V}(s+h)$, which is located at a different spot with respect to $s$, separated from $\mathbf{V}(s)$ by $h$. So to perform the subtraction we first have to parallel transport $\mathbf{V}(s)$ to where $\mathbf{V}(s+h)$ is located, as discussed above. We conclude, then, that the derivative is sensitive to parallel transport and eqn (2.51) is only correct (in components) if the vector is defined in Cartesian coordinates. Taking derivatives of vectors in other coordinate systems requires understanding parallel transport in those systems first. The same applies to taking derivatives of tensors, which, as the reader may guess, becomes terribly messy. We must then find a way to make it straightforwardly possible to do derivatives in any coordinate system; in other words, a *covariant* way of performing derivatives.

To further clarify this issue recall that in three-dimensional Cartesian coordinates the del or nabla operator $\vec{\nabla}$ is defined by

$$\vec{\nabla} = \hat{\mathbf{e}}_1 \frac{\partial}{\partial x} + \hat{\mathbf{e}}_2 \frac{\partial}{\partial y} + \hat{\mathbf{e}}_3 \frac{\partial}{\partial z} = \delta^{ij} \hat{\mathbf{e}}_i \partial_j \tag{2.52}$$

The divergence of a vector $\mathbf{A} = A^1 \hat{\mathbf{e}}_1 + A^2 \hat{\mathbf{e}}_2 + A^3 \hat{\mathbf{e}}_3 = A^i \hat{\mathbf{e}}_i$ is the dot product of $\vec{\nabla}$ with $\mathbf{A}$:

$$\begin{aligned}
\vec{\nabla} \cdot \mathbf{A} &= \left( \delta^{ij} \hat{\mathbf{e}}_i \partial_j \right) \cdot \left( A^k \hat{\mathbf{e}}_k \right) = \delta^{ij} \left( \partial_j A^k \right) \left( \hat{\mathbf{e}}_i \cdot \hat{\mathbf{e}}_k \right) \\
&= \delta^{ij} \left( \partial_j A^k \right) \delta_{ik} = \delta_k^j \left( \partial_j A^k \right) = \left( \partial_k A^k \right) \\
&= \frac{\partial A^1}{\partial x} + \frac{\partial A^2}{\partial y} + \frac{\partial A^3}{\partial z}.
\end{aligned} \tag{2.53}$$

Now consider a vector $\mathbf{B} = B^\rho \hat{\mathbf{g}}_\rho + B^\varphi \hat{\mathbf{g}}_\varphi + B^z \hat{\mathbf{g}}_\mathbf{z} = B^i \hat{\mathbf{g}}_i$ in cylindrical coordinates (the index $i$ here counts over $\rho$, $\varphi$, and $z$); the divergence *cannot* be written in the same form; that is

$$\vec{\nabla} \cdot \mathbf{B} \neq \left( \partial_i B^i \right) = \frac{\partial B^\rho}{\partial \rho} + \frac{\partial B^\varphi}{\partial \varphi} + \frac{\partial B^z}{\partial z}. \tag{2.54}$$

In other words, generally speaking the ordinary partial derivative $\partial_i$ is *not* covariant as defined. Another way of understanding this is noting that the curvilinear natural basis vectors $\hat{\mathbf{g}}$ (as well as their physical counterparts $\hat{\mathbf{v}}$ for that matter) are *not* constants, and their own derivatives have to be taken into account. In fact, as one can easily check by looking inside the cover of any book on electromagnetism or advanced mechanics,[11] the divergence in cylindrical coordinates has the following form:

$$\vec{\nabla} \cdot \mathbf{B} = \frac{\partial B^\rho}{\partial \rho} + \frac{B^\rho}{\rho} + \frac{1}{\rho} \frac{\partial B^\varphi}{\partial \varphi} + \frac{\partial B^z}{\partial z}. \tag{2.55}$$

Similarly, for a vector $\mathbf{A} = A^r \hat{\mathbf{g}}_r + A^\theta \hat{\mathbf{g}}_\theta + A^\varphi \hat{\mathbf{g}}_\varphi = A^i \hat{\mathbf{g}}_i$ in spherical coordinates the divergence is:

$$\vec{\nabla} \cdot \mathbf{A} = \frac{\partial A^r}{\partial r} + \frac{2}{r} A^r + \frac{1}{r} \frac{\partial A^\theta}{\partial \theta} + \frac{\cot \theta}{r} A^\theta + \frac{1}{r \sin \theta} \frac{\partial A^\varphi}{\partial \varphi}. \tag{2.56}$$

So, how can one write a covariant (coordinate-invariant) derivative in the index/tensor notation? This is our objective in this section. One possible way is to

---

[11] As noted in the paragraph preceding eqn (1.133), in this book applications of the del operator are defined using the *physical* basis. Most mechanics and electromagnetism textbooks use the *natural* basis. This results in slightly different expressions for divergence, curl, etc.

do it the way it is normally done in courses on multivariable calculus. One starts with the relationships between coordinate systems—for example (1.1)—takes the appropriate chain rule derivatives, and substitutes in the Cartesian divergence (2.53). This is not the route we will take here. It turns out that there is a more systematic way of figuring out how the derivative in a specific coordinate system looks using the metric itself. Starting with a given metric that defines the coordinate system of interest, one can go through a serious of predefined calculations to reach expressions such as (2.55) and (2.56). It does require us, however, to look very closely at the issue of parallel transport, as discussed earlier.

Consider a vector **V** with components $V^i$. The vector is originally defined to start at a point that has coordinates $x^k$ in a given coordinate system. If one wishes to move the vector to a new point with coordinates $x^k + \Delta x^k$, where $\Delta x^k$ is some change in each coordinate, then clearly a specific value has to be added or subtracted to each component of the vector. In other words,

$$V^i \left(x^k\right)^{\text{parallel transported}} = V^i \left(x^k\right) - \delta V^i, \tag{2.57}$$

where we have chosen a subtraction rather than an addition as a matter of convention. We guess that the change $\delta V^i$ in the vector's components must depend on the change in position $\Delta x^k$ as well as on the original components $V^i$ of the vector, since different vectors will change differently. In short, we can write

$$\delta V^i \propto V^1 \Delta x^1 + \cdots V^1 \Delta x^2 + \cdots V^2 \Delta x^1 + \cdots, \tag{2.58}$$

or symbolically

$$\delta V^i \propto V^j \Delta x^k. \tag{2.59}$$

In order for the indices in expression (2.59) to balance out, there must be a *three-index* quantity on the right-hand side as follows:

$$\delta V^i = \Gamma^i_{jk} V^j \Delta x^k, \tag{2.60}$$

or explicitly:

$$\delta V^i = \Gamma^i_{11} V^1 \Delta x^1 + \Gamma^i_{12} V^1 \Delta x^2 + \Gamma^i_{21} V^2 \Delta x^1 + \cdots \tag{2.61}$$

The quantities $\Gamma^i_{jk}$ are called the **Christoffel symbols,**[12] sometimes also referred to as the **connection coefficients** or simply the **connection,** as they *do* connect the vector and its change of location to the amount that is needed to correct the vector itself upon parallel transport.[13] We can now write the parallel transport eqns (2.57) as follows:

$$V^i \left(x^k\right)^{\text{parallel transported}} = V^i \left(x^k\right) - \Gamma^i_{jk} V^j \Delta x^k. \tag{2.62}$$

---

[12]Elwin Bruno Christoffel: German mathematician and physicist (1829–1900).

[13]Technically the Christoffel symbols are a special case of the more general mathematical notion of "metric connection," but a discussion of the difference will take us too far from our intended purposes here, so we postpone it until Chapter 7.

In other words, the components of a vector transported from a given point to a new point must be corrected by an expression involving these Christoffel symbols. A detailed example of how to calculate these symbols for the case of polar coordinates using geometry and trigonometry is given at the beginning of the next section. But, as we will see, it is most certainly a very painful way of finding $\Gamma^i_{jk}$. Our task then reduces to figuring out how to calculate these quantities with the least amount of pain, which we will do in §2.4.

Before we get there, let us go back to the question of the derivative and see how the Christoffel symbols play a role there. In ordinary non-covariant Cartesian coordinates the definition of the derivative of any vector **V** is given by eqn (2.51). Hence for each component $V^i$ of the vector **V** we write:

$$\frac{\partial V^i}{\partial x^k} = \lim_{\Delta x^k \to 0} \frac{V^i\left(x^k + \Delta x^k\right) - V^i\left(x^k\right)}{\Delta x^k}. \tag{2.63}$$

This definition of the ordinary partial derivative is not covariant; it works only in Cartesian coordinates. So we define a new type of derivative, the **covariant derivative**, which takes into account the appropriate parallel transport in any system of coordinates as intended. The difference would be in the use of $V^i\left(x^k\right)^{\text{parallel transported}}$ instead of $V^i\left(x^k\right)$:

$$\left(\frac{\partial V^i}{\partial x^k}\right)^{\text{covariant}} = \lim_{\Delta x^k \to 0} \frac{V^i\left(x^k + \Delta x^k\right) - V^i\left(x^k\right)^{\text{parallel transported}}}{\Delta x^k}. \tag{2.64}$$

Using (2.62) and simplifying,

$$\left(\frac{\partial V^i}{\partial x^k}\right)^{\text{covariant}} = \lim_{\Delta x^k \to 0} \frac{V^i\left(x^k + \Delta x^k\right) - V^i\left(x^k\right) + \Gamma^i_{jm}V^j\Delta x^m}{\Delta x^k}$$

$$= \lim_{\Delta x^k \to 0} \frac{V^i\left(x^k + \Delta x^k\right) - V^i\left(x^k\right)}{\Delta x^k} + \lim_{\Delta x^k \to 0} \Gamma^i_{jm}V^j\left(\frac{\Delta x^m}{\Delta x^k}\right). \tag{2.65}$$

The first term is just (2.63) and the limit $\Delta x^k \to 0$ changes the $\Delta$ ratio in the second term into a derivative, so

$$\left(\frac{\partial V^i}{\partial x^k}\right)^{\text{covariant}} = \frac{\partial V^i}{\partial x^k} + \Gamma^i_{jm}V^j\left(\frac{\partial x^m}{\partial x^k}\right), \tag{2.66}$$

or equivalently

$$\left(\partial_k V^i\right)^{\text{covariant}} = \partial_k V^i + \Gamma^i_{jm}V^j\left(\partial_k x^m\right). \tag{2.67}$$

Applying (2.21), we end up with

$$\left(\partial_k V^i\right)^{\text{covariant}} = \partial_k V^i + \Gamma^i_{jk} V^j. \tag{2.68}$$

Finally, rather than writing the word "covariant" every time we wish to denote a covariant derivative, we choose a new symbol for it:

$$\nabla_k = \left(\frac{\partial}{\partial x^k}\right)^{\text{covariant}}. \tag{2.69}$$

Hence

$$\nabla_k V^i = \partial_k V^i + \Gamma^i_{jk} V^j. \tag{2.70}$$

The covariant partial derivative (2.70) is finally independent of the choice of coordinates and looks the same in any coordinate system.[14] Of course, the actual values of the Christoffel symbols, which we are yet to calculate, will vary from one coordinate system to another. If we choose to work in Cartesian coordinates, then the Christoffel symbols should vanish, giving $\nabla_k V^i = \partial_k V^i$ as expected.

Now, we can use (2.70) to define a divergence that is truly covariant. This is easily done by setting $i = k$ in (2.70) to get

$$\vec{\nabla} \cdot \mathbf{V} = \nabla_i V^i = \partial_i V^i + \Gamma^i_{ji} V^j, \tag{2.71}$$

which goes back to being $\partial_i V^i$ in Cartesian coordinates.

This is all nice and fine, but (2.70) and (2.71) work only for upper-index vectors. To define covariant derivatives for a lower-index vector we can either start from scratch or we can make the following somewhat clever argument: Consider the norm of a vector: $V^i V_i$. Being a scalar quantity, it remains invariant upon parallel transport, i.e.

$$\left(V^i V_i\right)^{\text{parallel transported}} = V^i V_i - \delta\left(V^i V_i\right) = V^i V_i. \tag{2.72}$$

Since $\delta$ is an incremental change, we can apply the product rule on $\delta\left(V^i V_i\right)$ to get

$$\delta\left(V^i V_i\right) = \left(\delta V^i\right) V_i + \left(\delta V_i\right) V^i = 0, \tag{2.73}$$

then use (2.60) to get

$$\Gamma^i_{jk} V^j \Delta x^k V_i + \left(\delta V_i\right) V^i = 0. \tag{2.74}$$

In the second term, the index $i$ is a dummy index, so we rename it $i \to j$:

---

[14]Some textbooks, especially older ones, use the comma notation $A_{,k} \equiv \partial_k A$ and the semicolon $A_{;k} \equiv \nabla_k A$, which we find potentially confusing.

$$\Gamma^i_{jk} V^j \Delta x^k V_i + (\delta V_j) V^j = 0 \tag{2.75}$$

and take $V^j$ as a common factor:

$$\left[ \Gamma^i_{jk} \Delta x^k V_i + (\delta V_j) \right] V^j = 0. \tag{2.76}$$

If the product of two quantities vanishes, then one of them must be zero. Since the components of the vector $V^j$ are arbitrary and generally non-vanishing, then the quantity in brackets must vanish, leading to:

$$\delta V_j = -\Gamma^i_{jk} \Delta x^k V_i \tag{2.77}$$

as the analogous equation to (2.60). Substituting in the definition of the covariant derivative finally gets us:

$$\nabla_i V_j = \partial_i V_j - \Gamma^k_{ij} V_k. \tag{2.78}$$

Similar manipulations lead to a variety of equations for the covariant derivatives of all types of upper, lower, or mixed index tensors. For example, the covariant derivative of a rank $(0,2)$ tensor $A_{ij}$ is

$$\nabla_i A_{jk} = \partial_i A_{jk} - \Gamma^n_{ij} A_{nk} - \Gamma^n_{ik} A_{nj}, \tag{2.79}$$

while for ranks $(2,0)$ and $(1,1)$ they are as follows:

$$\nabla_i A^{jk} = \partial_i A^{jk} + \Gamma^j_{in} A^{nk} + \Gamma^k_{in} A^{jn} \tag{2.80}$$

$$\nabla_i A^j_k = \partial_i A^j_k + \Gamma^j_{in} A^n_k - \Gamma^n_{ik} A^j_n. \tag{2.81}$$

For a tensor of rank $(0,0)$, i.e. a scalar function $f$, we simply have

$$\nabla_i f = \partial_i f, \tag{2.82}$$

since scalar quantities do not change on parallel transport. There is no sense in defining "parallel" transport for scalars in the first place; parallel to what? Scalars have no direction. It is a rare occurrence indeed when one needs to perform the covariant derivative of tensors of more than second rank, so the explicit examples above should be sufficient for most purposes. However, if need be, one can read the patterns off of the following general expression to find the covariant derivative of any arbitrary tensor of rank $(p, q)$:

$$\begin{aligned}
\nabla_i T^{j_1 j_2 \cdots j_p}{}_{k_1 k_2 \cdots k_q} =\ & \partial_i T^{j_1 j_2 \cdots j_p}{}_{k_1 k_2 \cdots k_q} \\
& + \Gamma^{j_1}_{im} T^{m j_2 \cdots j_p}{}_{k_1 k_2 \cdots k_q} + \Gamma^{j_2}_{im} T^{j_1 m \cdots j_p}{}_{k_1 k_2 \cdots k_q} + \cdots \\
& - \Gamma^m_{ik_1} T^{j_1 j_2 \cdots j_p}{}_{m k_2 \cdots k_q} - \Gamma^m_{ik_2} T^{j_1 j_2 \cdots j_p}{}_{k_1 m \cdots k_q} - \cdots
\end{aligned} \tag{2.83}$$

Presumably these expressions are all that is needed to find the correct derivatives/divergence of any tensor in any coordinate system. The only thing that is left for us to do is figure out exactly how to compute the Christoffel symbols, $\Gamma^i_{jk}$, which we will do in the next section.

**Exercise 2.16** Using (2.83) deduce:

1. $\nabla_k T^{ij}{}_n$.
2. $\nabla_k T^i{}_{nm}$.
3. $\nabla_k T^{ij}{}_{nm}$.

## 2.4 Calculating the Christoffel Symbols

It is possible to calculate the Christoffel symbols from the geometry of parallel transport. Consider as an example the vector **V** in Fig. [2.7]. Its components in polar coordinates are:

$$V^\rho = V \cos\alpha$$
$$V^\varphi = \frac{V}{\rho} \sin\alpha. \tag{2.84}$$

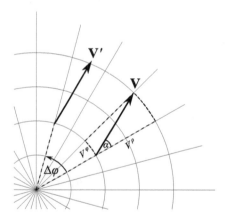

Fig. 2.7: Attempting to calculate the Christoffel symbols from the geometry of parallel transport.

It is not immediately obvious where the factor $1/\rho$ in $V^\varphi$ came from. It arises from the fact that the further one moves radially away from the origin, the narrower the component $V^\varphi$ becomes for the *same* angle $\alpha$—as the reader may check by sketching several vectors further and further away. In fact, at $\rho$ becoming so large as to be infinite, one sees that the vector becomes so close to the radial that its angular component vanishes. On the other hand, if we bring the vector to the origin $\rho = 0$, its angular component becomes undefined. This makes sense because of the fact that polar coordinates are defective at the origin, as we discussed before. The factor $1/\rho$ can also be deduced from writing the vector in natural and physical basis and noting that the physical components are related to the natural components by a factor of $\rho$, arising from $\hat{\mathbf{g}}_\varphi = \rho \hat{\mathbf{v}}_\varphi$.

Now imagine parallel transporting $\mathbf{V}$ through the angle $\Delta\varphi$ to the new location $\mathbf{V}'$. Its components are now

$$
\begin{aligned}
V'^\rho &= V \cos(\alpha - \Delta\varphi) \\
V'^\varphi &= \frac{V}{\rho} \sin(\alpha - \Delta\varphi).
\end{aligned}
\tag{2.85}
$$

Let's apply the trigonometric difference identities:

$$
\begin{aligned}
\cos(\alpha - \Delta\varphi) &= \cos\alpha \cos\Delta\varphi + \sin\alpha \sin\Delta\varphi \\
\sin(\alpha - \Delta\varphi) &= \sin\alpha \cos\Delta\varphi - \cos\alpha \sin\Delta\varphi,
\end{aligned}
\tag{2.86}
$$

which are true for any two angles $\alpha$ and $\Delta\varphi$. However, let us also take $\Delta\varphi$ to be infinitesimally small: $\Delta\varphi \to d\varphi$. Since it is always true that for small angles $\cos d\varphi \approx 1$ and $\sin d\varphi \approx d\varphi$, we end up with:

$$
\begin{aligned}
V'^\rho &= V (\cos\alpha + \sin\alpha \, d\varphi) \\
V'^\varphi &= \frac{V}{\rho} (\sin\alpha - \cos\alpha \, d\varphi),
\end{aligned}
\tag{2.87}
$$

or, using (2.84),

$$
\begin{aligned}
V'^\rho &= V^\rho + \rho V^\varphi d\varphi \\
V'^\varphi &= V^\varphi - \frac{1}{\rho} V^\rho d\varphi.
\end{aligned}
\tag{2.88}
$$

Now, comparing with (2.57), we conclude that

$$
\begin{aligned}
\delta V^\rho &= -\rho V^\varphi d\varphi \\
\delta V^\varphi &= \frac{1}{\rho} V^\rho d\varphi.
\end{aligned}
\tag{2.89}
$$

Comparing these with (2.60) by setting $x^k = (x^\rho, x^\varphi) = (\rho, \varphi)$ and taking the limit $\Delta x^k \to dx^k$, we identify the Christoffel symbols:

$$
\begin{aligned}
\Gamma^\rho_{\varphi\varphi} &= -\rho \\
\Gamma^\varphi_{\rho\varphi} &= \frac{1}{\rho}.
\end{aligned}
\tag{2.90}
$$

Notice that we have found only two out of eight possible Christoffel symbols (three indices, two dimensions each, so $2^3 = 8$). But we have only parallel transported $\mathbf{V}$ through an angle while keeping $\rho$ fixed! To find the remaining six Christoffel symbols we have to perform a similar calculation for parallel transporting along $\rho$ while keeping $\varphi$ fixed, then along both simultaneously.

**Exercise 2.17** For polar coordinates, parallel transport the vector **V** shown in Fig. [2.7] radially while keeping the angle fixed. See which Christoffel symbols you can find. If you give this case some thought you should be able to guess at *all* the remaining Christoffel symbols for polar coordinates. *And*, since cylindrical coordinates are a straightforward generalization, you should also be able to guess the Christoffel symbols for cylindrical coordinates in full.

**Exercise 2.18** If you really *really* want to do this, find the complete set of Christoffel symbols for spherical coordinates by parallel transporting a vector **V** in various directions and combinations of directions.

The method we just outlined is clearly quite tedious and obviously gets worse for other more complex curvilinear coordinate systems. It is also dependent on our ability to plot and visualize the coordinates. In other words, it won't work in dimensions higher than three. Luckily there is a much more elegant, as well as systematic, method of finding the Christoffel symbols (a method that also generalizes quite nicely to higher dimensions as well as curved spaces). Our target, as usual, is to refer everything to the metric. So, we ask, is there a way to calculate the Christoffel symbols directly from a given line element? It turns out there is; however, in order to get to it we have to make certain assumptions about the metric and its connection.[15] To begin with, the following property of the Christoffel symbols is assumed:

$$\Gamma^i_{jk} = \Gamma^i_{kj}, \tag{2.91}$$

meaning it is symmetric on exchange of the two lower indices. For reasons that we will discuss later, this is called the **torsion-free condition**, which is always true in flat Euclidean spaces. However, in the geometry of curved spaces, which we will get to eventually, it is only *assumed* based on the observed physics of our universe. The other property we are assuming is known as **metric compatibility**; it arises from the requirement that, upon parallel transport, lengths and angles are to remain invariant. This can be shown (which we will do in Chapter 7) to be satisfied by requiring that the covariant derivative of the metric *itself* vanishes:

$$\nabla_i g_{jk} = 0. \tag{2.92}$$

It turns out that these two properties are enough to fully pin down the Christoffel symbols. One proceeds as follows: First we write (2.92) in detail using (2.79):

$$\nabla_i g_{jk} = \partial_i g_{jk} - \Gamma^m_{ij} g_{mk} - \Gamma^m_{ik} g_{mj} = 0. \tag{2.93}$$

The idea is to solve for $\Gamma$ using (2.93) and write it purely in terms of the metric. We can do this by rewriting (2.93) in two other ways, differing only by the order of the indices:

---

[15]Recall that for our purposes the term "metric connection" is another name for the Christoffel symbols.

$$\nabla_j g_{ki} = \partial_j g_{ki} - \Gamma^m_{jk} g_{mi} - \Gamma^m_{ji} g_{mk} = 0 \tag{2.94}$$

$$\nabla_k g_{ij} = \partial_k g_{ij} - \Gamma^m_{ki} g_{mj} - \Gamma^m_{kj} g_{mi} = 0. \tag{2.95}$$

Now subtract (2.94) and (2.95) from (2.93):

$$\partial_i g_{jk} - \partial_j g_{ki} - \partial_k g_{ij}$$
$$+ \left( \Gamma^m_{ji} g_{mk} - \Gamma^m_{ij} g_{mk} \right) + \left( \Gamma^m_{ki} g_{mj} - \Gamma^m_{ik} g_{mj} \right) + \left( \Gamma^m_{kj} g_{mi} + \Gamma^m_{jk} g_{mi} \right) = 0. \tag{2.96}$$

Using the condition (2.91), the terms $\left( \Gamma^m_{ji} g_{mk} - \Gamma^m_{ij} g_{mk} \right)$ and $\left( \Gamma^m_{ki} g_{mj} - \Gamma^m_{ik} g_{mj} \right)$ cancel, while $\left( \Gamma^m_{kj} g_{mi} + \Gamma^m_{jk} g_{mi} \right)$ adds up:

$$2 \Gamma^m_{kj} g_{mi} = \partial_j g_{ki} + \partial_k g_{ij} - \partial_i g_{jk}. \tag{2.97}$$

Finally, to get rid of $g_{mi}$ on the left-hand side we contract both sides with $g^{ni}$ and use

$$g^{ni} g_{mi} = \delta^n_m \qquad \text{and} \qquad \delta^n_m \Gamma^m_{kj} = \Gamma^n_{kj}, \tag{2.98}$$

to end up with the formula defining the Christoffel symbols from the metric itself:

$$\Gamma^n_{jk} = \frac{1}{2} g^{ni} \left( \partial_j g_{ki} + \partial_k g_{ij} - \partial_i g_{jk} \right). \tag{2.99}$$

This is a particularly useful expression. It provides us with a systematic way of calculating the Christoffel symbols for any specific coordinate system directly from the metric. Clearly, if the metric components happen to be those for Cartesian coordinates, i.e. $g_{ij} = \delta_{ij}$, then all derivatives are zeros and the Christoffel symbols vanish, as expected.

As an example, lets calculate the non-vanishing Christoffel symbols for cylindrical polar coordinates using the metric (1.26). Since cylindrical polar coordinates include planar polar coordinates as a special case, we would then be able to compare with (2.90). Turns out there are only three non-vanishing components: $\Gamma^\rho_{\varphi\varphi}$, $\Gamma^\varphi_{\rho\varphi}$, and $\Gamma^\varphi_{\varphi\rho}$. However, the torsion-free property (2.91) requires that $\Gamma^\varphi_{\rho\varphi} = \Gamma^\varphi_{\varphi\rho}$; hence we need only calculate one of them. The explicit calculation is

$$g_{\rho\rho} = g^{\rho\rho} = g_{zz} = g^{zz} = 1, \quad g_{\varphi\varphi} = \rho^2, \quad g^{\varphi\varphi} = \frac{1}{\rho^2}$$

$$\Gamma^\rho_{\varphi\varphi} = \frac{1}{2} g^{\rho\rho} \left[ (\partial_\varphi g_{\rho\varphi}) + (\partial_\varphi g_{\rho\varphi}) - (\partial_\rho g_{\varphi\varphi}) \right]$$

$$= -\frac{1}{2} \left( \partial_\rho \rho^2 \right) = -\rho$$

$$\Gamma^\varphi_{\rho\varphi} = \Gamma^\varphi_{\varphi\rho} \frac{1}{2} g^{\varphi\varphi} \left[ (\partial_\rho g_{\varphi\varphi}) + (\partial_\varphi g_{\rho\varphi}) - (\partial_\varphi g_{\rho\varphi}) \right]$$

$$= \frac{1}{2} \frac{1}{\rho^2} \left( \partial_\rho \rho^2 \right) = \frac{1}{\rho}, \tag{2.100}$$

in complete agreement with (2.90) but without the pain of doing intense geometry and trigonometry.

One important point to emphasize: The Christoffel symbol by itself is *not* a tensor! In other words, it does not transform in the appropriate way, so

$$\Gamma^{k'}_{i'j'} \neq \lambda^{k'}_l \lambda^m_{i'} \lambda^n_{j'} \Gamma^l_{mn}, \tag{2.101}$$

as one can (easily?) verify. This is no surprise; the Christoffel symbols were *invented* to define a covariant partial derivative. In other words, since $\partial_i$ is not a tensor, we "add" to it certain quantities that are also non-tensors but specifically designed to make the result a tensor; namely $\nabla_i$. So the anatomy of an equation such as (2.70), for example, is simply: Tensor = non-tensor + non-tensor.

**Exercise 2.19** Prove that the Christoffel symbols are not the components of a tensor, i.e. prove (2.101). Show that the correct transformation law is:

$$\Gamma^{k'}_{i'j'} = \lambda^{k'}_l \lambda^m_{i'} \lambda^n_{j'} \Gamma^l_{mn} + \lambda^{k'}_m \frac{\partial^2 x^m}{\partial x^{i'} \partial x^{j'}}. \tag{2.102}$$

**Exercise 2.20** Calculate the complete set of Christoffel symbols for the following two-dimensional metric:

$$dl^2 = ydx^2 + xdy^2. \tag{2.103}$$

**Exercise 2.21** Show that the metric of Exercise 2.20 satisfies (2.92).

**Exercise 2.22** Calculate the complete set of Christoffel symbols for each of the following coordinate systems. Use the results of Exercise 1.23 where needed.

1. Spherical coordinates (1.27).
2. Elliptic cylindrical coordinates (1.6).
3. Parabolic cylindrical coordinates (1.7).
4. Parabolic coordinates (1.8).
5. Paraboloidal coordinates (1.9).
6. Oblate spheroidal coordinates (1.10).
7. Toroidal coordinates (1.11).
8. The skew coordinates of Exercise 1.2.
9. The two-dimensional metric $dl^2 = e^{-f(r)}dr^2 + r^2 d\theta^2$, where $f$ is an arbitrary function in $r$. You may use the notation $f'$ to signify the unknown derivative of $f$ in your results. Not that it matters here, but a metric such as this does not generally describe a flat plane, but rather a curved one.
10. The two-dimensional metric $dl^2 = 2dudv$. *Reminder*: The metric tensor $g_{ij}$ is a *symmetric* matrix.

**Exercise 2.23** Show that (2.92) is true for each of the cases in Exercise 2.22.

Finally, it is important to note, like we have done more than once so far,[16] that the formulae involving derivatives that we write here are based on the natural basis vectors **g** which means that they will look somewhat different from what you would find in most standard physics textbooks. Let's explore this briefly. Applying (2.82) to cylindrical polar coordinates and remembering that all our vectors are in the natural basis, we find

$$\vec{\nabla} f = \left(\frac{\partial f}{\partial \rho}\right)\hat{\mathbf{g}}_\rho + \left(\frac{\partial f}{\partial \varphi}\right)\hat{\mathbf{g}}_\varphi + \left(\frac{\partial f}{\partial z}\right)\hat{\mathbf{g}}_z. \tag{2.104}$$

However, if the reader checks the back or front covers of most books on classical mechanics or electromagnetism they are most likely to find something that looks like this:

$$\vec{\nabla} f = \left(\frac{\partial f}{\partial \rho}\right)\hat{\rho} + \frac{1}{\rho}\left(\frac{\partial f}{\partial \varphi}\right)\hat{\phi} + \left(\frac{\partial f}{\partial z}\right)\hat{\mathbf{z}}. \tag{2.105}$$

The difference is in the middle term; where did this $1/\rho$ come from? It arises from the fact that the basis vectors used for (2.105) are physical; i.e. they are our **v**'s: $\hat{\mathbf{v}}_\rho = \hat{\rho}$, $\hat{\mathbf{v}}_\varphi = \hat{\phi}$, and $\hat{\mathbf{v}}_z = \hat{\mathbf{z}}$. But we know (as discussed back in §1.6) that $\hat{\mathbf{g}}_\rho = \hat{\mathbf{v}}_\rho$, $\hat{\mathbf{g}}_\varphi = \frac{1}{\rho}\hat{\mathbf{v}}_\varphi$, and $\hat{\mathbf{v}}_z = \hat{\mathbf{z}}$. Hence we see that (2.104) is equivalent to (2.105), just written in a different basis. We will explore this difference further in Chapter 3.

---

**Exercise 2.24** Write the gradient $\vec{\nabla} f$ of a scalar function $f$ in spherical coordinates in the natural then the physical basis vectors. Check that the latter is what you find in any standard treatment on classical mechanics or electromagnetism. Which book did you check?

---

## 2.5  More on the Divergence and Laplacian Operators

There is a particularly simple way of calculating the divergence directly from the metric without having to first calculate the Christoffel symbols. Also, since the Laplacian operator is, by definition, the divergence of a gradient, i.e. $\nabla^2 = \nabla_i \nabla^i = \nabla^i \nabla_i$, this method also leads to an easy (*ish*) expression for $\nabla^2$. The starting point is the formula for a Christoffel symbol with its upper index contracted with one of its lower indices (it doesn't matter which one, since $\Gamma$ is symmetric in its lower indices):

$$\Gamma^j_{ji} = \frac{1}{\sqrt{|g|}}\partial_i\sqrt{|g|}, \tag{2.106}$$

where the quantity $|g|$ is the determinant of the metric $g_{ij}$.[17] The proof of this expression would take us too far afield into linear algebra, so we will skip it.

---

[16]For example, in the paragraph preceding eqn (1.133).

[17]The determinant of the metric $|g|$ is of course positive definite; however, in relativistic theories it is *not*, as we will see. Hence, when adapting any of the equations we are deriving here for relativity, one simply inserts a minus sign *by hand*, making $\sqrt{|g|} \to \sqrt{-|g|}$. We will remind the reader of this once we get to it.

**Exercise 2.25** Verify that (2.106) is true for each of the cases in Exercise 2.22 by comparing it with what you found using (2.99).

Now the divergence contains a contracted Christoffel symbol (2.71):

$$\nabla_i V^i = \partial_i V^i + \Gamma^j_{ji} V^i. \tag{2.107}$$

Substituting (2.106) into (2.107),

$$\nabla_i V^i = \partial_i V^i + \frac{1}{\sqrt{|g|}} \left( \partial_i \sqrt{|g|} \right) V^i. \tag{2.108}$$

Finally, we note that the right-hand side of (2.108) is actually equivalent to

$$\nabla_i V^i = \frac{1}{\sqrt{|g|}} \partial_i \left( \sqrt{|g|} V^i \right), \tag{2.109}$$

which can be verified by applying the product rule on the derivative in (2.109). This is the promised simplified expression for the divergence. It is clearly a lot simpler to use than (2.107), as it is much easier to calculate the determinant of the metric than it is to calculate the Christoffel symbols. It also leads to a simple expression for the Laplacian. But before we do that we should note a certain property of the Laplacian that would make our lives a lot easier. The Laplacian, being the dot product of two "vectors," is automatically a covariant operator, just as a dot product should be. In other words, it behaves just like a scalar. This means that applying the Laplacian to *any* tensor component, no matter how many indices it carries, should have the exact same form. To clarify, think back to the covariant derivative applied on different rank tensors: For example, for rank $(1,0)$ tensors the covariant derivative was defined by (2.70), having a single term containing the Christoffel symbols, while for rank $(0,2)$ the covariant derivative contained two terms with Christoffel symbols. In the case of the Laplacian being a "scalar" there should not be such a distinction, so expressions such as $\nabla^2 f$, $\nabla^2 V^i$, $\nabla^2 V_i$, $\nabla^2 A^{ij}$, and $\nabla^2 A_{ij}$ are all perfectly "legal."

It is then clear that the Laplacian can be applied directly on the components of tensors, not on the tensor as a whole. Since all tensor components are themselves scalars, we can just consider the Laplacian acting on any scalar function $f$:

$$\nabla^2 f = \nabla_i \nabla^i f. \tag{2.110}$$

Now, since the Laplacian is the divergence of a gradient, we can apply (2.109):

$$\nabla^2 f = \nabla_i \nabla^i f = \frac{1}{\sqrt{|g|}} \partial_i \left( \sqrt{|g|} \nabla^i f \right). \tag{2.111}$$

Noting that $\nabla^i f = \partial^i f$, we end up with

$$\nabla^2 f = \nabla_i \nabla^i f = \frac{1}{\sqrt{|g|}} \partial_i \left( \sqrt{|g|} \partial^i f \right). \tag{2.112}$$

This expression for the covariant Laplacian can now be applied to any component of any tensor; for example,

$$\nabla^2 V^k = \frac{1}{\sqrt{|g|}} \partial_i \left( \sqrt{|g|} \partial^i V^k \right),$$

$$\nabla^2 V_k = \frac{1}{\sqrt{|g|}} \partial_i \left( \sqrt{|g|} \partial^i V_k \right),$$

$$\nabla^2 A^k_n = \frac{1}{\sqrt{|g|}} \partial_i \left( \sqrt{|g|} \partial^i A^k_n \right), \tag{2.113}$$

and so on. This concludes our discussion of the various types of covariant first and second derivatives. There exist other formulae—divergences of tensors, for example—but they are rarely useful. The only remaining type of derivative common in three dimensions is the curl, which we'll briefly discuss in the next section.

---

**Exercise 2.26** Prove that the covariant derivative commutes with itself, i.e. $\nabla_i \nabla_j = \nabla_j \nabla_i$.

**Exercise 2.27** Derive:

1. The divergence in cylindrical and spherical coordinates applied to an arbitrary vector **A**.
2. The Laplacian in cylindrical and spherical coordinates applied to an arbitrary function $f$.

   By noting the difference between the natural basis and the physical basis, confirm that what you found agrees with the standard divergence and Laplacian found in the front or back cover of any standard textbook on classical mechanics or electromagnetism. Which book did you check?

**Exercise 2.28** Derive:

1. The divergence for each of the coordinate systems of Exercise 2.22.
2. The Laplacian for each of the coordinate systems of Exercise 2.22.

   Do not do spherical coordinates if you have already done Exercise 2.27.

**Exercise 2.29** Calculate the divergence of the following vectors. Use the results of Exercise 2.27 where needed.

1. $\mathbf{A} = \left( 5 \sin \theta, 2 \cos \varphi, 3r^2 \right)$ in spherical coordinates.
2. $\mathbf{B} = \left( -v, 2u^2, -2z + u \right)$ in elliptic cylindrical coordinates.
3. $\mathbf{C} = (2\xi, -\eta, 3z)$ in parabolic cylindrical coordinates.
4. $\mathbf{D} = (-u, 4v, 2\theta)$ in parabolic coordinates.
5. $\mathbf{E} = \left( 3\varphi\mu^2, 2v, -v + 3\mu \right)$ in oblate spheroidal coordinates.
6. $\mathbf{F} = \left( 2\sigma\tau, -\sigma^3, 3\tau \right)$ in toroidal coordinates.

**Exercise 2.30** Calculate the Laplacian of the following functions. Use the results of Exercise 2.27 where needed.

1. $A(r, \theta) = r^3 \tan \theta$ in spherical coordinates.
2. $B(u, v) = v \ln u$ in elliptic cylindrical coordinates.
3. $C(\xi, \eta, z) = \xi^2 e^{-\eta} + z$ in parabolic cylindrical coordinates.
4. $D(u, v, \theta) = uv \cosh \theta$ in parabolic coordinates.
5. $E(\mu, \nu, \varphi) = \mu \nu^2 \sin \varphi$ in oblate spheroidal coordinates.
6. $F(\sigma, \tau) = 2\sigma + \frac{\tau}{\sigma}$ in toroidal coordinates.

**Exercise 2.31** Prove the following identities:

1. $g_{ij} \left( \partial_k g^{jn} \right) = -g^{jn} \left( \partial_k g_{ij} \right)$.
2. $\left( \partial_i g^{jk} \right) = -\Gamma^j_{ni} g^{nk} - \Gamma^k_{ni} g^{nj}$.

## 2.6 Integrals and the Levi-Civita Tensor

So far we have defined covariant vectors, tensors, and derivatives using the index notation. The only thing that is left to consider is integrals. As a bonus, we will also be able to define a covariant Levi-Civita tensor, since the reader will recall that the ordinary totally antisymmetric symbol (1.94) is only valid in Cartesian coordinates. Given a specific set of coordinates defined by the metric tensor, how does one write down single, double, and triple integrals independently of the choice of coordinates; i.e. covariant integrals? The pertinent item to consider is the integrals' infinitesimal elements: $dx$ for single integrals, the area element $da$ for double integrals, and the volume element $dV$ for triple integrals. Consider first an ordinary one-dimensional integral,

$$I = \int f(x) \, dx, \qquad (2.114)$$

where $f(x)$ is some arbitrary integrand (smooth, continuous, etc.). Transforming this integral into a different coordinate parameter $x'$ gives

$$I = \int f(x') \, J dx', \qquad (2.115)$$

where we have replaced $x$ with $x'$ in the function $f$ and we have replaced $dx$ with $J dx'$. The factor $J$ is found by the usual substitution techniques. Let's consider an example: If we knew that $x' = x^2$ ($x$ squared, *not* an index), then

$$dx' = 2x dx$$
$$dx = \frac{dx'}{2x} = \frac{dx'}{2\sqrt{x'}}, \qquad (2.116)$$

leading to

$$I = \int f\left(x'\right) \frac{dx'}{2\sqrt{x'}}, \qquad (2.117)$$

so in this case

$$J = \frac{1}{2\sqrt{x'}}. \qquad (2.118)$$

In other words, the factor $J$, known in mathematical circles as the **Jacobian**,[18] is simply found by

$$J = \frac{dx}{dx'}. \qquad (2.119)$$

In the example we just considered, $x = \sqrt{x'}$, so (2.119) gives (2.118). Clearly, then, knowledge of how $x'$ is related to $x$ is the key to transforming an integral from one set of variables into another or one set of *coordinates* into another. Now consider double integrals in two dimensions:

$$I = \int \int f\left(x^1, x^2\right) da, \qquad (2.120)$$

where if $x^1$ and $x^2$ are the ordinary Cartesian coordinates, then the element of area $da$ is simply the area of the infinitesimal square $da = dx^1 dx^2$. If one wishes to transform such an integral to another coordinate system, say planar polar coordinates ($x^{1'} = \rho$ and $x^{2'} = \varphi$), then all one needs to do is replace $x^1$ and $x^2$ with the appropriate relations, specifically (1.24), which lead to

$$da = dx^1 dx^2 = J dx^{1'} dx^{2'} = \rho d\rho d\varphi, \qquad (2.121)$$

where the Jacobian is $J = \rho$ in this case. The integral becomes

$$I = \int \int f\left(\rho, \varphi\right) \rho d\rho d\varphi. \qquad (2.122)$$

Analogously to (2.119), the two-dimensional Jacobian can analytically be found as follows:

$$J = \left(\frac{\partial x^1}{\partial x^{1'}}\right)\left(\frac{\partial x^2}{\partial x^{2'}}\right) - \left(\frac{\partial x^2}{\partial x^{1'}}\right)\left(\frac{\partial x^1}{\partial x^{2'}}\right), \qquad (2.123)$$

where for polar coordinates this gives:

$$J = \rho \cos^2 \varphi + \rho \sin^2 \varphi = \rho, \qquad (2.124)$$

as needed. The expression (2.123) is in fact the determinant:

$$J = \begin{vmatrix} \left(\dfrac{\partial x^1}{\partial x^{1'}}\right) & \left(\dfrac{\partial x^1}{\partial x^{2'}}\right) \\[2ex] \left(\dfrac{\partial x^2}{\partial x^{1'}}\right) & \left(\dfrac{\partial x^2}{\partial x^{2'}}\right) \end{vmatrix} \qquad (2.125)$$

[18]Named after the German mathematician Carl Gustav Jacob Jacobi (1804–1851).

where $(x^1, x^2)$ are the original coordinates (not necessarily Cartesian) and $\left(x^{1'}, x^{2'}\right)$ are the new coordinates. It follows, then, that generally

$$da = J dx^1 dx^2 \tag{2.126}$$

is a covariant element of area. In connection to our previous study of tensor transformations, we note that the components of the Jacobian determinant are in fact the components of the transformation tensor $\lambda^i_{j'}$ (2.19). In other words, thinking of $\lambda$ as a matrix, the Jacobian is simply the determinant:

$$J = |\lambda| . \tag{2.127}$$

The same procedure can be easily generalized to three dimensions. Given a triple integral in Cartesian coordinates,

$$I = \int \int \int f\left(x^1, x^2, x^3\right) dV, \tag{2.128}$$

where now the volume element is $dV = dx^1 dx^2 dx^2$, the Jacobian is the straightforward generalization

$$J = \begin{vmatrix} \left(\frac{\partial x^1}{\partial x^{1'}}\right) & \left(\frac{\partial x^1}{\partial x^{2'}}\right) & \left(\frac{\partial x^1}{\partial x^{3'}}\right) \\ \left(\frac{\partial x^2}{\partial x^{1'}}\right) & \left(\frac{\partial x^2}{\partial x^{2'}}\right) & \left(\frac{\partial x^2}{\partial x^{3'}}\right) \\ \left(\frac{\partial x^3}{\partial x^{1'}}\right) & \left(\frac{\partial x^3}{\partial x^{2'}}\right) & \left(\frac{\partial x^3}{\partial x^{3'}}\right) \end{vmatrix}, \tag{2.129}$$

leading to

$$dV = dx^1 dx^2 dx^3 = J dx^{1'} dx^{2'} dx^{3'}. \tag{2.130}$$

We can check that this leads to the correct expressions for cylindrical and spherical coordinates respectively:

$$\begin{aligned} I &= \int \int \int f\left(\rho, \varphi, z\right) \rho d\rho d\varphi dz \\ &= \int \int \int f\left(r, \theta, \varphi\right) r^2 \sin\theta dr d\theta d\varphi, \end{aligned} \tag{2.131}$$

In a sense the Jacobian is to integrals what the Christoffel symbols are to derivatives. It guarantees the covariance of $dx$, $da$, and $dV$. Now, just as we related the Christoffel symbols to the metric, can we also relate the Jacobian to the metric? This can indeed be done. Begin by applying the tensor transformation rule (2.32) to the metric tensor:

$$g_{i'j'} = \lambda_{i'}^m \lambda_{j'}^n g_{mn}. \tag{2.132}$$

Treated as a product of matrices, we can take the determinant of both sides:

$$|g'| = |\lambda \lambda g|. \tag{2.133}$$

In linear algebra, we learn that if $[A]$, $[B]$, and $[C]$ are matrices, then the determinant of their product is equal to the product of their determinants $|ABC| = |A|\,|B|\,|C|$. Applying this to (2.133),

$$|g'| = |\lambda|^2\,|g|. \tag{2.134}$$

Equation (2.134) is true for any metric $g_{ij}$ and any transformed metric $g_{i'j'}$. Let's choose the original metric to be Cartesian, i.e. $g_{ij} = \delta_{ij}$; then (2.134) becomes

$$|g| = |\lambda|^2\,|\delta| \tag{2.135}$$

and the prime on $g$ is no longer necessary. Noting that the determinant of the Kronecker delta is just $|\delta| = 1$ and that $|\lambda| = J$, as observed earlier, we end up with the following simple rule:

$$J = \sqrt{|g|}. \tag{2.136}$$

In conclusion, then, a triple integral is covariant if and only if it looks like

$$\int \int \int \sqrt{|g|} f\left(x^i\right) dx^1 dx^2 dx^3, \tag{2.137}$$

where $\left(x^1, x^2, x^3\right)$ define any system of coordinates. Similar definitions can be written for single and double integrals.[19]

---

**Exercise 2.32** Using (2.136), calculate the Jacobian for the coordinate systems of Exercise 2.22.

---

The metric's determinant $|g|$ is a quantity with magnitude but no direction; however, it is not, strictly speaking, a scalar. If it were, then it would transform following the scalar (rank 0 tensors) transformation rule, which would simply be $|g'|=|g|$, but we already know that this is not true for $|g|$, as it transforms following (2.134). Quantities that transform like this are *not* tensors, they are a new class of objects known as **tensor densities**. Generally speaking, a tensor density is a quantity $\mathscr{A}$ (that may

---

[19]Once again these equations can be adapted to higher-dimensional metrics in relativity theory where the determinant $|g|$ is no longer positive definite and a minus sign is inserted by hand into the square root.

or may not carry indices) that transforms following a similar rule to that of tensors, except that there is a multiplicative factor of $|\lambda|^{\omega}$, where $\omega$ is some real number known as the **weight** of the tensor density. For example, a tensor density of rank $(0, 2)$, i.e. $\mathscr{A}_{ij}$, would transform as follows

$$
\begin{aligned}
\mathscr{A}_{i'j'} &= |\lambda|^{\omega} \, \lambda_{i'}^{m} \lambda_{j'}^{n} \mathscr{A}_{mn} \\
&= |\lambda|^{\omega} \left( \partial_{i'} x^{m} \right) \left( \partial_{j'} x^{n} \right) \mathscr{A}_{mn},
\end{aligned}
\tag{2.138}
$$

compared to the tensor transformation rule (2.32). One says that $\mathscr{A}_{ij}$ is a rank $(0, 2)$ tensor density of weight $\omega$. Clearly, then, $|g|$ is a rank $(0, 0)$ tensor density of weight 2, also sometimes known as a **scalar density**.

As a bonus to this discussion, we can use this concept of tensor density to finally define a covariant Levi-Civita symbol. Recall that, as defined in §1.5, the Levi-Civita symbol is only valid in Cartesian coordinates; in other words, it does not transform following the appropriate version of (2.36). It turns out that its covariant version is not a tensor, but rather a tensor density. Let us call the sought quantity $\mathscr{E}_{ijk}$ and write it as follows:

$$
\mathscr{E}_{ijk} = |g|^{\alpha} \, \varepsilon_{ijk}.
\tag{2.139}
$$

The object $\mathscr{E}_{ijk}$ clearly has the necessary antisymmetry, but is it a true tensor density? And if it is, what is the value of the power $\alpha$? To answer these questions let us first note that the determinant of any given matrix $A_{j}^{i}$ can be written as follows (the proof is in linear algebra):

$$
|A| = \frac{1}{3!} \varepsilon^{ijk} \varepsilon_{lmn} A_{i}^{l} A_{j}^{m} A_{k}^{n},
\tag{2.140}
$$

as the reader may explicitly check. Starting with (2.140) it is straightforward, but rather tedious, to derive the identity

$$
A_{i}^{l} A_{j}^{m} A_{k}^{n} \varepsilon_{lmn} = |A| \, \varepsilon_{ijk}.
\tag{2.141}
$$

**Exercise 2.33** Explicitly check that (2.140) is correct.

**Exercise 2.34** Starting from (2.141), derive (2.140).

Now, assuming it is a tensor, let us transform $\mathscr{E}_{lmn}$ to a different coordinate system using the appropriate transformation rule,

$$
\mathscr{E}_{i'j'k'} = \lambda_{i'}^{l} \lambda_{j'}^{m} \lambda_{k'}^{n} \mathscr{E}_{lmn},
\tag{2.142}
$$

then use (2.139)

$$
\mathscr{E}_{i'j'k'} = \lambda_{i'}^{l} \lambda_{j'}^{m} \lambda_{k'}^{n} |g|^{\alpha} \, \varepsilon_{lmn}.
\tag{2.143}
$$

The aim is to find the value of $\alpha$ that would guarantee that $\mathscr{E}_{i'j'k'}$ has exactly the same form as (2.139) in the new coordinate system. Now, apply (2.141) on $\lambda$ to get

$$\lambda_{i'}^l \lambda_{j'}^m \lambda_{k'}^n \varepsilon_{lmn} = |\lambda| \, \varepsilon_{i'j'k'}, \tag{2.144}$$

such that (2.143) becomes

$$\mathscr{E}_{i'j'k'} = |\lambda| \, |g|^\alpha \, \varepsilon_{i'j'k'}. \tag{2.145}$$

Using (2.134) we get

$$\mathscr{E}_{i'j'k'} = \sqrt{\frac{|g'|}{|g|}} \, |g|^\alpha \, \varepsilon_{i'j'k'}, \tag{2.146}$$

where it is clear that the unprimed $|g|$ cancels out only for $\alpha = 1/2$ and we get both (2.139) and (2.146) to look exactly the same as required, i.e.

$$\mathscr{E}_{ijk} = \sqrt{|g|}\varepsilon_{ijk}$$
$$\mathscr{E}_{i'j'k'} = \sqrt{|g'|}\varepsilon_{i'j'k'}. \tag{2.147}$$

We can thus conclude that $\mathscr{E}_{ijk}$ is indeed covariant. It can be shown that raising all the indices on $\mathscr{E}_{ijk}$ gives

$$\mathscr{E}^{ijk} = g^{il} g^{jm} g^{kn} \mathscr{E}_{lmn} = \frac{1}{\sqrt{|g|}} \varepsilon^{ijk}. \tag{2.148}$$

The relations (2.147) and (2.148) demonstrate that $\mathscr{E}_{ijk}$ is not quite a bona-fide tensor; comparison with (2.138) and using $|g| = |\lambda|^2$ indicates that it is technically a tensor density of weight $+1$. We will, however, continue to refer to it as a "tensor," since it is covariant and for our purposes that is all that is needed. It is straightforward to show that $\mathscr{E}_{ijk}$ satisfies similar relations to (1.105), (1.106), and (1.107), provided that it is understood that the indices are now raised and lowered by $g_{ij}$.

The Levi-Civita tensor is a useful quantity if one wishes to define covariant cross products and curls of vectors. Consider two vectors $\mathbf{A} = A^i \hat{\mathbf{g}}_i$ and $\mathbf{B} = B^i \hat{\mathbf{g}}_i$ in any arbitrary coordinate system; their cross product can now be written as

$$\mathbf{C} = \mathbf{A} \times \mathbf{B} = \mathscr{E}^n{}_{jk} A^j B^k \hat{\mathbf{g}}_n, \tag{2.149}$$

or in component form

$$C^n = g^{ni} \mathscr{E}_{ijk} A^j B^k. \tag{2.150}$$

We can also use it to define the covariant curl of a vector. One way of doing this is

$$\vec{\nabla} \times \mathbf{V} = \mathscr{E}^{ijk} \left[ \nabla_j V_k \right] \hat{\mathbf{g}}_i. \tag{2.151}$$

Notice that if we use (2.78) in (2.151),

$$\vec{\nabla} \times \mathbf{V} = \mathscr{E}^{ijk} \left[ (\partial_j V_k) - \Gamma^n_{jk} V_n \right] \hat{\mathbf{g}}_i$$
$$= \left[ \mathscr{E}^{ijk} (\partial_j V_k) - \mathscr{E}^{ijk} \Gamma^n_{jk} V_n \right] \hat{\mathbf{g}}_i, \tag{2.152}$$

the second term actually vanishes, since $\mathscr{E}^{ijk}$ is antisymmetric while $\Gamma^n_{jk}$ is symmetric in $j$ and $k$, so they cancel each other exactly. Hence a covariant curl does not really require the covariant derivative; the ordinary partial derivative suffices:

$$\vec{\nabla} \times \mathbf{V} = \mathscr{E}^{ijk} (\partial_j V_k) \hat{\mathbf{g}}_i. \tag{2.153}$$

We have combined the discussion on integrals *and* the discussion on the Levi-Civita tensor in one section simply because both "happen" to be dependent on tensor densities; in fact, on the same tensor density: $\sqrt{|g|}$. This is, in fact, *not* a coincidence; there *does* exist a deep relation between integrals and the Levi-Civita symbol. It is a rather unexpected one, but it requires knowledge of something called differential forms: a somewhat advanced technique that we will discuss in Chapter 9.

**Exercise 2.35** Using (2.153), derive the curl in cylindrical and spherical coordinates applied to an arbitrary vector **A**. By noting the difference between the natural basis and the physical basis, confirm that what you found agrees with the standard curl found in the front or back cover of any standard textbook on classical mechanics or electromagnetism. Which book did you check?

**Exercise 2.36** Using (2.153), derive the curl in each of the coordinate systems of Exercise 2.22. Do not do spherical coordinates if you have already done Exercise 2.35. You may use the results of Exercise 2.32.

**Exercise 2.37** Express the following laws of physics in covariant form. You may use the additional notation $\partial_t = \partial/\partial t$. You may also wish to refer to the results of Exercises 1.12 and 1.18.

1. The wave equation $\nabla^2 f - \frac{1}{v^2} \frac{\partial^2 f}{\partial t^2} = 0$. The quantity $v$ is a constant signifying the speed of the wave.
2. Maxwell's eqns (4.168), (4.169), (4.170), and (4.171).

**Exercise 2.38** For the following, find the components of $A$ in terms of the given arbitrary quantities. Indices are raised and lowered by an arbitrary metric tensor.

1. $kA^i + \mathscr{E}^{ijk} A_j B_k = C^i$.
2. $\mathscr{E}_{ijk} A^j B^k = C_i$.
3. $A^{ij} = -A^{ji}$, $A^{ij} B_i = C^j$, $A^{ij} C_i = 0$, and $B^i C_i = 0$.

**Exercise 2.39** Using the properties of the Levi-Civita and the metric tensor, prove the following identities for any coordinate system, where **r** is the position vector, $r$ is its magnitude, $d\mathbf{r}$ is the displacement vector, and **J** is an arbitrary vector field:

1. $\vec{\nabla} \times \left( \frac{\mathbf{r}}{r} \times \mathbf{J} \right) = \frac{\mathbf{J}}{r^3} - \frac{3(\mathbf{J} \cdot \mathbf{r}) \mathbf{r}}{r^5}$.   2. $d\mathbf{r} \times \vec{\nabla} \left( \frac{1}{r} \right) = \frac{\mathbf{r}}{r^3} \times d\mathbf{r}$.

# 3

# Classical Covariance

*The description of right lines and circles, upon which geometry is founded, belongs to mechanics. Geometry does not teach us to draw these lines, but requires them to be drawn.*

Sir Isaac Newton

## Introduction

In the first two chapters, we reformulated our understanding of vectors in terms of the concept of covariance. In the process, we introduced tensors as covariant multi-directional quantities as opposed to the single directional vectors. All of this was done using the so-called index or component notation. To apply these concepts to a physics problem, two pieces of information are needed: the first is what coordinate system is being used and the second is what type of covariance is in effect—both of which may be fully specified by the form of the metric. In this sense, non-relativistic physics is completely covariant in three-dimensional Euclidean space. We will consider Newtonian mechanics and reformulate its laws in coordinate covariant form. In addition, we will discuss another type of covariance: the invariance of the laws of nature between coordinate systems moving with a constant speed with respect to each other (so-called inertial frames of reference). This can be called Galilean covariance, as it was first introduced by none other than the famous Galileo himself. We will see that this type of covariance introduces time as the fourth dimension (in a certain sense) and paves the way to a more precise definition of this notion in the special theory of relativity. The other edifice of classical physics, Maxwell's theory of electrodynamics, is also coordinate covariant; however, it is *not* Galilean covariant. Hence we will postpone discussing it until the next chapter.

## 3.1 Point Particle Mechanics

We can systematically rebuild classical mechanics from the ground up using a principle that is sometimes referred to as **minimal substitution**. It goes like this:

1. Begin with any law of physics in its usual traditional form.

*Covariant Physics: From Classical Mechanics to General Relativity and Beyond.* Moataz H. Emam,
Oxford University Press (2021). © Moataz H. Emam.
DOI: 10.1093/oso/9780198864899.003.0003

2. Rewrite the formula using the index notation in *Cartesian* coordinates.

3. Perform the substitutions

$$\delta_{ij} \;\rightarrow\; g_{ij} \quad \text{and} \quad \delta^{ij} \rightarrow g^{ij} \quad (\delta^i_j \text{ remains unchanged})$$
$$\hat{\mathbf{e}}_i \;\rightarrow\; \hat{\mathbf{g}}_i$$
$$\partial_i \;\rightarrow\; \nabla_i$$
$$\varepsilon_{ijk} \;\rightarrow\; \mathscr{E}_{ijk}$$

Insert $\sqrt{|g|}$ in integrals as appropriate. $\hspace{2cm}$ (3.1)

The equations that result are *guaranteed* to be covariant, i.e. guaranteed to work in any coordinate system specified by $g_{ij}$.[1]

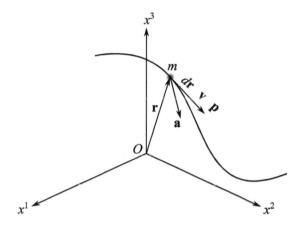

Fig. 3.1: The fundamental kinematical vectors of a point particle of mass $m$ with respect to the particle's trajectory.

Let's now define the basic kinematical quantities of point particle mechanics and write them in the index notation in Cartesian coordinates. These are the position, displacement, velocity, acceleration, and momentum vectors respectively (as shown in Fig. [3.1]):

$$\mathbf{r}\,(t) = x^i\,(t)\,\hat{\mathbf{e}}_i$$
$$d\mathbf{r}\,(t) = dx^i\,(t)\,\hat{\mathbf{e}}_i$$
$$\mathbf{v}\,(t) = v^i\,(t)\,\hat{\mathbf{e}}_i$$
$$\mathbf{a}\,(t) = a^i\,(t)\,\hat{\mathbf{e}}_i$$
$$\mathbf{p}\,(t) = mv^i\,(t)\,\hat{\mathbf{e}}_i, \hspace{2cm} (3.2)$$

where $m$ is the mass of the object under study, assumed a point particle for now, and an over dot is standard notation for a time derivative ($\dot{x} = \frac{dx}{dt}$, $\ddot{x} = \frac{d^2x}{dt^2}$). Using

---

[1]Although there are some subtleties, keep reading.

minimal substitution, we just replace $\hat{\mathbf{e}}_i \to \hat{\mathbf{g}}_i$ to get their covariant form in the natural basis:

$$\mathbf{r}(t) = x^i(t)\,\hat{\mathbf{g}}_i$$
$$d\mathbf{r}(t) = dx^i(t)\,\hat{\mathbf{g}}_i$$
$$\mathbf{v}(t) = v^i(t)\,\hat{\mathbf{g}}_i$$
$$\mathbf{a}(t) = a^i(t)\,\hat{\mathbf{g}}_i$$
$$\mathbf{p}(t) = mv^i(t)\,\hat{\mathbf{g}}_i, \tag{3.3}$$

where now the indices count over *any* three-dimensional coordinate system defined by the metric. As mentioned, these vectors are *kinematic*, in the sense that they describe the motion without reference to the *cause* of the motion. This last would be the realm of *dynamics*, which begins by defining the concept of "force" as the cause of acceleration, from which the other quantities can be derived. The connection between cause and effect is given by Newton's second law,

$$F^i = ma^i. \tag{3.4}$$

The left-hand side of (3.4) is the $i^{\text{th}}$ component of the *net* force acting on the object, in the sense that, given several forces, $F^i$ is their *resultant* sum. The standard problem of point particle mechanics is then defined as follows: Given a set of forces acting on a mass $m$, write (3.4) in terms of the time derivatives of $x^i$. This usually gives a set of second-order differential equations in $x^i$ referred to as the **equations of motion** of $m$. Finding these equations is typically the easy part; solving them for $x^i(t)$ is usually another story altogether.[2]

Before we proceed, let us confirm that the vectors in (3.3), found by minimal substitution, are in fact covariant, and whether or not we need to "fix" their components in any way to make them so. The position vector transforms straightforwardly. Defining $\mathbf{r}' = x^{i'}\hat{\mathbf{g}}_{i'}$ in arbitrary coordinates, and using (1.129), (2.18), and (2.23), we write:

$$\mathbf{r}' = x^{i'}\hat{\mathbf{g}}_{i'} = \lambda^{i'}_j x^j \frac{\partial \mathbf{r}}{\partial x^{i'}} = \lambda^{i'}_j x^j \frac{\partial x^k}{\partial x^{i'}}\hat{\mathbf{e}}_k$$
$$= \lambda^{i'}_j \lambda^k_{i'} x^j \hat{\mathbf{e}}_k = \delta^k_j x^j \hat{\mathbf{e}}_k = x^k \hat{\mathbf{e}}_k = \mathbf{r}. \tag{3.5}$$

Hence $\mathbf{r}' = x^{i'}\hat{\mathbf{g}}_{i'} = x^i\hat{\mathbf{g}}_i = \mathbf{r}$, or in component form: $x^{i'} = \lambda^{i'}_j x^j$. The displacement vector $d\mathbf{r}$ transforms as follows:

$$d\mathbf{r}' = d\left(x^{i'}\hat{\mathbf{g}}_{i'}\right) = x^{i'}d\hat{\mathbf{g}}_{i'} + \hat{\mathbf{g}}_{i'}dx^{i'}$$
$$= \lambda^{i'}_j x^j \hat{\mathbf{e}}_k d\lambda^k_{i'} + \lambda^k_{i'} x^j \hat{\mathbf{e}}_k d\lambda^{i'}_j + dx^{i'}\lambda^{i'}_j \lambda^k_{i'}\hat{\mathbf{e}}_k. \tag{3.6}$$

---

[2] As students of advanced classical mechanics know quite well there are other formulations of classical mechanics that do not refer to Newton's laws explicitly. The most famous example of these is the so-called "Lagrangian mechanics," which we will briefly discuss in Chapter 8.

The dependence on $d\lambda$ in the first two terms vanishes as follows:

$$d\lambda^k_{i'} = \frac{\partial \lambda^k_{i'}}{\partial x^j} dx^j = \frac{\partial}{\partial x^{i'}}\left(\frac{\partial x^k}{\partial x^j}\right) dx^j = \frac{\partial \delta^k_j}{\partial x^{i'}} dx^j = 0. \tag{3.7}$$

This works similarly for $d\lambda^{i'}_j$. Hence (3.6) reduces to

$$d\mathbf{r}' = dx^{i'}\,\hat{\mathbf{g}}_{i'} = dx^i \hat{\mathbf{e}}_i = d\mathbf{r}, \tag{3.8}$$

or in component form: $dx^{i'} = \lambda^{i'}_j\, dx^j$, as required. Since $\mathbf{v} = \frac{d\mathbf{r}}{dt}$, we also know that the velocity transforms correctly: $\mathbf{v}' = v^{i'}\,\hat{\mathbf{g}}_{i'} = v^i \hat{\mathbf{e}}_i = \mathbf{v}$, or $v^{i'} = \lambda^{i'}_j v^j$. A problem arises, however, when we get to the second derivative, i.e. acceleration:

$$\begin{aligned}\mathbf{a}' &= \frac{d\mathbf{v}'}{dt} = \frac{d}{dt}\left(v^{i'}\,\hat{\mathbf{g}}_{i'}\right) = \frac{dv^{i'}}{dt}\hat{\mathbf{g}}_{i'} + v^{i'}\frac{d\hat{\mathbf{g}}_{i'}}{dt}\\[2mm] &= \lambda^j_{i'}a^{i'}\hat{\mathbf{e}}_j + v^{i'}\frac{d\lambda^j_{i'}}{dt}\hat{\mathbf{e}}_j = a^j\hat{\mathbf{e}}_j + v^{i'}\frac{d\lambda^j_{i'}}{dt}\hat{\mathbf{e}}_j = \mathbf{a} + v^{i'}\frac{d\lambda^j_{i'}}{dt}\hat{\mathbf{e}}_j.\end{aligned} \tag{3.9}$$

The required transformation $\mathbf{a}' = \mathbf{a}$ is ruined by the presence of the second term, which in this case doesn't vanish:

$$\frac{d\lambda^j_{i'}}{dt} = \frac{\partial \lambda^j_{i'}}{\partial x^{k'}}\frac{\partial x^{k'}}{\partial t} = \frac{\partial^2 x^j}{\partial x^{k'}\partial x^{i'}}\dot{x}^{k'}. \tag{3.10}$$

Hence (3.9) becomes

$$\mathbf{a}' = \mathbf{a} + \frac{\partial^2 x^j}{\partial x^{k'}\partial x^{i'}}v^{i'}v^{k'}\hat{\mathbf{e}}_j, \tag{3.11}$$

or in component form:

$$a^{i'} = \lambda^{i'}_j a^j + \frac{\partial^2 x^{i'}}{\partial x^{j'}\partial x^{k'}}v^{j'}v^{k'}. \tag{3.12}$$

We conclude, then, that the vector $a^i = \dot{v}^i = \ddot{x}^i$ is not covariant; an extra term proportional to $\frac{\partial^2 x^i}{\partial x^j \partial x^k}v^j v^{k'}$ must be added to guarantee covariance. It turns out that this is exactly the Christoffel symbols $\Gamma^i_{jk}$; in other words, the definition

$$\mathbf{a} = a^i\hat{\mathbf{g}}_i = \left(\dot{v}^i + \Gamma^i_{jk}v^j v^k\right)\hat{\mathbf{g}}_i \tag{3.13}$$

*is* covariant. The proof that $\frac{\partial^2 x^i}{\partial x^{j'}\partial x^{k'}}$ leads to the Christoffel symbols is somewhat tricky; the curious student may consult, for example, appendix 1 of [4]. There does exist, however, a rather "handwavy" trick that further confirms (3.13). First write the components of the acceleration explicitly:

$$a^i(t) = \frac{dv^i}{dt}.$$
(3.14)

Secondly, apply the chain rule

$$a^i = \frac{dv^i}{dt} = \frac{\partial v^i}{\partial x^j}\frac{dx^j}{dt} = \left(\partial_j v^i\right)\frac{dx^j}{dt} = \left(\partial_j v^i\right)v^j,$$
(3.15)

*then* perform the minimal substitution $\partial_i \to \nabla_i$:

$$a^i = \left(\nabla_j v^i\right)v^j,$$
(3.16)

which is explicitly

$$a^i = \left(\partial_j v^i + \Gamma^i_{jk}v^k\right)v^j$$
$$= \left[\left(\partial_j v^i\right)v^j + \Gamma^i_{jk}v^k v^j\right].$$
(3.17)

In the last expression the first term is exactly the chain rule of $dv/dt$, hence we end up with

$$a^i = \dot{v}^i + \Gamma^i_{jk}v^k v^j$$
$$= \ddot{x}^i + \Gamma^i_{jk}\dot{x}^k \dot{x}^j,$$
(3.18)

as needed.

**Exercise 3.1** Explicitly verify that the acceleration (3.18) is covariant, i.e. transforms as a tensor.

We can now establish the covariant form of Newton's second law as:

$$F^i = m\ddot{x}^i + m\Gamma^i_{jk}\dot{x}^k\dot{x}^j.$$
(3.19)

This is exactly the expression that straightforwardly leads to the desired equations of motion in *any* coordinate system. For example, it gives (0.6) in polar coordinates (as is detailed from eqn [3.22] onwards below). All one needs is knowledge of the Christoffel symbols, which are found from the metric using (2.99). Of course, in Cartesian coordinates the Christoffel symbols vanish and the equations of motion reduce to

$$F^i = m\ddot{x}^i.$$
(3.20)

Now, it is worth noting that the term containing the connection in (3.19) does in fact have a physical meaning (in the context of Newton's law) beyond what we have already learned. If we move it to the left-hand side we get

$$F^i - m\Gamma^i_{jk}\dot{x}^k\dot{x}^j = m\ddot{x}^i \qquad (3.21)$$

$$\underbrace{\phantom{F^i - m\Gamma^i_{jk}\dot{x}^k\dot{x}^j}}_{F} \quad \underbrace{\phantom{m\ddot{x}^i}}_{ma}$$

which implies that $m\Gamma^i_{jk}\dot{x}^k\dot{x}^j$ may be viewed as extra "forces" that appear *only* in non-Cartesian coordinate systems. For an observer moving along the non-straight paths of a curvilinear coordinate system, the quantities $m\Gamma^i_{jk}\dot{x}^k\dot{x}^j$ are "perceived" as a number of extra "forces." But clearly these are not real forces. They don't arise from, say, ordinary push and pull, gravity, or any other external agent; they are a manifestation of the fact that the observer is accelerating along non-straight paths (circles of constant $\rho$ in polar coordinates, for example). In other words, they are a virtual construction of the coordinate system *itself*. These are in fact the fictitious pseudo forces everyone feels when they are in an accelerating reference frame, known in mechanics as **inertial forces**. This surprising interpretation should convince the reader that the Christoffel symbols are not just abstract mathematical constructions invented for theoretical purposes; they do have a physical effect, once viewed in context.

As an example, and at long last, let us show how the polar form of Newton's second law (0.6) arises. From (2.100) we know that the Christoffel symbols for polar coordinates (in other words, cylindrical coordinates without the $z$ direction) are $\Gamma^\rho_{\varphi\varphi} = -\rho$ and $\Gamma^\varphi_{\rho\varphi} = \Gamma^\varphi_{\varphi\rho} = \frac{1}{\rho}$. Identifying $x^1 = \rho$ in (3.19) gives

$$F^\rho = m\ddot{\rho} + m\Gamma^\rho_{\varphi\varphi}\dot{\varphi}^2 = m\ddot{\rho} - m\rho\dot{\varphi}^2, \qquad (3.22)$$

which is exactly the first equation of (0.6). Using $x^2 = \varphi$ we find

$$F^\varphi = m\ddot{\varphi} + m\Gamma^\varphi_{\rho\varphi}\dot{\rho}\dot{\varphi} + m\Gamma^\varphi_{\varphi\rho}\dot{\rho}\dot{\varphi} = m\ddot{\varphi} + 2m\frac{1}{\rho}\dot{\rho}\dot{\varphi}. \qquad (3.23)$$

Equation (3.23) is not quite the same as the second equation in (0.6). This "discrepancy" is due to the difference between using the physical basis vectors $\hat{\mathbf{v}}$, which are used to find (0.6) in most books on mechanics, and the natural basis vectors $\hat{\mathbf{g}}$, which we have been using throughout this book.[3] The confusion clears if we write the full vectorial form of Newton's second law in polar coordinates in the two systems of basis vectors:

$$\mathbf{F} = \left(m\ddot{\rho} - m\rho\dot{\varphi}^2\right)\hat{\mathbf{g}}_\rho + \left(m\ddot{\varphi} + 2m\frac{1}{\rho}\dot{\rho}\dot{\varphi}\right)\hat{\mathbf{g}}_\varphi \qquad (3.24)$$

$$\mathbf{F} = \left(m\ddot{\rho} - m\rho\dot{\varphi}^2\right)\hat{\mathbf{v}}_\rho + \left(m\rho\ddot{\varphi} + 2m\dot{\rho}\dot{\varphi}\right)\hat{\mathbf{v}}_\varphi. \qquad (3.25)$$

Equation (3.24) is what we found here, while (3.25) is what is normally found in standard treatments. Both equations are the same because $\hat{\mathbf{g}}_\rho = \hat{\mathbf{v}}_\rho$, while $\hat{\mathbf{g}}_\varphi = \rho\hat{\mathbf{v}}_\varphi$, as we found in §1.6. Now if we move the "extra" terms of eqn (3.25) to the left-hand

---

[3]We have noted this before in our discussion concerning the divergence in §2.4 as well as the paragraph preceding eqn (1.133).

side, following (3.21), the student who has studied a bit of advanced classical mechanics should recognize the term $m\rho\dot{\varphi}^2$ as representing the **centrifugal force**, while $-2m\dot{\rho}\dot{\varphi}$ is exactly the so-called **Coriolis force**. Both of these are the inertial forces one would expect to feel in moving along non-straight curves in the plane. We will further explore this in Exercise 3.3.

**Exercise 3.2**

1. Using (3.19), write the components of Newton's second law in the first eight coordinate systems of Exercise 2.22.
2. Write the full vectorial form of Newton's second law using your results from part 1. This would be in the natural basis.
3. Using the results of Exercise 1.21 along with (1.122), rewrite the full vectorial form of Newton's second law in these coordinate systems in the physical basis.

**Exercise 3.3** In traditional advanced texts in classical mechanics, the analysis leading to the inertial forces is different from, but equivalent to, what we have done here. Three-dimensionally the centrifugal force is found to be $m\vec{\omega} \times \left(\vec{\omega} \times \mathbf{r}\right)$ and the Coriolis force is $2m\vec{\omega} \times \dot{\mathbf{r}}$, where $\mathbf{r}$ is the position vector and $\vec{\omega}$ is the angular velocity vector. In polar coordinates in the plane, these are just $\mathbf{r} = \rho\hat{\mathbf{v}}_\rho$ and $\vec{\omega} = \dot{\varphi}\hat{\mathbf{v}}_\varphi$. Show that these expressions lead to exactly the ones we found in (3.25), i.e. $m\rho\dot{\varphi}^2\hat{\mathbf{v}}_\rho$ and $2m\dot{\rho}\dot{\varphi}\hat{\mathbf{v}}_\varphi$ respectively.

**Exercise 3.4** Redo Exercise 3.3 for spherical coordinates. Compare your results with the expressions of the centrifugal and Coriolis forces in spherical coordinates, as may be found in any standard text in classical mechanics.

### 3.1.1 The Geodesic Equation: Take One

Now let's consider the case where the forces acting on an object vanish, i.e. $F^i = 0$ in eqn (3.19). The result is vanishing acceleration $a^i = 0$, or

$$\ddot{x}^i + \Gamma^i_{jk}\dot{x}^k\dot{x}^j = 0. \tag{3.26}$$

Equations (3.26) are the equations of motion for a **free particle**, defined as a particle moving under the influence of zero forces. We know from Newton's first law that the solution is constant velocity along a straight line. In Cartesian coordinates the Christoffel symbols vanish and eqn (3.26) becomes the familiar zero acceleration expression

$$\ddot{x}^i = 0. \tag{3.27}$$

Solving (3.27) is straightforward: Integrating once gives

$$\dot{x}^i = v_0^i, \tag{3.28}$$

where $v_0^i$ are the constant components of the particle's initial velocity (at $t = 0$). Integrating again,

$$x^i(t) = v_0^i t + x_0^i, \tag{3.29}$$

where $x_0^i$ are the coordinates of the particle's initial starting point. Let's briefly switch back to standard notation, so that (3.29) becomes

$$
\begin{aligned}
x(t) &= v_0^x t + x_0 \\
y(t) &= v_0^y t + y_0 \\
z(t) &= v_0^z t + z_0.
\end{aligned} \tag{3.30}
$$

These are indeed the parametric equations of a straight line in three dimensions, as desired. If, for simplicity, we rotate our coordinate system such that the line described by (3.30) is made to live in the $x$–$y$ plane, then we need only consider the first two equations. Using the first to eliminate $t$, in the second we find

$$y(x) = mx + b, \tag{3.31}$$

as expected, where in this case the slope and intercept are

$$m = \frac{v_0^y}{v_0^x}$$

$$b = y_0 - \frac{v_0^y}{v_0^x} x_0 \tag{3.32}$$

respectively. In non-Cartesian coordinates, the free particle equations of motion (3.26) are not as simple to integrate. However, they still give the equation of a straight line, expressed in the used coordinates. For instance, the reader may find it interesting to check that in planar polar coordinates, (3.26) (which now has the form $\ddot{\rho} - \rho\dot{\varphi}^2 = 0$ and $\rho\ddot{\varphi} + 2\dot{\rho}\dot{\varphi} = 0$) has the solution

$$\rho = \frac{b}{\cos(\varphi - \chi)}, \tag{3.33}$$

where $b$ and $\chi$ are constants. It is another interesting challenge to show that this is in fact the equation of a straight line in the plane.

---

### Exercise 3.5

1. Show that (3.33) is a solution of the Geodesic equations in their polar form, $\ddot{\rho} - \rho\dot{\varphi}^2 = 0$ and $\rho\ddot{\varphi} + 2\dot{\rho}\dot{\varphi} = 0$.
2. Show analytically that (3.33) *is* the polar equation for a straight line in the $x$–$y$ plane.
3. Plot $\rho(\varphi)$ defined by (3.33) using the computer and demonstrate that it is indeed a straight line. You may use any appropriate values for the constants $b$ and $\chi$.

Because eqn (3.26) has the straight line as its *only* solution it is usually known as the **geodesic equation**, where the term "geodesic" literally means "the shortest distance." But why would one give a name to an equation whose solution is as trivial as a straight line? Why not just leave it simply at "the zero acceleration" equation or something similar? The reason is that (3.26) has the straight line as its solution *only* in flat spaces. When we start our discussion of general relativity, we will quickly realize that the relativistic analogue to (3.26) no longer gives a straight line, but rather gives the curve describing the "shortest distance between two points" (i.e. a geodesic) in whatever curved space we are in.

It is possible, however, to demonstrate this concept of non-straight geodesic motion while still in classical non-relativistic mechanics. Imagine the situation in Fig. [3.2], for instance. A particle is constrained to move on the surface of a sphere; i.e. it is not allowed to leave the surface. Now the constraint itself *is* a force and the full three-dimensional eqns (3.26) cannot be used since they refer only to force-free motion. *However*, imagine that an observer lives on the surface of the sphere and is unaware of, or intentionally ignoring, the constraint on the particle. To this observer, the particle is moving in a two-dimensional world, the sphere's surface, and is *exactly* force-free, since the only true force on it is orthogonal to the surface and hence does not belong to its two-dimensional life. This is not as unrealistic as you may imagine. In fact you, dear reader, are such an observer right now (assuming that you are reading this on Earth and are not currently aboard the International Space Station, for example). A marble sliding on a *frictionless* floor before you is such a force-free object! Gravity is the constraint keeping it on the floor, but all forces parallel to the floor are vanishing. Now, as long as you keep your experiment local, you can just ignore the spherical nature of Earth and treat the motion of the marble as if it were in a plane. But if you enlarge the domain of the experiment and imagine the need to plot the motion of said marble on the entire surface of the Earth, then you can no longer ignore that the motion of the marble cannot be a straight line!

To write the equations of motion of this object we first begin by writing the metric. Since the space in question is a sphere, we begin with the metric of spherical coordinates (1.27), then remove the third dimension, in this case the radial, by setting $r = R$ to be the constant radius of the sphere. This leads to $dr = 0$, and we end up with the metric of a sphere with radius $R$:

$$dl^2 = R^2 d\theta^2 + R^2 \sin^2\theta d\varphi^2. \tag{3.34}$$

It is conventional (and would be of future usefulness to us) to first define the metric of the **unit sphere**, i.e. a sphere of radius unity:[4]

$$d\Omega^2 = d\theta^2 + \sin^2\theta d\varphi^2, \tag{3.35}$$

*then* use that to define the metric of a sphere of radius $R$:

---

[4]The quantity $d\Omega$ in eqn (3.35) is sometimes referred to, confusingly in my opinion, as the **solid angle**.

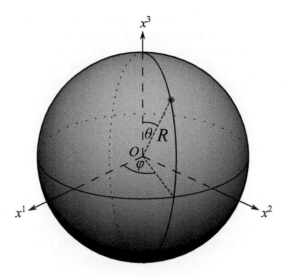

Fig. 3.2: The problem of force-free motion on the surface of a sphere.

$$dl^2 = R^2 d\Omega^2. \tag{3.36}$$

The components of the metric tensor on the sphere are hence $g_{11} = g_{\theta\theta} = R^2$ and $g_{22} = g_{\varphi\varphi} = R^2 \sin^2\theta$, with inverse components $g^{11} = g^{\theta\theta} = 1/R^2$ and $g^{22} = g^{\varphi\varphi} = 1/(R^2 \sin^2\theta)$. A straightforward application of (2.99) gives

$$\Gamma^\theta_{\varphi\varphi} = -\sin\theta\cos\theta$$
$$\Gamma^\varphi_{\theta\varphi} = \Gamma^\varphi_{\varphi\theta} = \cot\theta \tag{3.37}$$

as the only non-vanishing Christoffel symbols. The geodesic equations are hence

$$\ddot\theta - \dot\varphi^2 \sin\theta\cos\theta = 0$$
$$\ddot\varphi + 2\dot\theta\dot\varphi\cot\theta = 0. \tag{3.38}$$

These are the equations of motion of a free particle on the surface of the sphere. Mathematically, they are also the equations whose solutions parameterize the curve that describes the shortest distance between two points on the sphere. They are not at all easy to solve for $\theta(t)$ and $\varphi(t)$ generally; however, if we make the following choice of initial conditions, $\theta(0) = \frac{\pi}{2}$, $\varphi(0) = 0$, $\dot\theta(0) = 0$, and $\dot\varphi(0) = \varphi_0$=constant, then we can check that the solutions are

$$\theta(t) = \frac{\pi}{2}$$
$$\varphi(t) = \varphi_0 t. \tag{3.39}$$

For these initial conditions the particle is traveling along the equator $\theta = \frac{\pi}{2}$ with a constant angular speed of $\dot\varphi = \varphi_0$. Now because the sphere is a highly symmetric surface, the equator can be rotated to be *any* great circle (defined as any circle on the sphere whose radius is equal to the sphere's radius $R$). As such, we conclude that the motion of a free particle on a sphere is not, and cannot be, a straight line, but rather a segment of a great circle. An equivalent statement is: The shortest distance between two points on any sphere is necessarily a segment of a great circle. This correspondence between force-free motion and shortest distance is essential for understanding motion in curved spaces, be they actual surfaces, like that of our sphere here, or the curved spaces of general relativity.[5] Note that the statement "Force-free motion is always along the shortest distance between two points" in itself does require a rigorous proof. This can be done by showing that the geodesic eqn (3.26) which we extracted from Newton's second law can also be derived from the mathematical problem of calculating the shortest distance between two points on any given surface or space. This would require an excursion into a branch of mathematics known as the "calculus of variations." We will not do so here and just accept the correspondence between the two problems as axiomatic, but a proof does exist in case the reader is curious.

**Exercise 3.6** Compute the equations of motion of a free particle (i.e. the geodesic equations) on the surface described by constant $\rho$ in cylindrical coordinates (in other words, an infinite vertical cylinder). Solve them to find the parametric equations $\varphi(t)$ and $z(t)$ describing the path of a free particle on the cylinder. Using these, eliminate $t$ and find the equation $z(\varphi)$ that describes the geodesics on a cylinder. You should be able to find three different types of geodesics. What are they?

**Exercise 3.7** Compute the equations of motion of a free particle (i.e. the geodesic equations) for surfaces described by:

1. Constant $v$ in parabolic coordinates (1.8).
2. Constant $\nu$ in oblate spheroidal coordinates (1.10).
3. Constant $\tau$ in toroidal coordinates (1.11).

These should be found straightforwardly from the results of Exercise 3.2. You do not need to solve them. However, if you are so inclined, you might want to do so numerically using the computer, and assuming various sets of initial conditions. Analytical solutions might be way too complicated, if at all doable (but you can still try).

### 3.1.2   The Rest of Point Particle Mechanics

Curved surfaces aside, let us, for completeness' sake, list the most important equations of point particle mechanics that we haven't mentioned so far. For example, the study of free particle motion leads to one of the more important principles of classical mechanics: **conservation of linear momentum**. The argument goes like this: The original form of Newton's law is actually not (3.4), but rather

---

[5]Historically, this example is where the "geodesic" got its name: from "geodesy" the science of studying the shape and size of the Earth, where Geo is Greek for "Earth;" hence "Geology," "Geography," etc.

$$F^i = \frac{dp^i}{dt},\tag{3.40}$$

where the momentum $p^i$ is defined by (3.3). Equation (3.40) reduces to (3.4) only if the mass $m$ is a constant with respect to time. The covariant form of (3.40) would then be:

$$F^i = m\frac{dv^i}{dt} + v^i\frac{dm}{dt}$$
$$= m\ddot{x}^i + m\Gamma^i_{jk}\dot{x}^j\dot{x}^k + \dot{m}\dot{x}^i,\tag{3.41}$$

where as before, the two extra terms on the right-hand side can be moved to the left. The new "force" $\dot{m}\dot{x}^i$ can then be thought of as the force that the "missing" mass exerts on the main object. A typical example of this is the rocket problem, where $\dot{m}\dot{x}^i$ is taken to be the force that the rocket's exhaust applies to the rocket itself. A more detailed analysis of the physical implications of this can be found in any standard mechanics book.

Now consider a number of particles moving freely in space, possibly colliding with each other, and being influenced by a number of external forces. Equation (3.40) becomes

$$\sum F^i = \frac{d}{dt}\sum p^i.\tag{3.42}$$

In other words, the sum of external forces acting on a group of objects is equal to the time rate of change of the total sum of momenta of these objects.[6] Now our case of interest is when the sum of external forces cancels, leaving

$$\frac{d}{dt}\sum p^i = 0.\tag{3.43}$$

If the derivative of something vanishes, then the "something" must necessarily be a constant, so

$$\sum p^i = \text{constant}.\tag{3.44}$$

This is the famous law of conservation of linear momentum. It states that for a **closed system**, i.e. a system that does not exchange forces with its surroundings, the sum of the individual components for each particle's momentum stays constant, even if the particles collide with each other. In introductory courses the example of billiard balls sliding on a frictionless table (since friction would be an external force) is often quoted and provides for an excellent case study, but other examples are also common: the particles of gas in a specific specimen, astronauts and spaceships (in frictionless vacuum), and so on.

---

[6]The summations here are *not* over the index $i$ but over the number of forces and momenta. The index $i$ is a free index such that (3.42) represents the usual three equations.

**Exercise 3.8** As a refresher, and also for future comparison with the relativistic case, solve the following exercises using the principle of conservation of momentum:

1. On a frictionless surface, a 0.3kg object moving with a speed of 20m/s collides with a 0.4kg object initially at rest. After the collision, the two objects stick together and keep moving along the same straight line of the first object's velocity. What is the speed of the combined object after collision?

2. On a frictionless surface, a 0.3kg object moving with a speed of 20m/s collides with a 0.4kg object moving with a speed of 10m/s and both continue moving in the same direction. Find their speeds following the collision.

3. On a frictionless surface, a 0.3kg object moving with a speed of 20m/s collides with a 0.4kg object initially at rest. After the collision, each of these objects move along a direction making 30° on either side of the line of motion of the first object before the collision. Calculate their velocities.

The other famous conservation law of classical mechanics is the law of **conservation of mechanical energy**, which states that the sum of the kinetic and potential energies of a given closed system of objects is a constant, so

$$E = \sum (\text{K.E.} + V), \tag{3.45}$$

where $E$ is the constant **total energy**, $V$ is the **potential energy**, and K.E. is the **kinetic energy** of each object in the system defined by

$$\text{K.E.} = \frac{1}{2}mv^2 = \frac{1}{2}mv^i v_i = \frac{1}{2}mv^i v^j g_{ij}, \tag{3.46}$$

where $v^i$ are the components of the velocity and their indices are raised and lowered by $g_{ij}$ as usual. The concept of conservation of energy, while important, does not really lend itself as a good example of covariance. The reason is that energy is a scalar quantity and hence is automatically covariant. We simply include it here for completeness as well as for future comparison with the relativistic case.

A more interesting example of tensors in classical mechanics arises when formulating the laws of point particle motion in a "rotational" form. The so-called angular momentum **L** of a particle is defined as follows:

$$\mathbf{L} = \mathbf{r} \times \mathbf{p}. \tag{3.47}$$

Note that angular momentum is computed with respect to the origin of coordinates, since this is where the position vector **r** is defined. So the value of **L** for any given particle changes if one changes the location of the origin. Also note that even though we call this the "angular" or "rotational" version of mechanics, the particle may not even be rotating! Equation (3.47) is still well-defined for a particle traveling on a straight line. Now in the covariant index notation (3.47) becomes

$$L^i = \mathscr{E}^i{}_{jk} x^j p^k, \tag{3.48}$$

where minimal substitution is used. The next object to define is the torque that a given force **F** exerts on an object,

$$\tau = \mathbf{r} \times \mathbf{F}, \tag{3.49}$$

which becomes

$$\tau^i = \mathscr{E}^i{}_{jk} x^j F^k. \tag{3.50}$$

Finally, the rotational version of Newton's second law is

$$\sum \vec{\tau} = \frac{d}{dt} \sum \mathbf{L}, \tag{3.51}$$

where once again the summation is over the number of objects in the problem. The covariant form of this is just

$$\sum \tau^i = \frac{d}{dt} \sum L^i. \tag{3.52}$$

As before, if the system of particles in question experiences no external torques, eqn (3.52) leads to the law of conservation of angular momentum:

$$\sum L^i = \text{constant.} \tag{3.53}$$

Given the definitions (3.48) and (3.50), everything is guaranteed to work in any coordinate system defined by a given metric $g_{ij}$. Note that since the time derivative of **L** involves the time derivative of **p**, which in turn involves the time derivative of **v**, the Christoffel symbols are expected to arise in (3.52) just as they did in (3.19). We will not explore this any further, as it gets quite messy. The daring student may, however, attempt to reproduce the argument.

**Exercise 3.9** For the coordinate systems of Exercise 1.21, explicitly write the components $L^i$ of angular momentum using (3.48) and $\tau^i$ using (3.50). As an example, $L^\varphi = m(z\dot{\rho} - \rho\dot{z})$ in cylindrical coordinates.

**Exercise 3.10** Using the definition (3.48) and the properties of the $\delta$ and $\varepsilon$ (more precisely $g$ and $\mathscr{E}$), show that the norm of a given angular momentum vector is $L^2 = r^2 p^2 - (\mathbf{r} \cdot \mathbf{p})^2$.

## 3.2   Rigid Body Mechanics

The topic of rigid bodies, including their rotational properties, presents itself as a great application to covariance. This is because it contains one of the rare occurrences of second-rank tensors in undergraduate physics (other than the metric): the so-called **moment of inertia tensor**, or **inertia tensor** for short. This, however, requires the reader to have taken an upper-level classical mechanics class, so it might be a section to skip on first reading. Nevertheless, it is straightforward and self-explanatory if the reader decides to tackle it anyway.

A rigid body is defined as an object that does not deform in any way; that is, it is completely rigid. This is essentially only an approximation because any realistic object, no matter how tough, is ultimately deformable. However, it is still a good approximation if the object under study is robust enough. The starting point is pretty straightforward: rigid objects can be thought of as a large collection of point particles located at fixed points with respect to each other. Fig. [3.3] shows a rigid object as a collection of a number of $\mathcal{N}$ masses, not necessarily the same. We can assign an index to each mass to signify its number in the list:

$$m_1, m_2, m_3, \cdots, m_A, \cdots, m_{\mathcal{N}}. \tag{3.54}$$

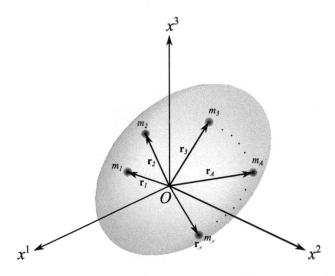

Fig. 3.3: A rigid body may be approximated by a series of point masses.

So an arbitrary mass in this list can be called $m_A$, where the index $A$ counts over the masses, *not* the spatial dimensions, as we are used to. Furthermore, when we are summing over the upper case indices we will *not* use the Einstein summation convention but will explicitly write the summation symbol, as an added precaution against confusion.[7] So the total mass of the object would be the sum of the individual masses

$$M = m_1 + m_2 + \cdots + m_{\mathscr{N}} = \sum_{A=1}^{\mathscr{N}} m_A. \tag{3.55}$$

On the other hand, if the number of masses is large enough, and the value of each mass is small enough then we can think of the rigid object as a continuum, in which case we use the rules of integral calculus to transform (3.55) to the continuum case. In other words, the limit of (3.55) leads to

$$M = \lim_{\substack{m_A \to 0 \\ \mathscr{N} \to \infty}} \sum_{A=1}^{\mathscr{N}} m_A = \int dm, \tag{3.56}$$

as shown in Fig. [3.4].

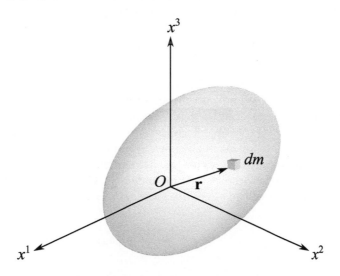

Fig. 3.4: A rigid body treated as a continuum.

The student who has seen such calculations before may recall that a good rule of thumb to allow a summation to become an integral is to perform the substitutions

$$m_A \to dm$$
$$\sum \to \int \tag{3.57}$$

---

[7]The summation convention would not work in this context anyway, as there will rarely be repeated upper-case indices, as you can see in eqn (3.55), for example.

which should work fine in most cases. It is also possible to write $dm$ in terms of how much volume $dV$ it occupies. This is accomplished by defining the concept of mass density $\rho$ in units of mass per volume:

$$dm = \rho\,(\mathbf{r})\,dV. \tag{3.58}$$

Generally the density may not be constant throughout the volume of the object. We allow for this by making it dependent on the position vector $\mathbf{r}$ of the volume element. Clearly in that context the integral (3.56) becomes a triple integral over the volume $dV$:

$$M = \int \int \int \rho\,(\mathbf{r})\,dV. \tag{3.59}$$

Now in Cartesian coordinates $dV$ is just $dV = dxdydz$, but we already know how to write (3.59) in any arbitrary system of coordinates. This is done by the method of the Jacobian, which we showed in §2.6 to be equal to the square root of the determinant of the metric, so the fully covariant form of (3.59) is

$$M = \int \int \int \rho\,(\mathbf{r})\,\sqrt{|g|}dx^1 dx^2 dx^3 = \int d^3x \sqrt{|g|}\rho\,(\mathbf{r}), \tag{3.60}$$

where $\left(x^1, x^2, x^3\right)$ are coordinates in any system defined by $g_{ij}$. It is typical notation to write $dx^1 dx^2 dx^3 = d^3x$ and just use one integration symbol for brevity. Now the stage is set and we can proceed to write any of the subsequent equations in covariant form. For example, any rigid object contains within it a certain point, and only one point, such that the distribution of the mass around it balances in every direction. This is the famous **center of mass**, or c.o.m. To find the location of this point, consider again the rigid object as a collection of masses in Cartesian coordinates. Then the coordinates $\left(\bar{x}^1, \bar{x}^2, \bar{x}^3\right)$ of the c.o.m are found via

$$\bar{x}^i = \frac{1}{M} \sum_{A=1}^{\mathcal{N}} m_A x_A^i, \tag{3.61}$$

where $x_A^i$ is read as the $i^{\text{th}}$ component $(1, 2, 3)$ of the position vector of the $A^{\text{th}}$ particle $(1, 2, \ldots, \mathcal{N})$. Once again, treating this like a continuum, we apply the substitution (3.57)

$$\bar{x}^i = \frac{1}{M} \int x^i dm \tag{3.62}$$

which straightforwardly leads to the covariant form

$$\bar{x}^i = \frac{1}{M} \int d^3x \sqrt{|g|} x^i \rho\,(\mathbf{r}). \tag{3.63}$$

The next major property of rigid bodies is the inertia tensor $I_j^i$, which is a tensor of rank $(1, 1)$ that appears when one considers the rotation of the body around some given axis. The inertia tensor in rotational mechanics plays the same role that mass plays in linear mechanics. So, if the amount of force one needs to impose on a given object to cause it to accelerate is proportional to $m$, in the sense that more massive objects need more force to achieve the same acceleration (essentially $F = ma$), then in rotations the amount of "torque" one needs to apply to an object to cause a change in its rotation is proportional to $I^i_j$. Intuitively, while it takes more "effort" to rotate massive objects than lighter ones, it is also true that certain shapes and sizes are "harder" to rotate. So the inertia tensor is dependent on *both* the mass *and* the dimensions of the object in question. Once again, if we assume that the object under study is made up of a number $\mathscr{N}$ of masses, the inertia tensor in Cartesian coordinates is defined as follows:

$$I^i_j = \sum_{A=1}^{\mathscr{N}} m_A \left( \delta^i_j x^k_A x_{A,k} - x^i_A x_{A,j} \right), \tag{3.64}$$

while as a continuum it becomes

$$I^i_j = \int dm \left( \delta^i_j x^k x_k - x^i x_j \right), \tag{3.65}$$

and finally, in fully covariant form,

$$I^i_j = \int d^3x \sqrt{|g|} \rho(\mathbf{r}) \left( \delta^i_j x^k x_k - x^i x_j \right). \tag{3.66}$$

As mentioned earlier, the inertia tensor is used to study the behavior of rigid objects under rotation. For example, given a rigid object with a known $I^i_j$ rotating about some axis with an angular velocity vector

$$\vec{\omega} = \omega^i \hat{\mathbf{g}}_i, \tag{3.67}$$

the components of its **angular momentum** vector are found by

$$L^i = I^i_j \omega^j. \tag{3.68}$$

Angular momentum is the rotational analogue to linear momentum, which we know as

$$p^i = mv^i. \tag{3.69}$$

The similarity between (3.68) and (3.69) is clear: linear/angular momentum is proportional to linear/angular velocity and the proportionality is $m/I^i_j$. This is the origin of the statement that in rotational mechanics $I^i_j$ plays the role that mass does in translational mechanics. The rotational analogue to Newton's second law (3.40) is also straightforward:

$$\tau^i = \frac{dL^i}{dt}, \tag{3.70}$$

where $\tau^i$ are the components of the torque vector needed to cause a rotational acceleration. In fact, if the inertia tensor is constant, one can write

$$\tau^i = \frac{d}{dt}\left(I_j^i \omega^j\right) = I_j^i \frac{d\omega^j}{dt}, \tag{3.71}$$

where the rate of change of angular velocity is the angular acceleration $\alpha^i$, leading to the rotational form of $F^i = ma^i$:

$$\tau^i = I_j^i \alpha^j. \tag{3.72}$$

**Exercise 3.11** Prove that (3.66) transforms like a tensor.

**Exercise 3.12** Consider a cube of dimensions $a$ and a uniformly distributed mass $M$. The origin of coordinates is located at the center of mass of the cube and the axes are parallel to its sides.

1. Calculate $I_j^i$ in Cartesian coordinates.
2. Calculate $I_j^i$ in cylindrical and spherical coordinates.
3. Write down the $I_{ij}$ and $I^{ij}$ components in all three coordinate systems.

**Exercise 3.13** If the cube of Exercise 3.12 is set to rotate with a constant angular velocity $\vec{\omega} = b\,(rad/s)\,\hat{\mathbf{e}}_x$, where $b$ is a constant:

1. Find the components $L^i$ of its angular momentum in Cartesian coordinates using eqn (3.68).
2. Find the components $L^i$ of its angular momentum in cylindrical and spherical coordinates.

**Exercise 3.14** Euler's equations for the generalized rotation of a rigid body are:

$$\frac{dL_1}{dt} - (\omega_3 L_2 - \omega_2 L_3) = \tau_1$$

$$\frac{dL_2}{dt} - (\omega_1 L_3 - \omega_3 L_1) = \tau_2$$

$$\frac{dL_3}{dt} - (\omega_2 L_1 - \omega_1 L_2) = \tau_3,$$

where $L_i$ are the components of the angular momentum vector, $\omega_i$ are the components of the angular velocity vector, and $\tau_i$ are the components of the torque vector. Rewrite these three equations as one coordinate covariant formula.

## 3.3   Motion in a Potential

The so-called **conservative forces** are defined as forces that can be derived as the negative gradient of a potential energy function $V(\mathbf{r})$ as follows:

$$\mathbf{F} = -\vec{\nabla}V. \tag{3.73}$$

They are called conservative because (3.73) is one of the requirements for a particle of mass $m$ moving under the influence of $\mathbf{F}$ to conserve energy. As in the theory of electrostatics, which the reader may be more familiar with, one can define the so-called **potential** $\Phi$ as a potential energy field per unit mass as follows:

$$\Phi = \frac{V}{m}.$$ (3.74)

Hence instead of a force one gets a vector field $\vec{g}$ filling all of space as a force per unit mass:

$$\vec{g} = -\vec{\nabla}\Phi.$$ (3.75)

An initial question that one may ask is this: Does the potential $\Phi$ for a given force actually exist? In other words, what conditions must be satisfied to guarantee that $\vec{g}$ is a conservative vector field? Let's first rewrite (3.75) in index form:

$$g^i = -\nabla^i \Phi.$$ (3.76)

Next, take the derivative of both sides and note that it can be written in either of the following forms:

$$\nabla^j g^i = -\nabla^j \nabla^i \Phi$$
$$\nabla^i g^j = -\nabla^i \nabla^j \Phi.$$ (3.77)

Since the covariant derivative commutes with itself, i.e. $\nabla^j \nabla^i = \nabla^i \nabla^j$ (see Exercise 2.26) then the right-hand sides of the two equations in (3.77) are equal, making the left-hand sides also equal. In other words, for $\Phi$ to exist and satisfy either equations in (3.77) it must be true that

$$\nabla^i g^j = \nabla^j g^i.$$ (3.78)

Equation (3.78) is sometimes referred to as an *integrability condition*, since it is the condition that *must* be satisfied for (3.76) to have a solution in $\Phi$ (i.e. be able to be integrated).

---

**Exercise 3.15** Show that (3.78) is also a consequence of (the possibly more familiar) $\vec{\nabla} \times \vec{g} = 0$, which is another way of saying that a solution of (3.76) must exist. This follows from the fifth identity in Exercise 1.19.

---

As a useful application, consider the motion in a potential $\Phi$ of two nearby particles. Let's discuss how these particles move with respect to each other. Denote the components of the position vectors of these particles by $x_1^i$ and $x_2^i$, where the $i$ index refers to the usual three coordinates, while the numbers 1 and 2 refer to the two particles. The equations of motion of such particles are found from Newton's second law (3.19) thus:

$$a_1^i = \ddot{x}_1^i + \Gamma_{jk}^i \dot{x}_1^k \dot{x}_1^j = -\nabla^i \Phi\left(x_1^k\right)$$
$$a_2^i = \ddot{x}_2^i + \Gamma_{jk}^i \dot{x}_2^k \dot{x}_1^j = -\nabla^i \Phi\left(x_2^k\right). \tag{3.79}$$

If we define the vector $s^i$ as the difference between the positions of the two particles, i.e. the vector pointing from one to the other indicating their separation:

$$s^i = x_2^i - x_1^i, \tag{3.80}$$

then eqns (3.79) can be combined to give the equation of motion of $s^i$:

$$A_s^i = a_2^i - a_1^i = -\nabla^i \Phi\left(x_2^k\right) + \nabla^i \Phi\left(x_1^k\right), \tag{3.81}$$

where $A_s^i = \ddot{s}^i + \Gamma_{jk}^i \dot{s}^k \dot{s}^j$ is the acceleration of $s^i$. Using (3.80) in the right-hand side of (3.81) and rearranging,

$$A_s^i = -\nabla^i \Phi\left(x_1^k + s^k\right) + \nabla^i \Phi\left(x_1^k\right)$$
$$= -\nabla^i \left[\Phi\left(x_1^k + s^k\right) - \Phi\left(x_1^k\right)\right]. \tag{3.82}$$

To evaluate this further, we can employ the use of the **Taylor series** theorem,[8] which states that any function $f\left(x+a\right)$, where $a$ is some change in $x$, can be written as an infinite sum of terms involving the derivatives of said function as follows:

$$f\left(x+a\right) = f\left(x\right) + \left(\frac{df}{dx}\right)a + \frac{1}{2!}\left(\frac{d^2 f}{dx^2}\right)a^2 + \frac{1}{3!}\left(\frac{d^3 f}{dx^3}\right)a^3 + \cdots. \tag{3.83}$$

If it happens that $a$ is a very small quantity, then higher powers of it get even smaller as one goes up the series. One may then ignore as many of the later terms as one wishes and write expressions like:

$$f\left(x+a\right) \approx f\left(x\right) + \left(\frac{df}{dx}\right)a. \tag{3.84}$$

For a vector-valued function $f\left(x^k + a^k\right)$, where $x^k$ are the Cartesian components of the position vector and $a^k$ are the components of some constant vector, the Taylor series becomes

$$f\left(x^k + a^k\right) = f\left(x^k\right) + \left(\partial_i f\right)a^i + \frac{1}{2!}\left(\partial_i \partial_j f\right)a^i a^j + \frac{1}{3!}\left(\partial_i \partial_j \partial_l f\right)a^i a^j a^l + \cdots. \tag{3.85}$$

---

[8]Although originally developed by the Scottish mathematician and astronomer James Gregory (1638–1675), the Taylor series is named after the English mathematician Brook Taylor (1685–1731) who formally introduced it to the mathematical community in 1715.

If the coordinates $x^k$ are non-Cartesian, one just performs the minimal substitution $\partial \to \nabla$ as usual:

$$f\left(x^k + a^k\right) = f\left(x^k\right) + \left(\nabla_i f\right) a^i + \frac{1}{2!}\left(\nabla_i \nabla_j f\right) a^i a^j + \frac{1}{3!}\left(\nabla_i \nabla_j \nabla_l f\right) a^i a^j a^l + \cdots . \quad (3.86)$$

Using this, let us expand $\Phi\left(x_1^k + s^k\right)$ in a Taylor series around $x_1^k$ and assume that the two particles are very close to each other, i.e. $s^k$ is a very small quantity. We may then write

$$\Phi\left(x^k + s^k\right) \approx \Phi\left(x^k\right) + \nabla_i \Phi\left(x^k\right) s^i, \quad (3.87)$$

where we have also dropped the subscript 1 on $x$, as it is no longer required. It is then true that $\Phi\left(x^k + s^k\right) - \Phi\left(x^k\right) \approx \nabla_i \Phi\left(x^k\right) s^i$, which we may use in (3.82):

$$A_s^i = -\nabla^i \nabla^j \Phi\left(x^k\right) s_j. \quad (3.88)$$

This equation of motion describes the behavior of two objects as they move in a potential $\Phi$, i.e. whether they get closer to or further away from each other, and at what rate, as described by their separation vector $s^i$. This behavior is not difficult to understand: two objects falling, say, toward Earth would start out at a larger separation from each other, then proceed to get closer as they followed the radial field lines of Earth's gravity. Sometimes such behavior is referred to as the **tidal effect**, and eqn (3.88) is hence known as the **tidal acceleration equation**. The quantity $\nabla^i \nabla^j \Phi\left(x^k\right)$ on the right-hand side of eqn (3.88) is what dictates this behavior; this can in fact be shown to transform as a second-rank tensor

$$R^{ij} = \nabla^i \nabla^j \Phi\left(x^k\right), \quad (3.89)$$

as will be explored in the exercises.

**Exercise 3.16** In the case of Newtonian gravity, the potential energy between two masses $m$ and $M$ is

$$V = -G\frac{mM}{r}, \quad (3.90)$$

where $G = 6.67408 \times 10^{-11} \mathrm{m}^3/\mathrm{kg\ s}^2$ is Newton's gravitational constant and $r$ is the distance between the centers of masses of $m$ and $M$. Applying (3.73) on (3.90) leads to **Newton's law of universal gravitation**,

$$\mathbf{F} = -G\frac{mM}{r^2}\hat{\mathbf{g}}_r, \quad (3.91)$$

in spherical coordinates. Following the arguments in this section, the Newtonian gravitational potential of a spherically symmetric mass distribution $M$ (say a planet or a star) is thus

$$\Phi = -G\frac{M}{r}, \quad (3.92)$$

leading to the vector field:

$$\vec{\mathfrak{g}} = -G\frac{M}{r^2}\hat{\mathbf{g}}_r, \quad (3.93)$$

which one calls the **gravitational field**. Derive $R^{ij}$ for Newtonian gravity and show that it transforms as a tensor.

**Exercise 3.17** Using the results of Exercise 3.16 along with eqn (3.88), show that masses falling radially toward the center of a spherical mass distribution, $M$, will behave in either of the following ways depending on their relative location with respect to each other:

1. If they are falling along the same radial, i.e. one mass on top of the other, the distance between them will increase with time.
2. If they are falling along two different radials, the distance between them will decrease.

In other words, an object falling in a radial gravitational field will tend to "elongate" along its length (defined as the direction of the radial from the center of $M$) and squeeze more tightly along its width (the plane perpendicular to the field lines). We don't normally observe this kind of behavior in mild gravitational fields such as those of Earth or even the Sun, but in truly strong gravitational fields such as black holes (as we will see), falling objects (or people) would visibly experience this "squishing" (and agonizing) phenomenon. This, for obvious reasons, is technically known as **spaghettification**.

**Exercise 3.18** Prove that $R^{ij} = \nabla^i \nabla^j \Phi \left( x^k \right)$ transforms as a tensor for *any* potential $\Phi$.

## 3.4   Continuum Mechanics

This branch of mechanics, concerned with the study of non-rigid materials, is filled with examples of higher-rank tensors, all the way up to rank 4. It is a vast field that further breaks down into different sub-categories—such as the theories of solid mechanics, fluid mechanics, elasticity, plasticity, and more—all of which study various types of real non-rigid objects. We will look at the simplest of cases, and in so doing we will introduce the tensors that describe the behavior of elastic materials, such as the one schematically shown in Fig. [2.1b]. It will not be obvious, but one of these tensors will make a comeback in Chapter 7 and will be extremely useful to us in the general theory of relativity. In the introduction to Chapter 2 we argued that Hooke's law (2.1) is not enough to describe the multidirectional relation between cause and effect in realistic matter, and can only be a special case of a more general formulation. Consider an object under the influence of some external forces. One would like to know how these forces "flow" inside the material and what deformations do they cause. In what follows we will assume that the deformations are not too large, and that the material therefore behaves elastically. If the external forces are removed, an elastic object springs back to its original shape and dimensions (this is known as a *linear* behavior). In contrast, so-called plastic materials do *not* return to their original form if the forces are removed and thus require a different treatment (non-linear behavior). Most realistic materials behave elastically at first, then break down into plasticity once the forces exceed a certain threshold; hence our assumption of small deformations.

Since we are looking at continuous matter as opposed to point masses, it is useful to define the concept of **stress**, which has units of force per unit area. In its simplest form, if a force $F$ is applied orthogonally on a flat surface with area $A$, then the stress $\sigma$ on that area is defined by[9]

---

[9] The reader may have noticed that this is exactly the definition of "pressure," used in fluid mechanics (with its various sub-branches of hydraulics, aerodynamics, etc.). This is correct: stress and

$$\sigma = \frac{F}{A}. \tag{3.94}$$

To generalize this concept consider an element of mass $dm = \rho(\mathbf{r})\,dV$ inside an elastic object, as shown in Fig. [3.4]. One would like to know which stresses are acting on this element, their directions, and the overall response of the material *due* to them. Firstly, we need to construct an object that contains not only the components of the internal stresses, but also the orientations of the surfaces that such stresses are acting on. Obviously we need a tensor—in this case one with *nine* components, each describing one of the three components of the stress as well as the surface it is acting on. Such is the second-rank **stress tensor** $\sigma_{ij}$, also known as the **Cauchy stress tensor**.[10] The components of the stress tensor describe the stresses on each of the surfaces of our mass element, as shown in Fig. [3.5]. The stresses perpendicular to the surfaces arise from forces perpendicular (orthogonal, or normal) to the surfaces, while the ones tangential to the surfaces arise from the tangential, or *shearing*, forces. As such $\sigma_{11}$, $\sigma_{22}$, and $\sigma_{33}$ can be termed the **normal stresses**, while the remaining components are called the **shearing stresses**. The numbering of the indices in the figure should be self-explanatory; for example, $\sigma_{23}$ is the *second* component of the stress acting on the surface perpendicular to the *third* axis.

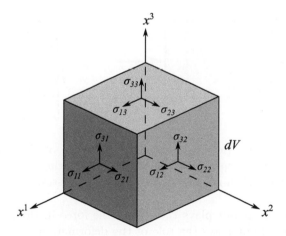

Fig. 3.5: The components of the stress tensor on an arbitrary volume element inside a specific object.

Once defined, there are several ways the stress tensor can be used. For example, consider a specific surface inside the object. Construct a vector $\hat{\mathbf{n}} = n^i \hat{\mathbf{g}}_i$ normal to that surface; then the stress vector $\mathbf{\Sigma} = \Sigma^i \hat{\mathbf{g}}_i$ acting on that surface is the projection

---

pressure are essentially the same thing; the former term more is commonly used for solids while the latter is used for fluids.

[10]In honor of Baron Augustin-Louis Cauchy, the French mathematician, engineer, and physicist who pioneered the study of continuum mechanics (1789–1857).

of the stress tensor on $\hat{n}$ thus:

$$\Sigma^i = \sigma^i{}_j n^j, \tag{3.95}$$

where the indices are raised and lowered by the appropriate metric. A proof that the stress tensor is symmetric exists. It is based, perhaps not too obviously, on the law of conservation of angular momentum. The interested student may look it up in any textbook on the subject. However, we can understand this property intuitively as follows: Consider that the shearing stress $\sigma_{12}$ is generating a component of torque (more precisely torque per unit area) that would tend to rotate the element around the $x^3$ axis in the clockwise direction as seen from above (the other component is around $x^2$). If the mass element is in equilibrium, this torque must be canceled by the opposing torque generated by $\sigma_{21}$. Since both stresses are equidistant to the $x^3$ axis, it must be that $\sigma_{12} = \sigma_{21}$, and similarly for all the others. Hence the components of the stress tensor are symmetric under the exchange $i \leftrightarrow j$:

$$\sigma_{ij} = \sigma_{ji}. \tag{3.96}$$

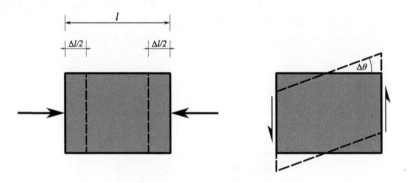

Fig. 3.6: Normal strain (left) is defined as the ratio of the deformation $\Delta l$ to the original length under normal stresses, while the shearing strain (right) is defined as the change of angle $\Delta\theta$ in radians under shearing stresses.

Now if the stress tensor plays the role of the force in Hooke's law (2.1), then the so-called **strain tensor** plays the role of the deformation $\mathbf{x}$, i.e. it is the response of the object to stress. The strain tensor $\epsilon_{ij}$ is usually defined in terms of infinitesimal deformations, and since its components are in direct response to the components of the stress tensor it is also a symmetric tensor: $\epsilon_{ij} = \epsilon_{ji}$. Generally the concept of strain is defined as a dimensionless ratio. For example, the diagonal components $\epsilon_{11}$, $\epsilon_{22}$, and $\epsilon_{33}$ describe the ratio of the change of length to the original length along the 1, 2, 3 directions respectively. They are the response to the normal stresses and are hence called the **normal strains**. In comparison, the off-diagonal shearing components can be thought of as the change of angle under shearing stresses. This is shown schematically in Fig. [3.6]. Finally, the relation between stress and strain is linear for elastic materials and involves the constant *fourth*-rank **elasticity tensor** $K_{ijkl}$ such that the **generalized Hooke's law** is

$$\sigma_{ij} = K_{ij}{}^{kl}\epsilon_{kl}. \tag{3.97}$$

Another name for $K_{ijkl}$ is the **stiffness tensor**, by analogy with the stiffness constant $k$ in the original form of Hooke's law (2.1).[11] The formula can be inverted:

$$\epsilon_{ij} = S_{ij}{}^{kl}\sigma_{kl}, \tag{3.98}$$

where $S_{ijkl}$ is known as the **compliance tensor**[12].

**Exercise 3.19** Show that $K_{ij}{}^{kl}S_{kl}{}^{ns} = \delta_i^n\delta_j^s = \delta_j^n\delta_i^s$.

**Exercise 3.20** Show that the stiffness tensor satisfies the symmetries:

$$\begin{aligned} K_{ijkl} &= K_{jikl} \\ K_{ijkl} &= K_{ijlk}. \end{aligned} \tag{3.99}$$

**Exercise 3.21** Prove that $K_{ijkl}$ transforms like a tensor if the stress $\sigma_{ij}$ and strain $\epsilon_{ij}$ components are assumed to transform like tensors.

A fourth-rank tensor in three dimensions has $3^4 = 81$ components; however, the symmetries (3.99) reduce them to only 36 independent components. Depending on the properties of the elastic material under study, this number can be further reduced. The components of these tensors are found experimentally by subjecting different materials to different stresses, measuring the strains, and computing the values of $K_{ijkl}$ and $S_{ijkl}$.

In fluid mechanics, which is a special case of the above, the stress tensor can be written for various types of fluids. For example, a non-viscous static fluid experiences only normal stress (conventionally renamed **pressure**, as noted earlier). This is intuitively true simply because a fluid by its very nature would not resist shearing forces; it would *flow* rather than deform. This means that in Cartesian basis the stress tensor is just

$$\sigma_{ij} = -p\delta_{ij}, \tag{3.100}$$

where $p$ is the magnitude of the pressure. It is taken to be the same in all directions by a simple argument of symmetry known as **Pascal's law**.[13] In a non-Cartesian

---

[11]It is interesting to note that the concept of tensors in its entirety was invented because of the study of the relation between stress and strain; *c.*1695–1705. The word "tensor" itself means "stretcher" in new Latin. It is equivalent to the old Latin word "tend;" which means "to stretch." Other commonly known words were also derived from the same origin, such as "tendon" in biology; namely the cords connecting muscles to bones.

[12]Some sources define (3.97) and (3.98) with a minus sign in analogy with (2.1).

[13]Blaise Pascal: French mathematician, physicist, inventor, writer and theologian (1623–1662).

coordinate system the minimal substitution $\delta \to g$ can be applied (the metric is still assumed to be diagonal):

$$\sigma_{ij} = -pg_{ij}. \tag{3.101}$$

On the other hand, fluids in motion experience shearing stresses which increase or decrease depending on the speed of the flow as well as the natural "stickiness," or "viscosity," of the fluid. Their stress tensor would have the general form

$$\sigma_{ij} = -pg_{ij} + d_{ij}, \tag{3.102}$$

where the so-called **deviatoric stress tensor** $d_{ij}$ is a consequence of the fluid's motion. It is perhaps intuitively true, as well as experimentally verifiable, that $d_{ij}$ is proportional to the velocity gradient of the fluid, i.e.

$$d_{ij} = A_{ij}{}^{kl} \left( \nabla_l v_k \right), \tag{3.103}$$

where the fourth-rank tensor $A_{ij}{}^{kl}$ is another constant tensor whose components may be found experimentally. Fluids that exhibit this relationship are sometimes referred to as **Newtonian fluids**, since Newton was the first to postulate a linear relationship between shearing stresses and velocity gradients. We can take this even further if we define the so-called **rate of strain tensor** $e_{ij}$ as follows:

$$e_{ij} = \frac{1}{2} \left( \nabla_i v_j + \nabla_j v_i \right). \tag{3.104}$$

It can then be shown that the most general form for the deviatoric stress tensor is

$$d_{ij} = 2\mu \left( e_{ij} - \frac{1}{3} e^k{}_k g_{ij} \right), \tag{3.105}$$

where $\mu(t, \mathbf{r})$ is a function referred to simply as the **viscosity**. This leads to the following general form for the stress tensor of a Newtonian fluid:

$$\sigma_{ij} = -pg_{ij} + 2\mu \left( e_{ij} - \frac{1}{3} e^k{}_k g_{ij} \right), \tag{3.106}$$

where the pressure is now assumed to be a non-constant scalar function $p(t, \mathbf{r})$ as well. The governing differential equation describing the general motion of a Newtonian fluid is the famous (perhaps infamous) **Navier–Stokes equation**;[14] by far one of the most interesting, as well as most challenging, equations in physics. We refer the interested reader to textbooks on the subject of fluid mechanics for more information.[15]

To conclude this section it is perhaps interesting to note that while our main focus was stresses and strains due to the "real" forces of push and pull inside matter, stress

---

[14]Claude-Louis Navier: French engineer and physicist (1785–1836) and Sir George Gabriel Stokes, first Baronet: Irish physicist and mathematician (1819–1903).

[15]For example, see [5].

tensors can be constructed for electromagnetic and other types of fields. It has long been known that electromagnetic fields carry energy and momentum and exert real and measurable forces on objects.[16] As such, it is perfectly reasonable to ask how one can write the stress tensor for an electromagnetic field, generally known as the **Maxwell stress tensor**. Consulting any textbook in standard electromagnetic theory, this is

$$\sigma_{ij} = \epsilon_0 E_i E_j + \frac{1}{\mu_0} B_i B_j - \frac{1}{2} \left( \epsilon_0 \mathbf{E} \cdot \mathbf{E} + \frac{1}{\mu_0} \mathbf{B} \cdot \mathbf{B} \right) g_{ij}, \qquad (3.107)$$

where $\mathbf{E} = E^i \hat{\mathbf{g}}_i$ and $\mathbf{B} = B^i \hat{\mathbf{g}}_i$ are the electric and magnetic field vectors and the constants $\epsilon_0$ and $\mu_0$ are the **permittivity of free space** and the **permeability of free space** respectively. The Maxwell stress tensor has units of energy per volume, which is the same as force per area.

There is more of course (there always is), and the entire curriculum of classical mechanics can be completely rewritten in the covariant index notation. However, in preparation for the next chapter we will need to shift our attention to a second type of covariance that classical mechanics is famous for.

**Exercise 3.22** Prove that the Maxwell stress tensor does indeed transform like a tensor, given that the electric and magnetic fields do.

## 3.5 Galilean Covariance

Coordinate covariance, i.e. the invariance of the laws of mechanics under rotations and translations, is supplemented by another type of covariance that is due to Galileo. Also known as the **principle of Galilean relativity**, it can be stated as follows:

*Any two observers moving at constant speed and direction with respect to one another will obtain the same results for all physical experiments.*

Basically what it says is this: Assume two observers; one we'll call $O$, represented by a set of coordinates $(x^1, x^2, x^3)$, and the other we'll call $O'$, whose point of view is represented by coordinates $\left(x^{1'}, x^{2'}, x^{3'}\right)$, as shown in Fig. [3.7]. Let's assume that $O'$ is moving to the right relative to $O$ with a *constant* velocity $v$; then the laws of physics measured by one are the *same* as those measured by the other. As in the case of rotational covariance, the particular numbers arising from some experiment

---

[16]It is tough to define what is truly "real" about forces. The Newtonian concept of "push and pull" forces is intuitive and easily understood, since this is exactly what we are used to in our everyday lives. However, a moment's thought would convince the reader that "push and pull" does not really exist, in the sense that nothing "truly" touches anything else. On an atomic scale there is always a "cushion" of electromagnetic fields between the molecules of, say, your hand and whatever it is you are "touching." So if you are surprised that electromagnetic fields exert measurable forces, consider that what you intuitively think of as a force is "really" electromagnetic in nature.

observed by both observers may be different, but the *form* of the laws of physics stays the same. This is clearly a case of covariance. Coordinate axes moving with constant velocities are called **inertial frames of reference**.

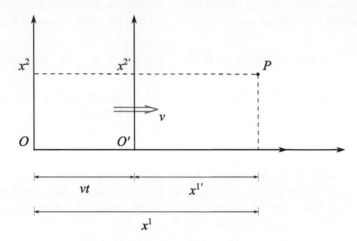

Fig. 3.7: Inertial reference frames.

Let's assume that both observers are looking at a specific event, $P$; for example, a light bulb that blinks once at a specific time. We further assume that at time $t = 0$ the points of origin in both frames coincided. Now we have chosen the axes to be parallel and the motion to be in the direction of $x$ for convenience. At any time $t$ the distance between $O$ and $O'$ is simply $X = vt$. The following equations relate the coordinates of the event $P$ at any time $t$ and are known as the **Galilean transformations**:

$$x^1 = x^{1'} + vt'$$
$$x^2 = x^{2'}, \qquad x^3 = x^{3'}. \tag{3.108}$$

Their inverse equations are as follows:

$$x^{1'} = x^1 - vt$$
$$x^{2'} = x^2, \qquad x^{3'} = x^3. \tag{3.109}$$

Intuitively, both observers also measure the same time for the event $P$; i.e.

$$t = t'. \tag{3.110}$$

This last is trivial and completely backed by our everyday experience; there just isn't any reason to think that time measurements made by $O$ and $O'$ are different. We only mention it because this will change in a non-trivial way once we delve into the special theory of relativity in the next chapter. A consequence of these transformations is the law of **Galilean addition of velocities**. Assume now that $P$ is an object moving with a constant speed $u'$ with respect to $O'$, as shown in Fig. [3.8]. Clearly $O$ would see it moving faster than $u'$; in fact $O$ would expect to see the object moving with speed $u$ such that

$$u = u' + v$$
$$u' = u - v. \qquad (3.111)$$

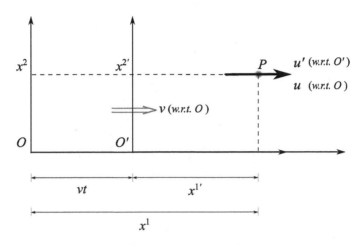

Fig. 3.8: Addition of velocities in inertial reference frames.

In fact, if $u'$ is exactly equal but opposite to $v$, then $u = 0$ and the object will not be moving with respect to $O$. Which child has not climbed *up* on a *downward* moving escalator or walked backwards on an airport's moving platform such that even though the child is moving with respect to the platform itself, they are really going nowhere with respect to their parents *outside* the platform; a direct consequence of (3.111). These two equations are the basics of the topic of relative velocity. They emphasize the fact that velocity is a relative concept and its value *only* makes sense with respect to a well-defined frame of reference.

**Exercise 3.23** Derive (3.111) from (3.108) and (3.109).

**Exercise 3.24** A car is traveling with a constant speed of 50m/s with respect to the sidewalk. A passenger inside the car is playing with a ball by throwing it vertically upward with an initial speed of 5m/s. Write the equations of motion of the ball with respect to the passenger (review the projectile problem in any introductory physics textbook). Now using the Galilean transformations derive the equations of motion with respect to an observer on the sidewalk. Sketch the motion of the ball with respect to both $O'$ and $O$. What curves do they describe, and why is this not a surprising result?

**Exercise 3.25** Show that Newton's second law in one dimension, $F = m\frac{d^2 x}{dt^2}$, is invariant under the Galilean transformations. In other words, it looks exactly the same in the primed and unprimed coordinates of Fig. [3.7]. This is easily generalized to three dimensions. Hence all of classical mechanics is Galilean covariant.

Now let's go back to (3.108) and (3.109). Since these are clearly coordinate transformation, we might choose to put them in the same form as (2.14). It is clear, however, that this should involve not only the spatial coordinates, but time as well, which means the addition of extra components to our transformation matrix $\lambda$. The reader can check that (3.108) and (3.109) are respectively equivalent to

$$
\begin{pmatrix} t \\ x^1 \\ x^2 \\ x^3 \end{pmatrix} = \begin{pmatrix} 1 & 0 & 0 & 0 \\ v & 1 & 0 & 0 \\ 0 & 0 & 1 & 0 \\ 0 & 0 & 0 & 1 \end{pmatrix} \begin{pmatrix} t' \\ x^{1'} \\ x^{2'} \\ x^{3'} \end{pmatrix}
\tag{3.112}
$$

and

$$
\begin{pmatrix} t' \\ x^{1'} \\ x^{2'} \\ x^{3'} \end{pmatrix} = \begin{pmatrix} 1 & 0 & 0 & 0 \\ -v & 1 & 0 & 0 \\ 0 & 0 & 1 & 0 \\ 0 & 0 & 0 & 1 \end{pmatrix} \begin{pmatrix} t \\ x^1 \\ x^2 \\ x^3 \end{pmatrix},
\tag{3.113}
$$

which also gives the trivial $t' = t$. Now, if one wishes to set this in the index form (2.14), it is clear that the indices can no longer count over just $1, 2, 3$. We will have to add an "extra" dimension. We choose to make the indices count over $0, 1, 2, 3$, where we define the zeroth dimension by $x^0 \equiv t$ and $x^{0'} \equiv t'$. To emphasize the difference we adopt Greek indices rather than Latin, so

$$
x^\mu = \lambda^\mu_{\nu'} x^{\nu'},
$$
$$
x^{\mu'} = \lambda^{\mu'}_{\nu} x^\nu, \qquad \mu, \nu = 0, 1, 2, 3.
\tag{3.114}
$$

The transformation matrix and its inverse have the non-vanishing components

$$
\lambda^0_{0'} = \lambda^1_{1'} = \lambda^2_{2'} = \lambda^3_{3'} = 1
$$
$$
\lambda^0_{1'} = v
$$
$$
\lambda^{0'}_0 = \lambda^{1'}_1 = \lambda^{2'}_2 = \lambda^{3'}_3 = 1
$$
$$
\lambda^{0'}_1 = -v.
\tag{3.115}
$$

Note that the relations (2.19) and (2.20) are still valid, except now upgraded to

$$
\lambda^\mu_{\nu'} = \frac{\partial x^\mu}{\partial x^{\nu'}} \quad \text{and} \quad \lambda^{\mu'}_\nu = \frac{\partial x^{\mu'}}{\partial x^\nu}.
\tag{3.116}
$$

The reader can easily check that the orthogonality condition $\lambda^\mu_{\mu'} \lambda^{\mu'}_\nu = \delta^\mu_\nu$ is satisfied:

$$\begin{pmatrix} 1 & 0 & 0 & 0 \\ +v & 1 & 0 & 0 \\ 0 & 0 & 1 & 0 \\ 0 & 0 & 0 & 1 \end{pmatrix} \begin{pmatrix} 1 & 0 & 0 & 0 \\ -v & 1 & 0 & 0 \\ 0 & 0 & 1 & 0 \\ 0 & 0 & 0 & 1 \end{pmatrix} = \begin{pmatrix} 1 & 0 & 0 & 0 \\ 0 & 1 & 0 & 0 \\ 0 & 0 & 1 & 0 \\ 0 & 0 & 0 & 1 \end{pmatrix}. \tag{3.117}$$

Note that the Galilean transformation matrix (3.115) is *not* symmetrical, which is representative of the fact that time and the spatial coordinates are *not* equivalent in classical mechanics. In contrast, we will see in the next chapter that time and space are treated equally in relativity. Now, describing the Galilean transformations using the index/tensor language is not something that is too popular in the literature. One of its main problems is that it doesn't quite lend itself to a physical understanding of what exactly the set of numbers $(x^0, x^1, x^2, x^3)$ represents. It cannot be a vector in the usual sense, because $x^0$ does not have the same units as the other $x$'s. One can also show that its "scalar product" $x^\mu x_\mu$, is difficult to define, as a four-dimensional covariant metric simply does not exist in the non-relativistic realm.[17] The fact that there are four "components" is not an issue: one can easily imagine a four-dimensional vector. However, there is no physical justification for considering $(x^0, x^1, x^2, x^3)$ to be a vector; they are just ordered as such in (3.112) as a matter of calculational convenience. Later when we start our discussion of relativity, we *will* be able to construct true four-dimensional vectors with time as the fourth dimension, and we will have plenty of justification for it.

Velocity addition can also be represented in this language. In reference to Fig. [3.8], write the four-dimensional velocity "vector" with respect to $O'$ as follows:

$$\mathbf{U}'_4 = (1, u', 0, 0), \tag{3.118}$$

where the zeroth "component" is just $U^{0'} = 1$. Next transform $\mathbf{U}'_4$ to $O$'s point of view

$$U^\nu = \lambda^\nu_{\mu'} U^{\mu'}, \tag{3.119}$$

or equivalently

$$\begin{pmatrix} U^0 \\ U^1 \\ U^2 \\ U^3 \end{pmatrix} = \begin{pmatrix} 1 & 0 & 0 & 0 \\ v & 1 & 0 & 0 \\ 0 & 0 & 1 & 0 \\ 0 & 0 & 0 & 1 \end{pmatrix} \begin{pmatrix} 1 \\ u' \\ 0 \\ 0 \end{pmatrix} = \begin{pmatrix} 1 \\ u'+v \\ 0 \\ 0 \end{pmatrix} = \begin{pmatrix} 1 \\ u \\ 0 \\ 0 \end{pmatrix}, \tag{3.120}$$

which is the correct answer. The transformation (3.119) and its inverse can be applied to momentum. We can define the four-dimensional momentum "vector"[18]

$$\begin{pmatrix} P^0 & P^1 & P^2 & P^3 \end{pmatrix} = m \begin{pmatrix} 1 & U^1 & U^2 & U^2 \end{pmatrix}, \tag{3.121}$$

that transforms following:

---

[17]This issue will be rectified in the next chapter.

[18]We continue to put the word "vector" in quotes to remind the reader that while it works, the four numbers $U^0$, $U^1$, $U^2$, and $U^3$ do not really constitute a vector . . . *yet!*

$$P^\nu = \lambda^\nu_{\mu'} P^{\mu'}. \tag{3.122}$$

However, in the case of acceleration (of the particle *not* of the reference frame; in other words, $v$ is still a constant), the four-dimensional "vector" approach breaks down. This is because taking the derivatives of (3.111) gives

$$a = \dot{u} = \dot{u}' = a'. \tag{3.123}$$

In other words, the acceleration of objects is exactly the same as measured by both $O$ and $O'$. The correct matrix of transformation is then just the identity matrix, not (3.115). Since force is proportional to acceleration, it is also true that different inertial observers will observe the same forces, as you will know if you have solved Exercise 3.25.

The transformation (3.112)/(3.114) can be useful in certain types of calculations, and for that purpose we will give it a name; let's call it the **Galilean boost**. While the reader may find this term in use in some sources, they are still warned that it is not very common. Another type of boost, the *relativistic* boost, is usually what most physicists will think of when they hear that word. But, since (3.112) plays the same role in Galilean covariance as the relativistic boost will play in relativistic covariance, we are fully justified in using a similar title for it.

The complete set of Galilean boost matrices are:

$$[\lambda]_{x'\to x} = \begin{pmatrix} 1 & 0 & 0 & 0 \\ v_x & 1 & 0 & 0 \\ 0 & 0 & 1 & 0 \\ 0 & 0 & 0 & 1 \end{pmatrix}, \qquad [\lambda]_{y'\to y} = \begin{pmatrix} 1 & 0 & 0 & 0 \\ 0 & 1 & 0 & 0 \\ v_y & 0 & 1 & 0 \\ 0 & 0 & 0 & 1 \end{pmatrix},$$

$$[\lambda]_{z'\to z} = \begin{pmatrix} 1 & 0 & 0 & 0 \\ 0 & 1 & 0 & 0 \\ 0 & 0 & 1 & 0 \\ v_z & 0 & 0 & 1 \end{pmatrix}, \tag{3.124}$$

where we have given the velocities subscripts to denote their direction. A generalized Galilean boost in an arbitrary direction for the velocity $\mathbf{v} = (v_x, v_y, v_z)$ can be found by multiplying the boost matrices together to give

$$[\lambda]_{O'\to O} = \begin{pmatrix} 1 & 0 & 0 & 0 \\ v_x & 1 & 0 & 0 \\ v_y & 0 & 1 & 0 \\ v_z & 0 & 0 & 1 \end{pmatrix}, \tag{3.125}$$

where the reader can check that the matrices (3.124) are commutative, i.e. that any order of multiplication gives (3.125).[19] The inverse boosts are obvious and easily found by a simple change of sign. The transformation (3.125) can be used for

---

[19]Recall that this was not true for rotational matrices.

the situation shown in Fig. [3.9]. The frame $O'$ is now traveling with velocity $\mathbf{v} = (v_x = v\cos\theta, v_y = v\sin\theta, v_z = 0)$ with respect to $O$ (which is two-dimensional for simplicity). It is easy to check by simple trigonometry that using (3.125) in the first equation of (3.114) gives the correct transformations.

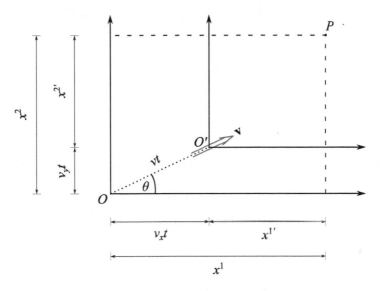

Fig. 3.9: General Galilean transformations.

**Exercise 3.26** Consider an airplane traveling at low altitude with a constant velocity $\mathbf{v} = 5\hat{\mathbf{x}} + 2\hat{\mathbf{y}} - 3\hat{\mathbf{z}}$ in m/s as measured by a coordinate system $O$ fixed with respect to the ground. The copilot is playing with a ball by throwing it vertically upwards with an initial speed of 5m/s. The person's rest frame $O'$, i.e. the coordinate system in which they are at rest, is parallel to $O$, similar to the one in Fig. [3.9] except in three dimensions. Write the equations of motion of the ball in $O'$, then use the Galilean transformation matrix (3.125) to write its equations of motion as seen by $O$. Since the airplane is at low altitude, the acceleration due to gravity is still approximately $g = 9.81$m/s$^2$ and the $x$–$y$ and $x'$–$y'$ planes are parallel to the Earth's surface. Sketch the motion of the ball with respect to both $O'$ and $O$.

It is also possible to combine boosts with rotations—useful if the $O'$ frame is rotated with respect to $O$, as in Fig. [3.10]. The "four-dimensional" rotation matrices are found by generalizing (2.10), (2.11), and (2.12) to

$$[\lambda]_{x^1} = \begin{pmatrix} 1 & 0 & 0 & 0 \\ 0 & 1 & 0 & 0 \\ 0 & 0 & \cos\theta & -\sin\theta \\ 0 & 0 & \sin\theta & \cos\theta \end{pmatrix}, \qquad [\lambda]_{x^2} = \begin{pmatrix} 1 & 0 & 0 & 0 \\ 0 & \cos\theta & 0 & -\sin\theta \\ 0 & 0 & 1 & 0 \\ 0 & \sin\theta & 0 & \cos\theta \end{pmatrix}$$

$$[\lambda]_{x^3} = \begin{pmatrix} 1 & 0 & 0 & 0 \\ 0 & \cos\theta & -\sin\theta & 0 \\ 0 & \sin\theta & \cos\theta & 0 \\ 0 & 0 & 0 & 1 \end{pmatrix}. \tag{3.126}$$

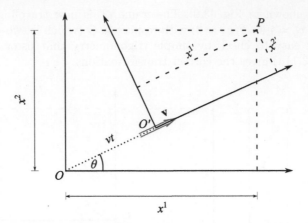

Fig. 3.10: A rotated boost. The boost is performed first, *then* the rotation.

Now the situation in Fig. [3.10] can be thought of in the following way: Both frames start together perfectly aligned and parallel at $t = 0$. A boost along the $x$ axis is done *first*,

$$\begin{pmatrix} 1 & 0 & 0 & 0 \\ v & 1 & 0 & 0 \\ 0 & 0 & 1 & 0 \\ 0 & 0 & 0 & 1 \end{pmatrix} \begin{pmatrix} t' \\ x' \\ y' \\ z' \end{pmatrix} = \begin{pmatrix} t' \\ x' + vt \\ y' \\ z' \end{pmatrix}, \tag{3.127}$$

followed by the counterclockwise rotation around the $z$ axis:

$$\begin{pmatrix} 1 & 0 & 0 & 0 \\ 0 & \cos\theta & -\sin\theta & 0 \\ 0 & \sin\theta & \cos\theta & 0 \\ 0 & 0 & 0 & 1 \end{pmatrix} \begin{pmatrix} t' \\ x' + vt \\ y' \\ z' \end{pmatrix} = \begin{pmatrix} t' \\ x' \cos\theta - y' \sin\theta + v\cos\theta t \\ x' \sin\theta + y' \cos\theta + v\sin\theta t \\ z' \end{pmatrix}$$

$$= \begin{pmatrix} t' \\ x' \cos\theta - y' \sin\theta + v_x t \\ x' \sin\theta + y' \cos\theta + v_y t \\ z' \end{pmatrix}. \tag{3.128}$$

This gives the correct answer, as can be checked with some (not too trivial) trigonometry. The overall matrix of transformations in this case is then

$$[\lambda] = [\lambda]_{x^3} \, [\lambda]_{x' \to x} \tag{3.129}$$

or

$$\begin{pmatrix} 1 & 0 & 0 & 0 \\ 0 & \cos\theta & -\sin\theta & 0 \\ 0 & \sin\theta & \cos\theta & 0 \\ 0 & 0 & 0 & 1 \end{pmatrix} \begin{pmatrix} 1 & 0 & 0 & 0 \\ v & 1 & 0 & 0 \\ 0 & 0 & 1 & 0 \\ 0 & 0 & 0 & 1 \end{pmatrix} = \begin{pmatrix} 1 & 0 & 0 & 0 \\ v\cos\theta & \cos\theta & -\sin\theta & 0 \\ v\sin\theta & \sin\theta & \cos\theta & 0 \\ 0 & 0 & 0 & 1 \end{pmatrix}$$

$$= \begin{pmatrix} 1 & 0 & 0 & 0 \\ v_x & \cos\theta & -\sin\theta & 0 \\ v_y & \sin\theta & \cos\theta & 0 \\ 0 & 0 & 0 & 1 \end{pmatrix}. \tag{3.130}$$

Note that the order of multiplying these matrices matters: reversing the order gives

$$\begin{pmatrix} 1 & 0 & 0 & 0 \\ v & 1 & 0 & 0 \\ 0 & 0 & 1 & 0 \\ 0 & 0 & 0 & 1 \end{pmatrix} \begin{pmatrix} 1 & 0 & 0 & 0 \\ 0 & \cos\theta & -\sin\theta & 0 \\ 0 & \sin\theta & \cos\theta & 0 \\ 0 & 0 & 0 & 1 \end{pmatrix} = \begin{pmatrix} 1 & 0 & 0 & 0 \\ v & \cos\theta & -\sin\theta & 0 \\ 0 & \sin\theta & \cos\theta & 0 \\ 0 & 0 & 0 & 0 \end{pmatrix}, \tag{3.131}$$

which does *not* correspond to the situation in Fig. [3.10]. What (3.131) *does* correspond to is a rotation *then* a boost along the $x$ axis, as shown in Fig. [3.11]. In other words, while Galilean boosts are commutative among themselves, rotations with rotations and rotations with boosts are not.

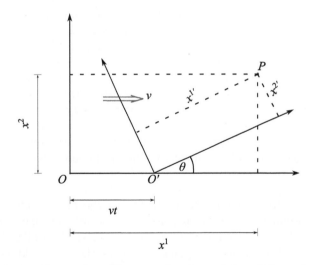

Fig. 3.11: A boosted rotation. The rotation is performed first, *then* the boost.

**Exercise 3.27** Consider the airplane of Exercise 3.26. Let the person's rest frame now be slanted with an angle such that its $x$ axis is along the direction of $\mathbf{v}$, similar to Fig. [3.10], except three-dimensionally. Write the equations of motion of the ball in $O'$, find the appropriate Galilean transformation matrix, and use it to write the equations of motion as seen by $O$. The $x$–$y$ plane is parallel to the Earth's surface. Sketch the motion of the ball with respect to both $O'$ and $O$.

Classical mechanics is entirely covariant under Galilean transformations, which together with rotations and constant translations may be termed the **classical covariance**.[20] The invariance of classical mechanics under boosts was explored in Exercise 3.25, where you showed that Newton's second law stays looking the same in any inertial reference frame. Hence one says that Newton's laws will be the same[21] whether measured by $O$ or $O'$. As such, another way of defining inertial frames of reference is that they are those frames that preserve the form of Newton's law.

To better understand the invariance of the laws of mechanics under the Galilean transformations, let us consider the situation where the velocity $v$ of $O'$ is no longer a constant. This would be the curious case of the **non-inertial observer**. Let us assume that $v$ is changing at a rate given by the acceleration of $O'$ such that $\ddot{X} = \dot{v} \neq 0$; then we have

$$F = m\frac{d^2x}{dt^2} = m\frac{d^2}{dt^2}(x' + X) = m\frac{d^2x'}{dt^2} + m\ddot{X}, \qquad (3.132)$$

which, upon rearranging, gives

$$F - m\ddot{X} = m\frac{d^2x'}{dt^2}, \qquad (3.133)$$

which is similar to what we observed in (3.21). Once again the term $m\ddot{X}$ is "felt" by the observer $O'$, the non-inertial observer, as an extra "force," unseen to observer $O$. In this case, $m\ddot{X}$ is the force one feels as one rides along in an accelerating car, $O'$, being pressed back against your seat as the car accelerates forward with acceleration $\ddot{X}$, or finding oneself pushed forward if the car decelerates (negative $\ddot{X}$). Notice that if the car is moving at a constant speed, no such forces exist. Another famous example of such **pseudo-forces** is the centrifugal force a rotating observer feels, which we discussed in more detail back in §3.1. Galilean covariance concerns itself only with inertial reference frames; the case of the non-inertial observer is usually studied separately in advanced mechanics courses. We will not have to concern ourselves with accelerated reference frames just yet.

Let us go back to inertial observers. Are all the laws of physics invariant under Galilean transformations? To the surprise of nineteenth-century physicists, the answer is no. Consider that other edifice of classical physics: Maxwell's theory of electrodynamics. Now, the foundational equations of this theory are known as the Maxwell equations and are well known as being correct. Ever since Maxwell presented them in the 1860s all kinds of experiments have been performed to validate them. The very existence of the computer I am using to type this and the television blaring in the room next to me is proof of the validity of Maxwell's electromagnetic field equations.

---

[20]Some sources assign the name "Galilean transformations" to boosts, rotations, and translations collectively.

[21]For any particular calculation like the different projectile problems explored here, the numerical results will differ for sure, but the *form* of the laws of mechanics will stay the same, which means that no new phenomena unobserved by $O$ will appear to $O'$. Read further for an example of what happens if $v$ is not a constant.

The surprising fact, however, is that the Maxwell equations are *not* invariant under the Galilean transformations! If one performs the same substitution we did in Exercise 3.25, the result will be equations describing *different* physics than do the original equations. Since we know that Maxwell's equations in my office work as well as they do in a speeding car (i.e. no "pseudo" fields appear), we have a problem. For years there were attempts to "fix" the Maxwell equations to make them covariant under Galilean transformations, but all of these attempts suffered from major problems: mostly they predicted physical phenomena that were not observed. That was the situation that Albert Einstein found and, building on the work of many others, finally fixed. It turns out that the problem was not in Maxwell's equations; it was in the Galilean transformations *themselves*. While they work fine with slow-moving objects, they fail if the speeds in question are considerable. In such a case the so-called relativistic effects manifest themselves and new transformations are needed. These were discovered by Lorentz and interpreted by Einstein and Minkowski. Furthermore, classical mechanics was modified to make it invariant under the new transformations. Hence began the era of the special theory of relativity.

**Exercise 3.28** Show that Maxwell's equations are not Galilean covariant. You may use the Cartesian form of the equations that you found in Exercise 1.18.

**Exercise 3.29** Show that the wave equation is not Galilean covariant. You may use the Cartesian form you found in Exercise 1.12. Since the wave equation can arise as a consequence of the Maxwell equations, this exercise also shows that Maxwell's equations are not Galilean covariant.

**Exercise 3.30** Using the transformation matrix $\lambda$ defined by (2.25), write down the generalized Galilean transformation matrix (3.125) in the cylindrical polar coordinate system.

**Exercise 3.31** If you have done Exercise 2.6 write down the generalized Galilean transformation matrix (3.125) in the coordinate systems of Exercise 2.5.

# 4

# Special Covariance

*Henceforth space by itself, and time by itself, are doomed to fade away into mere shadows, and only a kind of union of the two will preserve an independent reality.*

Hermann Minkowski

## Introduction

If one wished to define the special theory of relativity in the shortest possible terms using the language we have learned in the previous chapters, one could say the following: "Special Relativity is the statement that we live in a four-dimensional spacetime (ignoring gravity) with metric

$$ds^2 = -c^2 dt^2 + g_{ij}dx^i dx^j, \tag{4.1}$$

where $g_{ij}$ is any rotationally covariant (but not Galilean) three-dimensional metric, $t$ is time, and $c$ is the constant speed of light in vacuum in a given system of units." That's it! Everything else follows through using the techniques already developed: vectors, tensors, basis vectors, covariant derivatives, etc. However, the major difference is that all of these quantities will be four-dimensional, carrying Greek indices counting over 0, 1, 2, and 3. As we discussed earlier, the metric is a truly fundamental quantity. It tells us essentially all what we need to know. It also tells us what kind of covariance we have. The metric (4.1) has rotational covariance over the three-dimensional space $g_{ij}$, but the addition of an extra term with a *negative* sign introduces a new type of covariance in four dimensions— **special covariance**—where we now require vectors and tensors to transform *four*-dimensionally over the metric (4.1). The implications of this are vast, and they include all special relativistic phenomena that you may have heard of elsewhere. In this chapter we will explore special covariance from scratch, and our reference point will be the metric.

## 4.1 Special Relativity, the Basics

The usual way a student is exposed to special relativity is via Einstein's postulates, which state:

*Covariant Physics: From Classical Mechanics to General Relativity and Beyond.* Moataz H. Emam, Oxford University Press (2021). © Moataz H. Emam.
DOI: 10.1093/oso/9780198864899.003.0004

1. The laws of nature have the same form in all inertial (non-accelerating) reference frames.
2. The speed of light $c$ in vacuum is constant as measured by all inertial observers.

The first postulate we have already seen; it is exactly the Galilean postulate of relativity. The second one arose from the Maxwell equations and is well known to everyone nowadays; however, it is far from logical or obvious. In fact, based on classical covariance, the Galilean laws of addition of velocity (3.111) are in direct conflict with postulate 2. If an observer at rest in the moving frame $O'$ shines a beam of light forward with a velocity that they perceive as $c$, then according to (3.111) observer $O$ will measure the speed of the beam to be $c+v$. But postulate 2 says that this cannot be true! Einstein *had* to impose postulate 2, since all theory (Maxwell's equations, etc.) and all experiments up to that point had confirmed it. Consequently, it was Newton's mechanics that had to be fixed to agree with the second postulate.[1]

So how can one solve this dilemma? Since the speed of light is a universal constant, agreed upon by all observers, and since speed is distance over time, the inevitable conclusion here is that we have to change the way we make space and time measurements to *depend* on the observer in such a way that all observers measure the speed of light to be *exactly* the same. And since the term "observer" is another way of saying "coordinate system," what we are looking at here is a covariance of a new type.

Although special relativity is the work of many mathematicians and physicists, it is generally attributed to Albert Einstein because he was the first one to make sense of the mess and provide a solid foundation to these ideas. In terms of the concept of covariance, however, it was Hermann Minkowski who provided the bottom line: The metric of space, discussed in the previous chapters, needs to be amended. Since space and time work together, the "real" metric has to include *both* space *and* time in such a way as to give the observed phenomena when applied to physics. Incredibly, we can actually *derive* this **spacetime metric** using Einstein's postulates combined with the Pythagorean theorem! Consider a light signal sent in vacuum from point 1 to point 2 over a distance $\Delta l$ (Fig. [4.1]). Because of the second postulate, the signal's speed has to be exactly $c$. In other words, it has to satisfy the condition $\Delta l = c\Delta t$ for *all* observers. Now, working in Cartesian coordinates, $\Delta l$ satisfies the Pythagorean theorem as in (1.12). Hence for a beam of light it must be true that

$$\Delta l^2 = \Delta x^2 + \Delta y^2 + \Delta z^2 = c^2 \Delta t^2, \tag{4.2}$$

where we have generalized to three dimensions. Working with infinitesimals, this becomes

$$dl^2 = dx^2 + dy^2 + dz^2 = c^2 dt^2. \tag{4.3}$$

---

[1]In his "Autobiographical Notes" published in 1949, Einstein wrote a famous apology to Newton, "Newton, forgive me," he wrote, "You found the only way which, in your age, was just about possible for a man of highest thought and creative power."

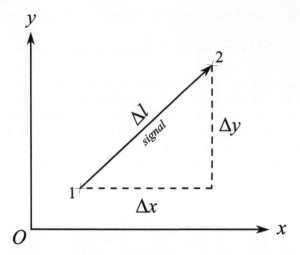

Fig. 4.1: Considering a signal sent from point 1 to point 2.

On the other hand, a signal sent via speeds slower than light must satisfy $\Delta l < c\Delta t$, or in differential form

$$dx^2 + dy^2 + dz^2 < c^2 dt^2, \qquad (4.4)$$

while a signal sent with speeds faster than light (which is physically impossible for reasons we will explain later) must satisfy

$$dx^2 + dy^2 + dz^2 > c^2 dt^2. \qquad (4.5)$$

In short, one can define a new quantity,

$$ds^2 = -c^2 dt^2 + dx^2 + dy^2 + dz^2, \qquad (4.6)$$

that can have three different values:

$$ds^2 < 0 \qquad \text{slower than light signal, called } \textit{timelike}, \qquad (4.7)$$
$$ds^2 > 0 \qquad \text{faster than light signal, called } \textit{spacelike}, \qquad (4.8)$$
$$ds^2 = 0 \qquad \text{signal traveling at light speed, called } \textit{lightlike}, \text{ or } \textit{null}. \qquad (4.9)$$

Minkowski realized that (4.6) describes a four-dimensional *metric*. Three of the components are the usual spatial Cartesian coordinates, and the fourth is something completely new. This is the origin of the statement that "time is the fourth dimension." If both of Einstein's postulates are required to be true, then $ds^2$ must be a quantity that *all* observers, no matter how fast they are traveling with respect to each other, must agree upon. They may *not* agree on the individual spatial measurements $dx$, $dy$, $dz$, or even on the time measurement $dt$, but they will agree on $ds^2$. This is the foundation of special relativity: The four-dimensional "spacetime" we live in has a metric defined by (4.6), known as the **Minkowski metric**. Other common names for $ds^2$ are the **interval** or **line element**.

**Exercise 4.1** A laser located 100km from an observer automatically fires a beam at exactly $t = 0$. At what later time does the observer see the light?

**Exercise 4.2** The following exercises can be solved by considering the finite time it takes a light signal to travel:

1. Consider a rod that has length $l_0$ when measured at rest with respect to an observer. Now the rod is made to move from left to right at a speed $v = 0.9c$, i.e. 90 percent of the speed of light. When the left end of the rod passes a camera, a picture is taken of the rod with a background of a stationary graded meterstick. The picture shows that the left end of the rod coincided with the 0 mark on the meterstick, while the right end coincided with the 0.8 mark. In other words, the moving length of the rod appears to be $l = 0.8$m. Find $l_0$, which may be termed as the actual, or *rest*, length, of the rod. This exercise demonstrates a phenomenon known as "length contraction," which we will come back to later.

2. Just as the length of moving objects appears different from the length of the same objects at rest, time also exhibits a similar phenomenon known as "time dilation," which we will also come back to later. Consider the following situation: Two events, say the explosions of two previously prepared firecrackers, occur at a given distance from each other on the $x$ axis. Maya stands exactly halfway between them and sees the light coming from the two events at the same instant in time. Hence she declares that the two explosions happened simultaneously. Now, her brother Mourad, in his own rest axis $x'$, is passing by with a speed $v$ as measured by Maya, such that $x'$ is parallel to $x$. Explain why Mourad will disagree with Maya on the simultaneity of the two explosions.

These two exercises explore two accepted results of special relativity. The first one demonstrates that two inertial observers will disagree on spatial distances, while the second demonstrates a disagreement in time differences. This is consistent with our claim that in the metric (4.6) different inertial observers will disagree on $dx$, $dy$, $dz$, and $dt$, while agreeing on $ds$ itself.

---

We can now proceed to define true four-dimensional vectors and tensors (as compared with the not-so-true Galilean four-dimensional vectors we discussed in the previous chapter). Based on (4.6), our four-dimensional spacetime is defined by the Cartesian coordinates $x^1$, $x^2$, $x^3$ as usual, with the addition of one more coordinate, defined by $x^0 \equiv ct$. Hence, we can now write

$$ds^2 = dx^\mu dx_\mu = dx_\mu dx^\mu = \eta_{\mu\nu} dx^\mu dx^\nu, \qquad (4.10)$$

where the Greek indices will count over $0, 1, 2, 3$. The quantity $\eta_{\mu\nu}$ is known as the **Minkowski metric tensor** and has components $\eta_{00} = -1$, $\eta_{11} = \eta_{22} = \eta_{33} = +1$. The metric (4.10) is in Cartesian coordinates, and $\eta_{\mu\nu}$ is a generalization of $\delta_{ij}$ to four dimensions (*not* $\delta_{\mu\nu}$; this would just be $\delta_{00} = \delta_{11} = \delta_{22} = \delta_{33} = +1$, which would not give the correct signs in $ds^2$).

---

**Exercise 4.3** Explicitly expand the double sum $\eta_{\mu\nu} dx^\mu dx^\nu$ to show that it gives (4.6). Also show that (4.10) implies that if $dx^0 = cdt$, then $dx_0 = -cdt$, while $dx^1 = dx_1$, $dx^3 = dx_3$, and $dx^3 = dx_3$.

If we wished, we could have written

$$ds^2 = -c^2 dt^2 + g_{ij} dx^i dx^j \tag{4.11}$$

or

$$ds^2 = g_{\mu\nu} dx^\mu dx^\nu, \tag{4.12}$$

where $g_{\mu\nu}$ has $1, 2, 3$ components that can be any of the usual curvilinear coordinates, with the extra pieces $g_{00} = -1$, and $g_{0i} = g_{i0} = 0$. For example,

$$ds^2 = -c^2 dt^2 + dr^2 + r^2 d\theta^2 + r^2 \sin^2 \theta d\varphi^2 \tag{4.13}$$

is the specially covariant metric in spherical coordinates. In the language of matrices, the specially covariant metric tensor is either

$$[\eta] = \begin{pmatrix} -1 & 0 \\ 0 & \delta_{ij} \end{pmatrix} = \begin{pmatrix} -1 & 0 & 0 & 0 \\ 0 & 1 & 0 & 0 \\ 0 & 0 & 1 & 0 \\ 0 & 0 & 0 & 1 \end{pmatrix} \tag{4.14}$$

or

$$[g] = \begin{pmatrix} -1 & 0 \\ 0 & g_{ij} \end{pmatrix} = \begin{pmatrix} -1 & 0 & 0 & 0 \\ 0 & g_{11} & 0 & 0 \\ 0 & 0 & g_{22} & 0 \\ 0 & 0 & 0 & g_{33} \end{pmatrix}. \tag{4.15}$$

Both (4.14) and (4.15) can be referred to as the Minkowski metric,[2] although it is somewhat conventional in the literature to exclusively work with Cartesian coordinates, i.e. with (4.6)/(4.10)/(4.14), just for the sake of simplicity. Most textbooks do not even mention (4.12)/(4.15). Whenever there is a possibility for confusion we will refer to (4.6)/(4.10)/(4.14) as the **Minkowski-Cartesian metric**.

Another warning to the reader: Because of the general approach used to derive the metric, it is possible to have defined (4.6) as

$$ds^2 = c^2 dt^2 - dx^2 - dy^2 - dz^2 \quad \text{(not one we will use)}. \tag{4.16}$$

Either form gives the correct physical behavior, although various later equations will change signs accordingly. The conventional choice of which terms are positive and which are negative is generally known as the **signature of the metric**. Usually books on relativity use the $(-, +, +, +)$ signature, while books on field theory or particle physics use the opposite signature $(+, -, -, -)$.[3] Usually the book in question will make it clear as early as possible which signature is being used, so be warned.

---

[2]We are once again assuming $g_{ij}$ to be diagonal for simplicity, but it doesn't have to be. One can just as easily define a Minkowskian metric tensor in skew coordinates where the metric will not be diagonal.

[3]While rare, some books on relativity use the particle physicist's signature; most notably [6].

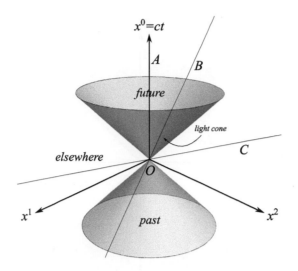

Fig. 4.2: A three-dimensional spacetime diagram.

Now four-dimensional spacetime may be "visualized" by considering the so-called **spacetime diagram** shown in Fig. [4.2]. First we *must* suppress at least one of the three spatial dimensions in order to be able to draw the diagram; in this case, we have chosen $x^3$. The origin point $t = 0$ is referred to as "now," which means that below the $x$–$y$ plane is the past and above it is the future. The path of a particle in spacetime diagrams is called its **world line**. Consider the following cases: A particle at rest at the origin $O$ will plot a perfectly vertical world line since it is not changing position at any time $t$; this is denoted by the letter $A$ in Fig. [4.2]. A timelike particle, i.e. one traveling with a constant speed $v < c$, will plot a straight world line with slope $c/v$, as denoted by $B$. Note that the slope of $B$ is *necessarily* greater than unity. A lightlike particle traveling at exactly the speed of light $v = c$ will have a slope of 1. The world lines of all possible lightlike particles form a cone with unit slope in the spacetime diagram. This is called the **light cone**. Finally, a spacelike particle traveling faster than light will have a world line outside of the light cone, here denoted by $C$. The different regions of past and future are labeled in the diagram: For timelike particles they are within the light cone while for the (unphysical) spacelike particles their past and future lie in the so-called "*elsewhere*" region, sometimes also referred to as "*elsewhen*." It is common to suppress two spatial directions and plot an even more easily visualizable spacetime diagram, as shown in Fig. [4.3].

With these definitions we can proceed and write all the rules of vectors and tensors "upgraded" to the new number of dimensions and the new metric. For example, a "4-vector" $\mathbf{V}_{4\text{d}}$ would have components $V^\mu = \left(V^0, V^1, V^2, V^3\right)$. There are two issues here. Firstly, visualization is lost: While we can sketch four-dimensional spacetime by ignoring one or two spatial directions, a full sketch is no longer possible. The second issue is the question of interpretation: While the $1, 2, 3$ components are the usual three-dimensional components, that extra "zeroth" component would have different

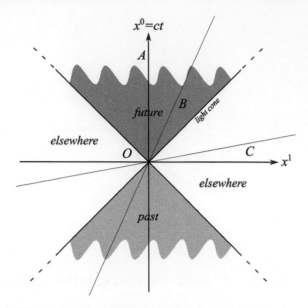

Fig. 4.3: A two-dimensional spacetime diagram.

interpretations depending on the physical meaning of the vector in question. We will get to that eventually. For now, let us review the mathematics of vectors and tensors in this new world. In Cartesian coordinates, 4-vectors can be written in terms of unit basis vectors as

$$\mathbf{V}_{4\mathrm{d}} = V^\mu \hat{\mathbf{e}}_\mu, \tag{4.17}$$

where $\hat{\mathbf{e}}_\mu$ are the spacetime basis vectors defined by

$$\hat{\mathbf{e}}_\mu \cdot \hat{\mathbf{e}}_\nu = \eta_{\mu\nu}. \tag{4.18}$$

The component $\hat{\mathbf{e}}_0$ points along the time axis in Figs. [4.2] and [4.3], and as such it is a timelike basis vector, while the $\hat{\mathbf{e}}_i$ are parallel to the spatial axis and are hence spacelike (these are just the usual Cartesian unit vectors in three dimensions). The scalar product of vectors is then

$$\mathbf{V} \cdot \mathbf{U} = V^\mu U^\nu \hat{\mathbf{e}}_\mu \cdot \hat{\mathbf{e}}_\nu = V^\mu U^\nu \eta_{\mu\nu}$$
$$= V^\mu U_\mu = V_\mu U^\mu = V_\mu U_\nu \eta^{\mu\nu}. \tag{4.19}$$

Indices are hence raised and lowered via

$$V^\mu \eta_{\mu\nu} = V_\nu \tag{4.20}$$

or

$$V_\mu \eta^{\mu\nu} = V^\nu, \tag{4.21}$$

where $\eta^{\mu\nu}$ is the inverse metric tensor, defined such that orthogonality is satisfied:

$$\eta_{\mu\alpha}\eta^{\alpha\nu} = \delta^{\nu}_{\mu}. \tag{4.22}$$

Hence

$$V^0 = -V_0, \quad V^1 = V_1, \quad V^2 = V_2, \quad V^3 = V_3. \tag{4.23}$$

The components of the inverse Minkowski tensor are then also $\eta^{00} = -1$, $\eta^{11} = \eta^{22} = \eta^{33} = +1$. In the language of matrices, (4.22) becomes

$$\begin{bmatrix} -1 & 0 & 0 & 0 \\ 0 & 1 & 0 & 0 \\ 0 & 0 & 1 & 0 \\ 0 & 0 & 0 & 1 \end{bmatrix} \begin{bmatrix} -1 & 0 & 0 & 0 \\ 0 & 1 & 0 & 0 \\ 0 & 0 & 1 & 0 \\ 0 & 0 & 0 & 1 \end{bmatrix} = \begin{bmatrix} 1 & 0 & 0 & 0 \\ 0 & 1 & 0 & 0 \\ 0 & 0 & 1 & 0 \\ 0 & 0 & 0 & 1 \end{bmatrix}, \tag{4.24}$$

as the reader can easily check. Note that the hybrid Kronecker delta, and *only* the hybrid delta, still exists in four dimensions and still plays the same role it did in three; specifically, it can be used to rename indices as in (1.63):

$$U^{\nu}\delta^{\mu}_{\nu} = U^{\mu}, \quad U_{\mu}\delta^{\mu}_{\nu} = U_{\nu}. \tag{4.25}$$

In Cartesian coordinates, then, the detailed scalar product of two 4-vectors $V^{\mu}$ and $U^{\mu}$ is

$$\begin{aligned} V^{\mu}U_{\nu} &= V^0U_0 + V^1U_1 + V^2U_2 + V^3U_3 \\ &= -V^0U^0 + V^1U^1 + V^2U^2 + V^3U^3 \\ &= -V_0U_0 + V_1U_1 + V_2U_2 + V_3U_3. \end{aligned} \tag{4.26}$$

Derivatives in Cartesian coordinates can still be defined by

$$\partial_{\mu} = \frac{\partial}{\partial x^{\mu}}, \tag{4.27}$$

where $\partial_1$, $\partial_2$, and $\partial_3$ (in other words, $\partial_i$) have the same interpretation as before; i.e. Cartesian derivatives with respect to the three spatial directions. The fourth derivative, $\partial_0$, is a time derivative:

$$\partial_0 = \frac{\partial}{\partial x^0} = \frac{1}{c}\frac{\partial}{\partial t}. \tag{4.28}$$

For the reader fluent in the language of multivariable calculus, we note that the four-dimensional Laplacian

$$\eta^{\mu\nu}\partial_{\mu}\partial_{\nu} = \eta_{\mu\nu}\partial^{\mu}\partial^{\nu} = \partial^{\mu}\partial_{\mu} \tag{4.29}$$

is in fact

$$\eta^{\mu\nu}\partial_\mu\partial_\nu = -\frac{1}{c^2}\frac{\partial^2}{\partial t^2} + \frac{\partial^2}{\partial x^2} + \frac{\partial^2}{\partial y^2} + \frac{\partial^2}{\partial z^2}$$

$$= -\frac{1}{c^2}\frac{\partial^2}{\partial t^2} + \nabla^2. \tag{4.30}$$

This is known as the **d'Alembertian operator**,[4] or the **wave operator**, since it appears in the wave equation of classical physics. A common notation for it is the so-called **Box operator**:

$$\Box f = -\frac{1}{c^2}\frac{\partial^2 f}{\partial t^2} + \nabla^2 f, \tag{4.31}$$

where $f(x^\alpha)$ is some scalar function. In this notation the wave equation is simply $\Box f(x^\alpha) = 0$, generalizing the **Laplace equation** $\nabla^2 f(x^i) = 0$. In fact, as far as the metric is concerned, the d'Alembertian *is* indeed just the four-dimensional Laplacian. Updating these equations to a curvilinear specially covariant metric is just a matter of minimal substitution, so

$$\hat{\mathbf{e}}_\mu \to \hat{\mathbf{g}}_\mu$$

$$\eta_{\mu\nu} \to g_{\mu\nu}$$

$$\partial_\mu \to \nabla_\mu \tag{4.32}$$

are guaranteed to make all equations applicable to any rotationally covariant coordinate system, where the Christoffel symbols are still given by

$$\Gamma^\rho_{\mu\nu} = \frac{1}{2}g^{\rho\sigma}\left(\partial_\mu g_{\nu\sigma} + \partial_\nu g_{\sigma\mu} - \partial_\sigma g_{\mu\nu}\right). \tag{4.33}$$

### Exercise 4.4

1. What are the *numerical* values of $\partial_\mu x_\nu$ and $\partial_\mu x^\nu$ for all the possible combinations of indices? How are these expressions related to the Minkowski tensor $\eta$ and the Kronecker $\delta$?

2. Using your result from the previous exercise find the numerical value of $\partial_\mu x^\mu$. Compare with Exercise 1.13. Based on this, can you guess the numerical result of $\partial_\mu x^\mu$ one would get for an $N$-dimensional spacetime?

3. Write the Minkowski metric for each of the coordinates of Exercise 1.21 using (4.12).

4. Using the metrics you found in the previous exercise, find the "missing" components of the Christoffel symbols that involve time. In other words, what are $\Gamma^0_{00}$, $\Gamma^0_{i0}$, $\Gamma^0_{ij}$, $\Gamma^i_{00}$, and so on? You don't need to write them in terms of the individual values of the spatial indices $i$ and $j$. In other words, there is no need to write $\Gamma^0_{12}$, $\Gamma^3_{01}$, etc.; just leave the spatial indices symbolic. The remaining components involving only the spatial directions—namely $\Gamma^i_{jk}$—would be the same as you found before in Exercise 2.22.

5. Calculate $\partial_\mu\left(\ln\sqrt{x^\alpha x_\alpha}\right)$ and simplify as much as possible. Compare your answer to that of Exercise 1.13.

---

[4] Jean-Baptiste le Rond d'Alembert: French mathematician (1717–1783).

**Exercise 4.5** In some theories of physics (e.g., superstring theory) it is useful to define the so-called **light cone coordinates**; this is a coordinate system that defines axis $x^+$ and $x^-$ along the light cone itself as follows:

$$x^+ = \frac{1}{\sqrt{2}} \left( x^0 + x^1 \right)$$

$$x^- = \frac{1}{\sqrt{2}} \left( x^0 - x^1 \right), \tag{4.34}$$

where $x^0 = ct$, $x^1 = x$. The remaining Cartesian coordinates $x^2 = y$ and $x^3 = z$ remain unchanged. Hence the complete set of light cone coordinates is simply $\left( x^+, x^-, x^2, x^3 \right)$. Careful inspection of these equations will convince the reader that $x^+$ is along the top right side of the light cone in Fig. [4.3] and the axis $x^-$ is along the top left side, hence the name. The factor of $1/\sqrt{2}$ is there because this is the value of both the sine and cosine of $45°$, the angle the light cone makes with the $x^0$ and $x^1$ axis. Notice that the new coordinates *mix* the spatial axis $x$ with the time axis, and as such neither $x^+$ nor $x^-$ are fully spatial or fully temporal. Perform the following calculations:

1. Find the line element $ds^2$ in light cone coordinates.
2. Write down the components of the light cone metric tensor; let's call it $\hat{\eta}_{\mu\nu}$, where the indices now count over $+, -, 2, 3$.
3. Find the inverse light cone metric tensor $\hat{\eta}^{\mu\nu}$ and verify orthogonality.
4. Given the spacetime vectors $a^\mu = \left( a^+, a^-, a^2, a^3 \right)$ and $b^\mu = \left( b^+, b^-, b^2, b^3 \right)$, write out the details of the scalar product $a^\mu b_\mu$, as seen in (4.26).
5. Find out how $a^\mu$ is related to $a_\mu$ in light cone coordinates, as seen in (4.23).

## 4.2  Four-Vectors and Four-Tensors

We are now in a position to ask a question similar to the one we asked in §2.1: What is a 4-vector? But the answer in this case will *define* special covariance, just as the answer for 3-vectors defined rotational (or more generally classical) covariance. Recall that we require that all the laws of non-relativistic physics be rotationally invariant, and that we have found that they are also invariant under Galilean boosts (at least mechanics). In relativity, rotational invariance is still true but because of the second postulate we require invariance under a new type of boosts: **relativistic boosts**. As before, different observers moving with different speeds may disagree on the components of vectors; however, they must still agree on the dot product of vectors. The vector transformation rules are still the same, just upgraded to four dimensions:

$$V^{\mu'} = \lambda^{\mu'}_\nu V^\nu. \tag{4.35}$$

The explicit form of the transformation tensor $\lambda$ is still

$$\lambda^{\mu'}_\nu = \frac{\partial x^{\mu'}}{\partial x^\nu}, \tag{4.36}$$

if we know how the transformed coordinates $x^{\mu'}$ are related to the original ones $x^{\nu}$. So what are the components of this four-dimensional $\lambda$? It must include rotations like its three-dimensional counterparts. In fact, the transformations (3.126) are still valid.

What about boosts? What are the relativistic counterparts to (3.112) and (3.113)? These are the so-called **Lorentz transformations**, or **Lorentz boosts**, which guarantee the invariance of the 4-vector dot product between moving coordinate systems while at the same time respecting the second postulate. Consider once again the two Cartesian coordinate systems shown in Fig. [3.7]. One is moving with velocity $v$ with respect to the other. There are several different methods of deriving the Lorentz transformations.[5] Our program is to derive everything from the metric, so that is what we will do. A major difference between the transformations we seek and the Galilean ones is that we can no longer expect $t' = t$; the times as measured by different inertial observers have to be related via a more complicated formula. Next we will also assume that the required transformations are linear, just as the Galilean transformations were. What this means is that $t'$ and $x'$ can only depend on unit powers of $t$ and/or $x$, so there can be no $t^3$ or $\sqrt{x}$ or any other non-linear dependence. This is an assumption required by physics, since we know that powers different from unity would lead to accelerating reference frames, which will necessarily generate non-inertial pseudo-forces, violating Einstein's first postulate. In short, we seek transformations of the form

$$t = At' + Bx'$$
$$x = Dt' + Ex', \tag{4.37}$$

where $A$, $B$, $D$, and $E$ are unknown constants to be determined. We require that (4.37) satisfy the invariance of the metric between the $O$ and $O'$ frames. This translates into

$$-c^2 dt^2 + dx^2 + dy^2 + dz^2 = -c^2 dt'^2 + dx'^2 + dy'^2 + dz'^2. \tag{4.38}$$

We are interested in a specific spacetime point rather than an interval (Point $P$ in Fig. [3.7]), so we can instead use

$$-c^2 t^2 + x^2 + y^2 + z^2 = -c^2 t'^2 + x'^2 + y'^2 + z'^2, \tag{4.39}$$

which is just the statement that the scalar product of the position vector is invariant, i.e. $x^{\mu} x_{\mu} = x^{\mu'} x_{\mu'}$. Furthermore, since we have chosen everything to be happening along the $x$ axis, we can assume $y = y'$ and $z = z'$, just like in the Galilean boosts. Equation (4.39) then reduces to

$$-c^2 t^2 + x^2 = -c^2 t'^2 + x'^2. \tag{4.40}$$

Next, substitute (4.37) into (4.40):

---

[5]These range from the very simple, such as the one proposed by Einstein in his popular science book [7], to the very complex, which includes the way Lorentz himself found them. He did so by asking: If the Maxwell equations are *not* invariant under the Galilean transformations, then what transformations are they invariant under?

$$-c^2 \left( At' + Bx' \right)^2 + \left( Dt' + Ex' \right)^2 = -c^2 t'^2 + x'^2. \tag{4.41}$$

Expanding and rearranging,

$$-c^2 t'^2 + x'^2 = \left( D^2 - c^2 A^2 \right) t'^2 + \left( E^2 - c^2 B^2 \right) x'^2 + 2 \left( ED - c^2 AB \right) x' t'. \tag{4.42}$$

The coefficients of $t'^2$, $x'^2$, and $x't'$ must match; hence

$$D^2 - c^2 A^2 = -c^2$$
$$E^2 - c^2 B^2 = 1$$
$$ED - c^2 AB = 0. \tag{4.43}$$

Being three equations in four unknowns, the set (4.43) is not enough; we need one more. Now, by definition the origin point of the unprimed frame $x = 0$ is always separated from the primed frame by the distance $vt$, as one clearly sees from Fig. [3.7]. Hence its location in the primed frame is always $x' = -vt'$. This can be used in the second equation of (4.37) as follows:

$$x = Dt' + Ex' = 0$$
$$= Dt' - Evt'; \tag{4.44}$$

hence the needed fourth equation is

$$D = vE. \tag{4.45}$$

Solving (4.43) and (4.45) together, we find

$$A = E = \frac{1}{\sqrt{1 - \frac{v^2}{c^2}}}$$

$$B = -\frac{v}{c^2 \sqrt{1 - \frac{v^2}{c^2}}}$$

$$D = \frac{v}{\sqrt{1 - \frac{v^2}{c^2}}}. \tag{4.46}$$

The ratio $v/c$ as well as the factor with the square root appears frequently in many of the subsequent equations, so for brevity's sake it is conventional to define the so-called **boost vector** (a 3-vector)

$$\vec{\beta} = \frac{\mathbf{v}}{c} \tag{4.47}$$

and the **Lorentz factor**

$$\gamma = \frac{1}{\sqrt{1-\beta^2}}, \tag{4.48}$$

where $\beta^2 = \vec{\beta} \cdot \vec{\beta} = v^2/c^2$. Putting everything back into (4.37), we find the Lorentz transformations,

$$x^0 = \gamma \left( x^{0'} + \beta x^{1'} \right)$$
$$x^1 = \gamma \left( x^{1'} + \beta x^{0'} \right)$$
$$x^2 = x^{2'}, \qquad x^3 = x^{3'}, \tag{4.49}$$

with their inverse relations

$$x^{0'} = \gamma \left( x^0 - \beta x^1 \right)$$
$$x^{1'} = \gamma \left( x^1 - \beta x^0 \right)$$
$$x^{2'} = x^2, \qquad x^{3'} = x^3. \tag{4.50}$$

Notice that in the low speed limit, i.e. $v << c$, we have $\beta \to 0$ and $\gamma \to 1$, which reduces (4.49) to exactly the Galilean transformations (3.108) plus the classically trivial $t' = t$. The factors $\beta$ and $\gamma$ run over the values

$$\beta = 0 \to 1 \tag{4.51}$$
$$\gamma = 1 \to \infty, \tag{4.52}$$

where the form of $\gamma$ as given by (4.48) constrains $\beta$ from becoming greater than 1; in other words, $v$ cannot be greater than $c$. This is the origin of the "rule" that the speed of light in vacuum $c$ is a cosmic speed limit: No object or signal can travel faster than light. But what about the case $v = c$ of light itself? How does one transform to a reference frame that is traveling *exactly* at the speed of light? As it turns out, the Lorentz transformations do not allow for this to happen, since $\gamma$ blows up at precisely that value. If $v = c$, then $O'$ becomes the frame of reference where light is not moving. But this is *not* allowed because it violates Einstein's second postulate. One then concludes that light *does not have a rest frame*! In other words, no inertial frame of reference can exist where light is at rest.[6] The Lorentz transformations apply *only* to reference frames traveling slower than light. Note that, although you can never get to exactly $v = c$, you are allowed to get arbitrarily close. There is no problem, in principle, with non-light particles achieving speeds of $v = 0.9999999c$ or higher, as long as it is not exactly $v = c$. In fact, subatomic particles have indeed been accelerated to such high speeds but, as we will see later, this is constrained by the amount of energy one has to pour into them.[7]

---

[6] Sorry if this feels a bit like cheating, but it is a fundamental property that has profound consequences.

[7] The Large Hadron Collider, or LHC, in Geneva, Switzerland, has accelerated electrons to about $v = 0.999999999988c$, while cosmic ray particles hitting our atmosphere from outer space have been clocked at roughly $v = 0.99999999999999999999973c$.

Now, we can write the Lorentz boost (4.49) in matrix form as follows, along the $x^1$, $x^2$ and $x^3$ axes respectively:

$$[\lambda]_{x^{1'}\to x^1} = \begin{pmatrix} \gamma & \beta\gamma & 0 & 0 \\ \beta\gamma & \gamma & 0 & 0 \\ 0 & 0 & 1 & 0 \\ 0 & 0 & 0 & 1 \end{pmatrix}, \qquad [\lambda]_{x^{2'}\to x^2} = \begin{pmatrix} \gamma & 0 & \beta\gamma & 0 \\ 0 & 1 & 0 & 0 \\ \beta\gamma & 0 & \gamma & 0 \\ 0 & 0 & 0 & 1 \end{pmatrix}$$

$$[\lambda]_{x^{3'}\to x^3} = \begin{pmatrix} \gamma & 0 & 0 & \beta\gamma \\ 0 & 1 & 0 & 0 \\ 0 & 0 & 1 & 0 \\ \beta\gamma & 0 & 0 & \gamma \end{pmatrix}, \tag{4.53}$$

along with their inverse,

$$[\lambda]_{x^1\to x^{1'}} = \begin{pmatrix} \gamma & -\beta\gamma & 0 & 0 \\ -\beta\gamma & \gamma & 0 & 0 \\ 0 & 0 & 1 & 0 \\ 0 & 0 & 0 & 1 \end{pmatrix}, \qquad [\lambda]_{x^2\to x^{2'}} = \begin{pmatrix} \gamma & 0 & -\beta\gamma & 0 \\ 0 & 1 & 0 & 0 \\ -\beta\gamma & 0 & \gamma & 0 \\ 0 & 0 & 0 & 1 \end{pmatrix}$$

$$[\lambda]_{x^3\to x^{3'}} = \begin{pmatrix} \gamma & 0 & 0 & -\beta\gamma \\ 0 & 1 & 0 & 0 \\ 0 & 0 & 1 & 0 \\ -\beta\gamma & 0 & 0 & \gamma \end{pmatrix}. \tag{4.54}$$

The algebraic expressions equivalent to (4.49) can be found by considering a boost along the $x^1$ direction as follows,

$$\begin{pmatrix} x^0 \\ x^1 \\ x^2 \\ x^3 \end{pmatrix} = \begin{pmatrix} \gamma & \beta\gamma & 0 & 0 \\ \beta\gamma & \gamma & 0 & 0 \\ 0 & 0 & 1 & 0 \\ 0 & 0 & 0 & 1 \end{pmatrix} \begin{pmatrix} x^{0'} \\ x^{1'} \\ x^{2'} \\ x^{3'} \end{pmatrix}, \tag{4.55}$$

making this the generalization of (3.112). One can check that the orthogonality relation (2.18) is still valid,

$$\lambda^{\mu}_{\nu'}\lambda^{\nu'}_{\rho} = \delta^{\mu}_{\rho}, \tag{4.56}$$

or equivalently

$$\begin{pmatrix} \gamma & \beta\gamma & 0 & 0 \\ \beta\gamma & \gamma & 0 & 0 \\ 0 & 0 & 1 & 0 \\ 0 & 0 & 0 & 1 \end{pmatrix} \begin{pmatrix} \gamma & -\beta\gamma & 0 & 0 \\ -\beta\gamma & \gamma & 0 & 0 \\ 0 & 0 & 1 & 0 \\ 0 & 0 & 0 & 1 \end{pmatrix} = \begin{pmatrix} 1 & 0 & 0 & 0 \\ 0 & 1 & 0 & 0 \\ 0 & 0 & 1 & 0 \\ 0 & 0 & 0 & 1 \end{pmatrix}. \tag{4.57}$$

Recall that the orthogonality relation is required for consistency: Transforming an event to a reference frame (whether a rotation, translation, or a boost) then transforming it back must return it to its original situation. We can now continue as we did in the case of the Galilean transformations by combining the Lorentzian boosts with each other or with rotations as needed. The arguments are exactly the same as the ones in the previous chapter. For example, a Lorentz boost in an arbitrary direction, corresponding to Fig. [3.9], is the product of the three boosts (4.54), giving[8]

$$
[\lambda]_{O \to O'} = \begin{bmatrix} \gamma & -\gamma\beta_x & -\gamma\beta_y & -\gamma\beta_z \\ -\gamma\beta_x & 1 + (\gamma - 1)\frac{\beta_x^2}{\beta^2} & (\gamma - 1)\frac{\beta_x\beta_y}{\beta^2} & (\gamma - 1)\frac{\beta_x\beta_z}{\beta^2} \\ -\gamma\beta_y & (\gamma - 1)\frac{\beta_x\beta_y}{\beta^2} & 1 + (\gamma - 1)\frac{\beta_y^2}{\beta^2} & (\gamma - 1)\frac{\beta_y\beta_z}{\beta^2} \\ -\gamma\beta_z & (\gamma - 1)\frac{\beta_x\beta_z}{\beta^2} & (\gamma - 1)\frac{\beta_y\beta_z}{\beta^2} & 1 + (\gamma - 1)\frac{\beta_z^2}{\beta^2} \end{bmatrix}. \tag{4.58}
$$

Notice that (4.58) is a symmetric transformation tensor, as opposed to the non-symmetric one (3.125). As mentioned before, this is due to the fact that in special relativity time and space are treated on a perfectly equal footing. As far as nature is concerned, time is just a dimension equivalent to space.[9] The other cases of combining boosts with rotations are the same as in the Galilean transformations discussion in the previous chapter, except that now we use the Lorentzian $\lambda$ tensors instead of the Galilean ones.

**Exercise 4.6** Explicitly verify (4.57). Find the inverse of the generalized Lorentz transformation matrix (4.58) and verify *its* orthogonality.

**Exercise 4.7**

1. A firecracker explodes in space. An observer $O$ measures the spacetime location of this event at $x = 10$km, $y = 100$km, $z = 3$km, and $t = 0$. Another observer $O'$ is traveling with a speed of $0.9c$ along the common $x$–$x'$ axis as in Fig. [3.7]. Find $x'$, $y'$, $z'$, and $t'$.

2. A firecracker explodes in space. An observer $O$ measures the spacetime location of this event at $x = 10$km, $y = 100$km, $z = 3$km, and $t = 0$. Another observer $O'$ is traveling with a velocity of $\mathbf{v} = (0.7c, 0.8c, 0.2c)$ as in (the three-dimensional version of) Fig. [3.9]. What are the components of $\lambda$ in this case? Find $x'$, $y'$, $z'$, and $t'$.

3. Redo Exercises 3.26 and 3.27 using the Lorentz transformation tensors instead of the Galilean ones. You may assume that the motion of the projectile in the rest frame of the airplane is still non-relativistic, i.e. just the usual physics 101 projectile situation. To make the numbers not too small, change the velocity of the airplane to $\mathbf{v} = 0.5c\hat{\mathbf{x}} + 0.2c\hat{\mathbf{y}} - 0.7c\hat{\mathbf{z}}$ (making it a futuristic spaceship). Can you still roughly sketch these motions?

---

[8]More precisely, this corresponds to the three-dimensional version of Fig. [3.9].

[9]Okay, so this is a bit too idealized. We all know that time, although a dimension in the sense of the metric, *does* hold a special place in physics. For example, consider that one may freely move in the positive or negative spatial directions, but one is not free to move *backwards* in time! This is the unsolved problem of the "arrow of time;" in other words why is it that time "flows" in one direction but not the other? It is true, though, that all the fundamental theories of physics, such as quantum mechanics, electrodynamics, and so on, do not really put any constraints on the arrow of time. The question only arises when one discusses large conglomerates of matter, as in thermodynamics and statistical mechanics. In this context the flow of time is tied to **entropy**, a thermodynamical property that changes in one direction only. Now here I find myself in a footnote that could easily turn into a book of its own. So I will stop and just refer the reader to the appropriate textbooks, such as [8].

Four-dimensional tensors are defined the same way as before: A tensor is a multi-index collection of objects of rank $(p, q)$ whose components $T^{\mu_1 \cdots \mu_p}_{\phantom{\mu_1 \cdots \mu_p} \nu_1 \cdots \nu_q}$ transform from one coordinate system to another, boosted, translated, or rotated, using the same exact rules as before, except upgraded to Greek indices to denote four dimensions. Here are some examples:

$$V_{\mu'} = \lambda^{\nu}_{\mu'} V_{\nu}. \tag{4.59}$$

$$A^{\mu' \nu'} = \lambda^{\mu'}_{\rho} \lambda^{\nu'}_{\sigma} A^{\rho \sigma}$$
$$A^{\mu' \nu'}_{\phantom{\mu' \nu'} \kappa' \pi' \rho'} = \lambda^{\mu'}_{\varphi} \lambda^{\nu'}_{\gamma} \lambda^{\tau}_{\kappa'} \lambda^{\chi}_{\pi'} \lambda^{\psi}_{\rho'} A^{\varphi \gamma}_{\phantom{\varphi \gamma} \tau \chi \psi}. \tag{4.60}$$

And, generally,

$$A^{\mu'_1 \cdots \mu'_p}_{\phantom{\mu'_1 \cdots \mu'_p} \nu'_1 \cdots \nu'_q} = \lambda^{\mu'_1}_{\sigma_1} \cdots \lambda^{\mu'_p}_{\sigma_1} \lambda^{\rho_1}_{\nu'_1} \cdots \lambda^{\rho_q}_{\nu'_q} A^{\sigma_1 \cdots \sigma_p}_{\phantom{\sigma_1 \cdots \sigma_p} \rho_1 \cdots \rho_q}. \tag{4.61}$$

Finally, one can also generalize the Levi-Civita totally antisymmetric symbol and tensor by just adding an extra index. Recall that the number of indices for $\varepsilon$ and $\mathscr{E}$ must match the number of dimensions. So we define the spacetime antisymmetric symbol

$$\varepsilon_{\mu\nu\rho\sigma} = \begin{cases} \text{zero} & \text{for any equal indices} \\ +1 & \text{for even permutations} \\ -1 & \text{for odd permutations} \end{cases} \tag{4.62}$$

and $\quad \varepsilon_{0123} = +1,$

which is covariant under (Cartesian) boosts but *not* rotations. The Levi-Civita *tensor*, covariant under everything, is

$$\mathscr{E}_{\mu\nu\rho\sigma} = \sqrt{-|g|} \varepsilon_{\mu\nu\rho\sigma}. \tag{4.63}$$

Note that the determinant of the Minkowski metric $g_{\mu\nu}$ is no longer positive definite! Hence we must insert a minus sign by hand to make the square root real (see the footnote in §2.5). If we are in Minkowski-Cartesian coordinates, then $\sqrt{-|g|} = \sqrt{-|\eta|} = 1$ and we just get $\mathscr{E}_{\mu\nu\rho\sigma} = \varepsilon_{\mu\nu\rho\sigma}$. Integrals can also be generalized to four dimensions in a straightforward way. For example, a four-dimensional integral looks like

$$\int \int \int \int \sqrt{-|g|} f(x^{\mu}) \, dx^0 dx^1 dx^2 dx^3, \tag{4.64}$$

or more compactly:

$$\int d^4 x \sqrt{-|g|} f(x^{\mu}). \tag{4.65}$$

**Exercise 4.8** Verify that the Minkowski-Cartesian metric is invariant under the Lorentz transformations.

**Exercise 4.9** If you have done Exercise 4.5, then continue to find the Lorentz transformations in light cone coordinates both in algebraic form, as in (4.49), and matrix form, as in (4.53).

**Exercise 4.10** The Levi-Civita identities we studied in §1.5 may be updated to four dimensions. Make a guess at the spacetime equivalent to (1.106), verify your guess component by component, then derive $\varepsilon_{\mu\nu\rho\sigma}\varepsilon^{\lambda\kappa\tau\sigma}$, $\varepsilon_{\mu\nu\rho\sigma}\varepsilon^{\lambda\kappa\rho\sigma}$, $\varepsilon_{\mu\nu\rho\sigma}\varepsilon^{\lambda\nu\rho\sigma}$, and $\varepsilon_{\mu\nu\rho\sigma}\varepsilon^{\mu\nu\rho\sigma}$ from it. Do the results depend on the signature of the Minkowski-Cartesian metric? In other words, if the components $\eta_{\mu\nu}$ were $(+,-,-,-)$, or even unphysical signatures such as $(+,+,+,+)$, or $(+,-,+,-)$, would your results be different? If so, how?

**Exercise 4.11**

1. If two events are separated by a spacelike interval, prove that there must exist a frame of reference in which they occur simultaneously, while a frame where they happen at the same point cannot exist.
2. Similarly, if two events are separated by a timelike interval, prove that there must exist a frame where they happen at the same point but no frames can exist in which they are simultaneous.

## 4.3 Lorentz Boosts as Rotations

The symmetry between time and space, i.e. the realization that time is "just" another axis in four dimensions, begs the interpretation of the Lorentz boosts as a "rotation" in four-dimensional spacetime. Consider a spacetime diagram like that in Fig. [4.3]. It is plotted from the point of view of some inertial observer $O$. Now can we boost $O$'s point of view to another observer's *and* plot it on the same diagram? What would it look like? Can we interpret it as a rotation like that of, say, Fig. [2.2]? Let's do so graphically first then analytically. In Fig. [4.4] observer $O$ is located at $x = 0$. At time $t = 0$ (i.e. the origin), $O$ points a flashlight in the $x$ direction and sends a light signal. The point $P$ has coordinates $x$ and $t$ representing the distance traveled by the signal over time $t$ as measured by $O$. Now if there is another observer $O'$ whose origin point coincides with that of $O$, meaning that at time $t' = t = 0$ we had $x' = x = 0$, moving with velocity $v$ with respect to $O$, what would the coordinates of the point $P$ be in their reference frame? This is given by the Lorentz transformations (4.50). Now we want to plot both the $x^{0'}$ and the $x^{1'}$ axes right on the same diagram, just like we did in Fig. [2.2]. The $x^{1'}$ axis is the line $x^{0'} = 0$, which upon substituting in the first of eqns (4.50) gives

$$x^0 = \beta x^1. \tag{4.66}$$

Similarly the $x^{0'}$ axis is the line $x^{1'} = 0$, which upon substituting in the second of eqns (4.50) gives

$$x^0 = \frac{1}{\beta}x^1. \qquad (4.67)$$

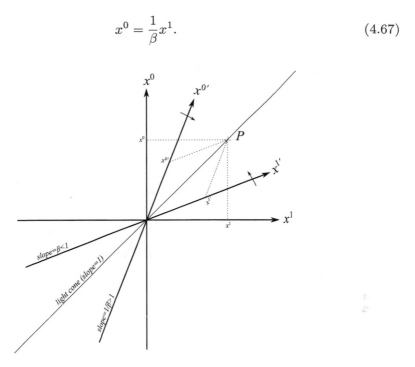

Fig. 4.4: Lorentz boosts as four-dimensional "rotations" in spacetime.

These two equations represent straight lines passing through the origin with slopes $\beta < 1$ and $\frac{1}{\beta} > 1$ respectively. As such, it seems that the $O'$ axes are *not* a rotation in the usual sense! This is called a hyperbolic rotation for reasons that we will explain shortly. As the speed $v$ of the $O'$ reference frame increases the slopes of the $O'$ axes continue to converge toward each other, with the light cone *always* bisecting the angle between them, such that at *exactly* $v = c$ ($\beta = 1$) they both collapse on top of the light cone. At $v = c$ time and space "coincide" and cease to have independent meanings.[10] This is just another way of saying that light has no rest frame. Fig. [4.4], then, is the spacetime analogue of Fig. [2.2]. Notice that the $O'$ coordinate system is a skew system of coordinates, i.e. the axes are not orthogonal.

What if we wanted to draw the inverse diagram, i.e. one where the $O'$ axis are orthogonal? In the sense of ordinary rotations such as Fig. [2.2], this would be easy to visualize: just rotate the whole thing clockwise by the angle $\theta$. But our spacetime "rotation" is not as simple. It is left as an exercise to the reader to show that Fig. [4.5] is the one we are seeking.

**Exercise 4.12** Convince yourself that Fig. [4.5] is the inverse spacetime rotation diagram via an argument like the one used to derive Fig. [4.4] in the text.

---

[10]I wonder how many science fiction stories were inspired by statements like this one.

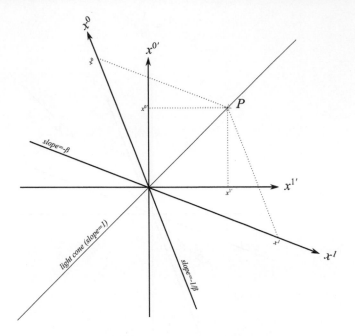

Fig. 4.5: Inverse Lorentz transformations.

These spacetime diagrams are also sometimes referred to as **Minkowski diagrams**. They represent the so-called hyperbolic geometry that special relativity follows, first explained by Minkowski himself. The analytical way to represent these "rotations" is as follows. First construct a quantity that we can call imaginary time, $w = ict$. In this new language the metric becomes

$$ds^2 = dw^2 + dx^2 + dy^2 + dz^2. \tag{4.68}$$

Notice that because the minus sign is no longer there[11] this is a *Euclidean* metric in four-dimensional space, a simple generalization of (1.15). Or, to put it another way, this is just a case of the $N = 4$ in eqn (1.17). Rotations in this space are *ordinary* rotations with cosines and sines in the usual way.[12] The position vector is just $\mathbf{r}_4 = (w, x, y, z)$. Suppose now we wish to rotate this vector in the $w$–$x$ plane; we can write something exactly analogous to (2.3):

[11] It can always be retrieved by returning to real time.

[12] Note that in four Euclidean dimensions, rotations do not happen about an axis, but rather about a *plane*. So for example, a rotation *in* the $w$–$x$ plane happens *about* the $y$–$z$ plane. This could be understood in the following way: In two dimensions you rotate around a point (dimension zero), in three dimensions you rotate around an axis (dimension one); hence in four dimensions you rotate about a plane, and so on. Don't worry if you can't visualize this; no one can.

$$
\begin{pmatrix} w' \\ x' \\ y' \\ z' \end{pmatrix} = \begin{pmatrix} \cos\theta & \sin\theta & 0 & 0 \\ -\sin\theta & \cos\theta & 0 & 0 \\ 0 & 0 & 1 & 0 \\ 0 & 0 & 0 & 1 \end{pmatrix} \begin{pmatrix} w \\ x \\ y \\ z \end{pmatrix}, \tag{4.69}
$$

which yields the equivalent of (2.2). Switching back to real time, $w = ict$, *and* also performing the (complex) variable redefinition $\theta = -i\psi$, we end up with

$$
ict' = ict\cos i\psi + x\sin i\psi
$$
$$
x' = -ict\sin i\psi + x\cos i\psi. \tag{4.70}
$$

A useful set of identities relating trigonometric functions of complex angles with the hyperbolic trigonometric functions are

$$
\sin(i\psi) = i\sinh\psi,
$$
$$
\cos i\psi = \cosh\psi. \tag{4.71}
$$

These identities are not so mysterious; they can be easily proven via the well-known relations

$$
\sinh\chi = \tfrac{1}{2}\left(e^{\chi} - e^{-\chi}\right) \qquad \cosh\chi = \frac{1}{2}\left(e^{\chi} + e^{-\chi}\right),
$$
$$
\sin\vartheta = \tfrac{1}{2i}\left(e^{i\vartheta} - e^{-i\vartheta}\right) \qquad \cos\vartheta = \frac{1}{2}\left(e^{i\vartheta} + e^{-i\vartheta}\right). \tag{4.72}
$$

Hence we can now write[13]

$$
\begin{pmatrix} x^{0'} \\ x^{1'} \\ x^{2'} \\ x^{3'} \end{pmatrix} = \begin{pmatrix} \cosh\psi & -\sinh\psi & 0 & 0 \\ -\sinh\psi & \cosh\psi & 0 & 0 \\ 0 & 0 & 1 & 0 \\ 0 & 0 & 0 & 1 \end{pmatrix} \begin{pmatrix} x^{0} \\ x^{1} \\ x^{2} \\ x^{3} \end{pmatrix}. \tag{4.73}
$$

Comparing with the inverse of (4.55), this leads to

$$
\cosh\psi = \gamma,
$$
$$
\sinh\psi = \beta\gamma, \tag{4.74}
$$

or equivalently

$$
\tanh\psi = \beta, \tag{4.75}
$$

where $\psi$ is usually referred to as the **rapidity**. Spacetime rotations are then hyperbolic, in the sense that rotated points do not trace circles as they would in an ordinary

---

[13]We also used the property $\sin(-\vartheta) = -\sin(\vartheta)$.

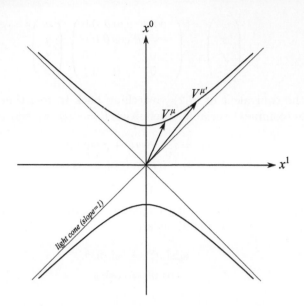

Fig. 4.6: Spacetime rotations (boosts) trace hyperbolas.

rotation such as Fig. [2.2]. Instead, they trace hyperbolas, whose asymptotes are exactly the 45° lines of the light cone. The tip of a rotated vector in ordinary space, such as in Fig. [2.3], traces a circle while the tip of a spacetime 4-vector traces a hyperbola. A boosted spacetime vector would "rotate," as shown in Fig. [4.6]. It may seem to the untrained eye that the norm of the vector has increased upon this boost; however, this is not true. Recall that the norm of the vector is no longer strictly Pythagorean; the minus sign in the metric makes the norm Minkowskian, which is an entirely different beast. In hyperbolic rotations the norm of a 4-vector stays invariant as seen by different (boosted) observers, as opposed to the norm of a Euclidean rotated vector, which stays invariant as seen by different (rotated) observers.[14] Using hyperbolic geometry is useful but not essential. It might make some problem-solving easier, especially when boosts are combined with (actual) rotations, but in general it isn't necessary.

**Exercise 4.13** An observer at rest in some orthogonal frame of reference $O$ sees the location of an event happening at $x = 1,000$m and $ct = 2,000$m. Another observer $O'$ on a frame of reference traveling at a speed $v = 0.87c$ with respect to $O$ sees the same event at $ct' = 2,291.64$m. Draw an accurate to-scale spacetime diagram to describe the situation and find $x'$ from it; i.e. the location of the event as measured by $O'$. Verify your result analytically using the Lorentz transformations. What is the rapidity of this boost?

---

[14]The choice of imaginary time that we did here was a preference of some early textbooks on relativity. It is very possible to just work with it from the start. In this case the metric stays Euclidean and scalar products have their usual forms without the minus sign. Although obsolete, this language is sometimes used in relativistic field theories, when a calculation in this complex space proves to be easier than its analogue in real spacetime.

**Exercise 4.14** Redraw the situation in Exercise 4.13 from $O'$'s point of view; in other words, the frame of reference $O'$ is now the orthogonal one.

**Exercise 4.15**

1. In part 1 of Exercise 4.2 we discussed the length contraction of a given rod as seen by a moving observer. Sketch a to-scale spacetime diagram for this situation and show that up to measurement errors your result for $l_0$ is the same as you found analytically.
2. In part 2 of the same exercise we briefly discussed the concept of simultaneity in special relativity. Let's demonstrate it again using spacetime diagrams. Consider Maya for whom two events $A$ and $B$ are simultaneous and occur $6,000$km apart. What is the time difference in seconds between those two events as determined by Mourad, who measures their spatial separation to be $12,000$km? Sketch a to-scale spacetime diagram for this situation and read the answer off it.

**Exercise 4.16** Research the following famous apparent paradoxes in relativity and sketch spacetime diagrams for each. State the paradox, then based on your diagrams explain in words why none of these are really paradoxes at all:

1. The twin paradox.
2. The pole-barn, also known as the ladder, paradox.

## 4.4   A Bit of Algebra

Putting aside 4-vectors and tensors for a while, it is instructive to briefly discuss special relativistic effects using ordinary algebra, which will allow us to further understand the differences between the Galilean and the relativistic views. This is the way special relativity is usually studied in introductory texts.

### 4.4.1   Addition of Velocities

Let us figure out how to add velocities in special relativity; in other words, the relativistic equivalent to (3.111). We can do so by simply differentiating (4.49), or equivalently, and perhaps more easily, find the ratio of $dx^{1'}$ to $dt'$. From (4.49)

$$dt' = \gamma \left( dt - \frac{v dx^1}{c^2} \right)$$
$$dx^{1'} = \gamma \left( dx^1 - v dt \right),$$
(4.76)

leading to

$$\frac{dx^{1'}}{dt'} = \frac{\gamma \left( dx^1 - v dt \right)}{\gamma \left( dt - \frac{v dx^1}{c^2} \right)}.$$
(4.77)

Simplifying, we get

$$\frac{dx^{1'}}{dt'} = \frac{\left(\frac{dx^1}{dt} - v\right)}{\left(1 - \frac{v}{c^2}\frac{dx^1}{dt}\right)} \tag{4.78}$$

and end up with

$$u' = \frac{u - v}{1 - \frac{uv}{c^2}}. \tag{4.79}$$

A similar argument leads to the inverse relation

$$u = \frac{u' + v}{1 + \frac{u'v}{c^2}}. \tag{4.80}$$

Clearly, eqns (4.79, 4.80) reduce to (3.111) for slow velocities; i.e. if $\frac{u'v}{c^2} \ll 1$. For high velocities where $\frac{u'v}{c^2}$ can no longer be ignored, one can check that (4.79, 4.80) respect Einstein's second postulate, in the sense that no matter how high the values of $u'$ or $v$ are (but still less than $c$), $u$ can never be *greater* than $c$. One can easily derive tensorial forms of these equations similar to (3.119), but a bit of work is needed first.

---

**Exercise 4.17** A sample radioactive material at rest in the lab, ejects two electrons in opposite directions. One of the electrons has a speed of $0.6c$, and the other has a speed of $0.7c$, as measured by Miriam, who is a laboratory observer.

1. According to the Galilean velocity addition formulae, what will the speed of one electron be with respect to the other electron?
2. Redo the exercise using the relativistic addition of velocities. Compare.

**Exercise 4.18** The starship "USS Enterprise," commanded by captain Jean-Luc Picard, is moving away from the "Deep Space 9" station with speed $v = 0.8c$. A shuttle craft, piloted by Lieutenant Commander Geordi La Forge, launches with velocity $v$ with respect to the Enterprise in the forward direction. Closely afterwards, La Forge fires a probe forward with a speed $v$ with respect to the shuttle (with the intent of exploring the nearby wormhole perhaps?). Find:

1. The speed of the shuttle craft relative to Deep Space 9.
2. The speed of the probe relative to Deep Space 9.

First compute the results symbolically in terms of $v$, then find the answer in terms of $c$.

---

### 4.4.2   Time Dilation

We know from the Lorentz transformations that different observers in different coordinate systems boosted with respect to each other will experience different clocks. This is a crucial point that did not exist in non-relativistic Newtonian mechanics, where $t'$ and $t$ were always the same thing and no two observers have ever measured different clocks. In high speed motion this is no longer the case. To figure out how different observers will measure different times it is convenient to start by defining the so-called **proper time**: the duration of an event as measured by an observer in the event's own rest frame.

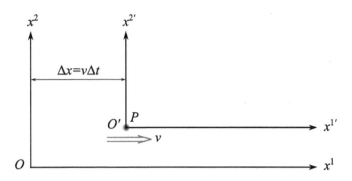

Fig. 4.7: The rest frame of a moving particle.

Given Fig. [4.7], consider a given physical phenomenon, say a point particle $P$ is moving to the right with a speed $v$ as measured by observer $O$. This observer will see the particle's displacement as $\Delta x$, as well as measure the time that elapses for the particle as $\Delta t$, such that

$$v = \frac{\Delta x}{\Delta t}. \tag{4.81}$$

If we want to be more precise, then we can work with infinitesimals instead; hence

$$v = \frac{dx}{dt}. \tag{4.82}$$

Now the particle's own **rest frame** is the one denoted by $O'$, moving along with the particle at the same speed $v$. The flow of time as measured by $O'$'s own clock in the particle's rest frame is what we call the proper time $\tau$. Furthermore, since $P$ is at rest with respect to $O'$, all changes in its position in that frame must vanish, so $dx' = dy' = dz' = 0$, such that the interval $ds^2$ as measured by $O'$ is simply

$$ds^2 = -c^2 d\tau^2. \tag{4.83}$$

This is an important equation; it gives us a physical interpretation of what the interval $ds^2$ really is: a quantity proportional to the proper time. Because of this correspondence, some books define the metric from the start using $\tau$ instead of $s$. Now recall that all observers will agree on the value of $ds^2$, which means that they will also agree on $d\tau^2$; in short, the proper time is a Lorentz invariant.

Now observer $O$ will measure time as $t$ and will see $P$ changing position as it moves to the right, i.e. for $O$ the value of $dx$ is not equal to zero but because the motion is along the $x$ axis, $dy$ and $dz$ vanish. Observer $O$ measures the interval as[15]

$$ds^2 = -c^2 dt^2 + dx^2. \tag{4.84}$$

But now recall that both observers will agree on $ds^2$, so using (4.83) and (4.84) together gives

$$-c^2 d\tau^2 = -c^2 dt^2 + dx^2. \tag{4.85}$$

Upon rearranging, we get

$$\left(\frac{d\tau}{dt}\right)^2 = 1 - \frac{1}{c^2}\left(\frac{dx}{dt}\right)^2. \tag{4.86}$$

Now using (4.82), we end up with

$$\left(\frac{d\tau}{dt}\right)^2 = 1 - \frac{v^2}{c^2}. \tag{4.87}$$

But this is just

$$\left(\frac{d\tau}{dt}\right)^2 = \frac{1}{\gamma^2}, \tag{4.88}$$

which finally leads to

$$dt = \gamma d\tau. \tag{4.89}$$

This is the famous **time dilation** equation that relates to how a moving observer measures the time $t$ of a given phenomenon, as opposed to the proper time measured in the event's rest frame. It is an essential formula, and we will use it extensively in what follows. If the time in question is finite, then we can easily rewrite (4.89):

$$\Delta t = \gamma \Delta \tau. \tag{4.90}$$

**Exercise 4.19** With respect to some observer, at what speed is a clock moving if it is observed to run at a rate half that of one at rest? Express your answer in terms of $c$.

**Exercise 4.20** The average lifetime of the particle known as the $\mu$-meson is approximately $6 \times 10^{-6}$s as measured by a given observer, with respect to whom the particle is speeding at a velocity of $0.95c$. Compute the average lifetime as measured by an observer at rest with respect to the particle.

---

[15]The name "proper time" is somewhat misleading, as it implies that $\tau$ has some sort of advantage over $t$. A student reading about these topics for the first time tends to think of $\tau$ as the "real" time and that $t$ is an illusion. This is incorrect in the relativistic world. There is no longer such a thing as "real time;" both $\tau$ and $t$ are "real" in the sense that time is no longer absolute but is rather a *relative* concept. In other words, the question of which time is real has no meaning; it essentially depends on whom you ask.

### 4.4.3   Length Contraction

Just as time in a moving rest frame "shrinks" with respect to a stationary observer, one can also show that lengths will "contract." This is probably best done by applying the Lorentz transformations. Consider an inertial observer $O$. A meter stick is moving with respect to $O$ with velocity $v$, as shown in Fig. [4.8]. Let $O'$ be the stick's own rest frame. We define the stick's length with respect to $O'$ as its **proper length** $l_0$:

$$l_0 = \Delta x' = x'_2 - x'_1. \tag{4.91}$$

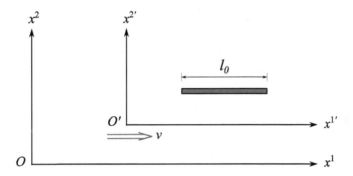

Fig. 4.8: Length contraction.

The length as measured by the stationary observer $O$ can be found by applying the transformations (4.49):

$$\begin{aligned} l_0 = \Delta x' &= \gamma \left(x_2 - vt_2\right) - \gamma \left(x_1 - vt_1\right) \\ &= \gamma \left(x_2 - x_1\right) - \gamma v \left(t_2 - t_1\right). \end{aligned} \tag{4.92}$$

If one assumes that the measurements $x_1$ and $x_2$ are done instantaneously in the frame $O$, then $t_1 = t_2$ and we end up with the length contraction formula

$$l_0 = \gamma l, \tag{4.93}$$

where $l = \Delta x = x_2 - x_1$. Different observers will then disagree on the length of the stick.

**Exercise 4.21** How fast (in terms of $c$) does a rocket ship have to go for its length to be contracted to 99 percent of its rest length?

**Exercise 4.22** A rod of rest length $l_0$ makes an angle $\theta_0$ with respect to the $x'$ axis of its rest frame $O'$. If it is moving with a speed $v$ with respect to an observer $O$, find its length $l$ in that frame and the angle $\theta$ it makes with the $x$ axis as measured by $O$. You should be able to show that the rod appears to $O$ not only contracted but also *rotated*.

## 4.5   Specially Covariant Mechanics

Let us generalize Newtonian mechanics to four dimensions. The recipe involves simply adding a zeroth component to the known Newtonian vectors. Our starting point here, just as in ordinary mechanics, is the position vector. We already know that it is

$$\mathbf{r}_{4d} = (ct, x, y, z).  \tag{4.94}$$

We choose the convention that the components of the vector (4.94) have upper indices, i.e. $x^0 = ct$, $x^1 = x$, $x^2 = y$, and $x^3 = z$. On the other hand, the lower index components of the "dual" position vector $\mathbf{r}_{4d}^*$ must be $x_0 = -ct$, $x_1 = x$, $x_2 = y$, and $x_3 = z$, such that

$$\begin{aligned} d\mathbf{r}_{4d} \cdot d\mathbf{r}_{4d}^* &= dx^\mu dx_\mu = \eta_{\mu\nu} dx^\mu dx^\nu \\ &= -c^2 dt^2 + dx^2 + dy^2 + dz^2, \end{aligned}  \tag{4.95}$$

as desired. Traditionally expressions such as "$\mathbf{r}_{4d}$" and "$\mathbf{r}_{4d}^*$" are rarely used. Despite an obvious abuse of notation, most authors just write

$$\begin{aligned} x^\mu &= (ct, x, y, z), \\ x_\mu &= (-ct, x, y, z) \end{aligned}  \tag{4.96}$$

to signify the same thing. One can also write expressions such as $x^\mu = \left(x^0, x^i\right)$ or $x^\mu = \left(x^0, \mathbf{r}\right)$. The position 4-vector contains more information than its three-dimensional counterpart $x^i$; it not only tells us the *location* of an event, it also contains the *time* when the event happened: $x^0 = ct$. It is the position of a *spacetime point*! Graphically, the 4-position is the vector that parameterizes the world line in a spacetime diagram of a moving particle, as shown in Fig. [4.9].[16] Just as in ordinary three-dimensional mechanics, the 4-position vector is a function in time. The question here, however, is "*whose* time?" As discussed before, observers will generally disagree on the time $t$, each measuring their own; however, they will always agree on the proper time $\tau$ (since they will agree on $ds$ which is proportional to $d\tau$); hence we choose the 4-position to be a function in $\tau$,

$$x^\mu(\tau) = [ct(\tau), x(\tau), y(\tau), z(\tau)],  \tag{4.97}$$

and similarly for $x_\mu(\tau)$. The next quantity we need to define is the 4-displacement $dx^\mu$. Similar to its three-dimensional counterpart, it is a vector tangential to the world line and pointing in the direction of motion. The 4-velocity $U^\mu$ follows naturally as the ratio of displacement to $d\tau$; i.e. the derivative of the position vector with respect to $\tau$. We can then use (4.89) to write it for any observer:

$$U^\mu = \frac{dx^\mu}{d\tau} = \left( c\frac{dt}{d\tau}, \frac{dx}{d\tau}, \frac{dy}{d\tau}, \frac{dz}{d\tau} \right)  \tag{4.98}$$

$$= \left( c\gamma, \gamma\frac{dx}{dt}, \gamma\frac{dy}{dt}, \gamma\frac{dz}{dt} \right)  \tag{4.99}$$

$$= \gamma(c, \mathbf{v}).  \tag{4.100}$$

---

[16]The other quantities in this figure will be explained shortly.

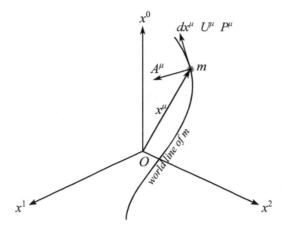

Fig. 4.9: The kinematical vectors of four-dimensional spacetime.

The reader is encouraged to compare Fig. [4.9] to Fig. [3.1] for further clarity. For future use, we note that the components of the 3-velocity **v** can be found from the components of the 4-velocity as follows:

$$v^i = c\frac{U^i}{U^0},\tag{4.101}$$

since $U^0 = \gamma c$ and $U^i = \gamma v^i$. Now the rest frame observer $O'$ sees $\mathbf{v} = 0$ and $\gamma = 1$, giving

$$U^\mu|_{\text{rest frame}} = (c, \mathbf{0}).\tag{4.102}$$

As expected, $O$ and $O'$ see different components to the 4-velocity; however, by the very definition of covariance they must both agree on the scalar product of $U^\mu$ with itself, whether measured in $O$ or in $O'$. In the particle's rest frame this is

$$U^\mu U_\mu = -c^2.\tag{4.103}$$

But since all observers must agree on the norm of any vector, we must conclude that (4.103) is a general result, even if $\mathbf{v} \neq 0$. We can easily check this by taking the scalar product of (4.100) with itself to find

$$U^\mu U_\mu = \gamma^2 \left(-c^2 + v^2\right)$$
$$= -c^2\gamma^2 \left(1 - \frac{v^2}{c^2}\right)$$
$$= -c^2,\tag{4.104}$$

as expected.[17] In (4.104) we have used $v^2 = \mathbf{v} \cdot \mathbf{v}$ and $\left(1 - \frac{v^2}{c^2}\right) = 1/\gamma^2$. It is very useful to note the following relationship between the norm of the 4-velocity and the

---

[17]The norm of the 4-velocity is one of those instances where the result on the right-hand side would have been positive had we chosen the particle physicist's signature (4.16).

metric:

$$U^\mu U_\mu = g_{\mu\nu}U^\mu U^\nu = g_{\mu\nu}\frac{dx^\mu}{d\tau}\frac{dx^\nu}{d\tau} = \frac{ds^2}{d\tau^2} = -c^2, \tag{4.105}$$

leading us back to $ds^2 = -c^2 d\tau^2$. Note that this is true for *massive* timelike particles *only*, since for *massless* lightlike particles (e.g. photons of light) we know that $ds^2 = 0$; hence the argument in (4.105) leads to

$$U^\mu U_\mu = 0. \tag{4.106}$$

In this case, one cannot define the "proper time" of a lightlike object, since these have no rest frames. Instead, we define the so-called "affine parameter" $\lambda$ such that

$$U^\mu = \frac{dx^\mu}{d\lambda}. \tag{4.107}$$

The exact definition of $\lambda$ will be given in the next chapter. For now it is sufficient to note that it is an arbitrary parameter that varies along the straight world lines of lightlike signals. It is not a very useful quantity in the special theory of relativity, and we postpone its usage to our discussion of the general theory. Also note that we have been implicitly assuming that massive objects are *necessarily* timelike (traveling at $v < c$), while massless objects are necessarily lightlike (traveling at $v = c$). There are many proofs to this assertion, and we will give one of them toward the end of the next section, but for now we will just accept it as an axiom. Spacelike particles, on the other hand, are unphysical, as discussed earlier; however, it is still possible to define $U^\mu U_\mu = +c^2$ for them.

**Exercise 4.23** Consider a particle traveling with constant speed about a circle of radius $R$ centered on the origin in the $x$–$y$ plane. From introductory physics we know that its 3-velocity in cylindrical coordinates is just $\mathbf{v} = R\omega\mathbf{v}_\varphi$, where $\omega$ is its angular speed and $\mathbf{v}_\varphi$ is the physical basis. In the natural basis $\mathbf{g}_\varphi = R\mathbf{v}_\varphi$ and consequently $\mathbf{v} = \omega\mathbf{g}_\varphi$. Construct the particle's 4-velocity $U^\mu$. Explicitly show that the normalization (4.105) is satisfied.

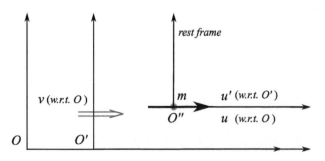

Fig. 4.10: Relativistic addition of velocities treated as a succession of boosts.

Finally, let us show how the relativistic addition of velocities (4.80) arises in the language of 4-vectors. The situation is as shown in Fig. [3.8] and can be thought of in the following way: There are in fact *three* reference frames in this diagram: the ones shown *plus* the particle's own rest frame. So we re-sketch the problem in Fig. [4.10] to include the particle's rest frame $O''$. With respect to $O''$, the particle has 4-velocity:

$$U^{\mu''} = (c, 0, 0, 0).\tag{4.108}$$

Now observer $O'$ sees the boosted 4-velocity

$$U^{\mu'} = \lambda^{\mu'}_{\nu''} U^{\mu''},\tag{4.109}$$

or in matrix notation

$$\begin{pmatrix} U^{0'} \\ U^{1'} \\ U^{2'} \\ U^{3'} \end{pmatrix} = \gamma_{u'} \begin{pmatrix} 1 & \frac{u'}{c} & 0 & 0 \\ \frac{u'}{c} & 1 & 0 & 0 \\ 0 & 0 & 1 & 0 \\ 0 & 0 & 0 & 1 \end{pmatrix} \begin{pmatrix} c \\ 0 \\ 0 \\ 0 \end{pmatrix} = \gamma_{u'} \begin{pmatrix} c \\ u' \\ 0 \\ 0 \end{pmatrix},\tag{4.110}$$

where

$$\gamma_{u'} = \frac{1}{\sqrt{1 - \frac{u'^2}{c^2}}}.\tag{4.111}$$

Hence $U^{\mu'} = \gamma_{u'}(c, u', 0, 0)$, exactly matching (4.100) as one would expect. Now, another boost gets us observer $O$'s point of view,

$$U^{\mu} = \lambda^{\mu}_{\nu'} U^{\nu'}.\tag{4.112}$$

Explicitly,

$$\begin{pmatrix} U^0 \\ U^1 \\ U^2 \\ U^3 \end{pmatrix} = \gamma_v \gamma_{u'} \begin{pmatrix} 1 & \frac{v}{c} & 0 & 0 \\ \frac{v}{c} & 1 & 0 & 0 \\ 0 & 0 & 1 & 0 \\ 0 & 0 & 0 & 1 \end{pmatrix} \begin{pmatrix} c \\ u' \\ 0 \\ 0 \end{pmatrix} = \gamma_v \gamma_{u'} \begin{pmatrix} c + \frac{vu'}{c} \\ u' + v \\ 0 \\ 0 \end{pmatrix},\tag{4.113}$$

where this time

$$\gamma_v = \frac{1}{\sqrt{1 - \frac{v^2}{c^2}}}.\tag{4.114}$$

Hence the non-trivial components of the 4-velocity in the $O$ frame are

$$U^0 = \gamma_v \gamma_{u'} c \left(1 + \frac{vu'}{c^2}\right),$$
$$U^1 = \gamma_v \gamma_{u'} (u' + v).\tag{4.115}$$

Using (4.101),

$$u = c\frac{U^1}{U^0} = \frac{u' + v}{1 + \frac{vu'}{c^2}}, \tag{4.116}$$

which is exactly (4.80) as desired.

**Exercise 4.24** Consider a third observer in Fig. [4.10], let's call her Miriam, with respect to whom $O$ is traveling at a speed of $V$ in the positive $x$ direction. Calculate the speed of $m$ with respect to Miriam. Explicitly show that the 4-velocity of $m$ as measured by her satisfies the normalization condition (4.105).

Now, with the knowledge of the 4-velocity, the 4-acceleration follows

$$A^\mu = \frac{dU^\mu}{d\tau} = \frac{dU^\mu}{dt}\frac{dt}{d\tau} = \gamma\frac{dU^\mu}{dt}$$

$$= \gamma\frac{d}{dt}\left[\gamma\left(c, \mathbf{v}\right)\right]$$

$$= \gamma\left[c\left(\frac{d\gamma}{dt}\right), \mathbf{a}\right], \tag{4.117}$$

where

$$\mathbf{a} = \frac{d}{dt}\left(\gamma\mathbf{v}\right) \tag{4.118}$$

is the relativistic 3-acceleration of the object. An interesting consequence is that it turns out that the 4-velocity and 4-acceleration vectors are orthogonal to each other. It is easy to see this by considering the rest frame, where $U^\mu = (c, \mathbf{0})$ and $A^\mu = (0, \mathbf{a})$. Consequently,

$$A^\mu U_\mu = 0. \tag{4.119}$$

Since (4.119) is invariant, it must be true in all frames of reference. The reader can explicitly check this using the general forms of $U^\mu$ and $A^\mu$.

Now that we have defined the most essential four-dimensional kinematical quantities, we can proceed along the lines of solving relativistic mechanical problems involving forces and accelerations. With some exceptions, for example the case of constant acceleration which we will study in Exercises 4.27 and 4.28, as well as in Chapter 8, these problems are not very interesting or useful. The case of vanishing 4-acceleration, i.e. the motion of a free particle, which we discuss next—is much more practical and a lot more informative.

**Exercise 4.25** In an accelerating situation, ($v \neq$ constant), show that

$$\frac{d\gamma}{dt} = \gamma^3\vec{\beta} \cdot \frac{d\vec{\beta}}{dt}. \tag{4.120}$$

Substitute in (4.117), then show that if we choose the $x$ axis to be the direction of $\vec{\beta}$, then

$$A^{\mu} = c\gamma^4 \frac{d\beta}{dt}\,(\beta, 1, 0, 0)\,. \tag{4.121}$$

**Exercise 4.26** Explicitly verify (4.119) using the non-rest forms of the 4-velocity (4.100) and the 4-acceleration (4.117).

**Exercise 4.27** Consider a particle of mass $m$ moving along the $x$ axis under the influence of a constant force $F$, which means a constant acceleration $a = F/m$. In the non-relativistic case this leads to the famous freshman physics equations

$$x(t) = x_0 + v_0 t + \frac{1}{2}at^2 \tag{4.122}$$

$$v(t) = \frac{dx}{dt} = v_0 + at, \tag{4.123}$$

where $x_0$ and $v_0$ are the initial position and speed of the particle respectively. The position of the particle $x(t)$ is described by the parabolic curve (4.122), while its speed $v(t)$ is the straight line curve (4.123). Note that this means that there is no upper limit on how fast the object can go; its speed will grow *indefinitely* for as long as it is under the influence of the constant force.

This situation is corrected if we consider the relativistic version of the 3-acceleration. Show that by considering eqn (4.118) in the $x$ direction and setting $a$ equal to a constant,

$$a = \frac{d}{dt}(\gamma v) = \alpha = \text{constant}, \tag{4.124}$$

the relativistic version of (4.123) sets the expected upper limit $v = c$. Also show that instead of a parabola, $x(t)$ would be described by a *hyperbola*. We will redo this problem in §8.1.3 using a different method. Fig. [8.5] shows a comparison between the non-relativistic parabolic path and the relativistic hyperbolic path.

**Exercise 4.28** In a $ct$–$x$ spacetime diagram the world line of the relativistically accelerating particle of Exercise 4.27 can be described by the **Rindler coordinates**[18]

$$t(\tau) = \frac{c}{a}\sinh\left(\frac{a\tau}{c}\right)$$
$$x(\tau) = \frac{c^2}{a}\left[\cosh\left(\frac{a\tau}{c}\right) - 1\right]. \tag{4.125}$$

Derive the metric that arises from these coordinates. It should be a particularly simple one. Does it surprise you? Interpret.

**Exercise 4.29** In the Rindler coordinates of Exercise 4.28, find the accelerating particle's 4-velocity as well as its 3-velocity.

[18]Introduced by the Austrian physicist Wolfgang Rindler (1924–2019). Rindler is also responsible for several famous contributions to the theory of relativity that include the so-called "ladder paradox," which we worked out in Exercise 4.16, as well as coining the term "event horizon" in connection with black holes; a concept that will play a central role in our discussions in later chapters.

## 4.6    Conservation of Four-Momentum

Cases of zero 4-acceleration lead to the exact same conclusion as in ordinary mechanics: momentum conservation. It does, however, lead to interesting relativistic effects. We define the **4-momentum** of a massive particle in exactly the same way that 3-momentum is defined: the product of mass with velocity:

$$P^\mu = mU^\mu. \tag{4.126}$$

Newton's second law may then be written as

$$f^\mu = \frac{dP^\mu}{d\tau} = \gamma\frac{dP^\mu}{dt}, \tag{4.127}$$

where $f^\mu$ is the 4-force. Now, if the sum of external 4-forces on a given mechanical system vanishes, then

$$\frac{dP^\mu}{d\tau} = \gamma\frac{dP^\mu}{dt} = 0, \tag{4.128}$$

leading to the statement that the sum of 4-momenta in that system is a constant:

$$\sum P^\mu = \text{constant}. \tag{4.129}$$

This is true in some *specific* reference frame. But, since $P^\mu$ is covariant, its norm is also conserved under transformations, so

$$P^\mu P_\mu = \text{constant}. \tag{4.130}$$

It is important to emphasize that (4.129) is only true in the *same* reference frame, while (4.130) is true *between* two different reference frames $O$ and $O'$:

$$\sum P^{\mu'} \neq \sum P^\mu \quad \text{while}$$
$$P^{\mu'} P_{\mu'} = P^\mu P_\mu. \tag{4.131}$$

Having said that, one can still find the sum of all momenta between frames by performing the usual transformations on $P^\mu$, i.e.

$$P^{\mu'} = \lambda^{\mu'}_\nu P^\nu. \tag{4.132}$$

Equations (4.129) and (4.130) (along with [4.132] if needed) provide extremely useful tools in analyzing relativistic collision problems. But before we do this, we need to first explore what the components of the 4-momentum are. Let's continue from (4.126):

$$P^\mu = m\gamma\left(c, \mathbf{v}\right) = \left(m\gamma c, m\gamma\mathbf{v}\right). \tag{4.133}$$

We interpret $m\gamma\mathbf{v}$ as just the ordinary 3-momentum corrected relativistically with a factor of $\gamma$, i.e.

$$\mathbf{p} = m\gamma\mathbf{v}. \tag{4.134}$$

What about the zeroth component of the 4-momentum $P^0 = m\gamma c$? What physical interpretation does it have? A useful approach to figuring this out is to look at the non-relativistic limit, since whatever $P^0$ is it should give something easily recognizable in the ordinary slow velocities world that we know so well. Now if we just set $\gamma \to 1$, $P^0$ just becomes $mc$, which is still not very informative. Perhaps we have made things *too* slow. A common trick is to expand $\gamma$ in an infinite series in $v^2/c^2$ and take as many terms as needed. We begin by writing

$$P^0 = m\gamma c = mc\left(1 - \frac{v^2}{c^2}\right)^{-\frac{1}{2}}, \tag{4.135}$$

and employ the use of the so-called **binomial theorem**[19] which states that any bracket of the form $(1 + k)^n$ can be expanded into a series as follows:

$$(1+k)^n = 1 + nk + \frac{n\left(n-1\right)}{2!}k^2 + \frac{n\left(n-1\right)\left(n-2\right)}{3!}k^3 + \cdots. \tag{4.136}$$

This series terminates only for positive integer values of $n$, otherwise it goes on forever. For example, the simplest case $n = 2$ gives

$$(1+k)^2 = 1 + 2k + k^2, \tag{4.137}$$

as one can check explicitly: $(1+k)^2 = (1+k)(1+k)$. Now here comes the trick: if the quantity $k$ happens to be very small compared to unity i.e. $k \ll 1$, then $k^2$ is much smaller and $k^3$ is even smaller still, and so on. So, for small enough values of $k$, we can simply ignore all the higher-order terms in (4.136) and write

$$\lim_{k \to 0} (1+k)^n \approx 1 + nk. \tag{4.138}$$

This is exactly what we need in the low speed limit of (4.135), where $k = -v^2/c^2$ and $n = -1/2$:

$$\lim_{\frac{v}{c} \to 0} P^0 \approx mc\left(1 + \frac{v^2}{2c^2}\right) = mc + \frac{1}{2c}mv^2, \tag{4.139}$$

---

[19]Discovered by Newton himself.

which may be starting to look familiar already. Now, multiplying both sides by $c$, we get:

$$cP^0 \approx mc^2 + \frac{1}{2}mv^2. \tag{4.140}$$

Remembering that we are in the Newtonian limit $v \ll c$, we immediately recognize the second term as the non-relativistic kinetic energy! It must then be true that $cP^0$ is the *total* energy of a free particle, where the second term is its kinetic energy and the first term is something new that has no Newtonian counterpart.[20] In fact, $mc^2$ is the famous **rest energy** of a particle of mass $m$. So, in a slow speed limit, the free particle's total energy is

$$E = cP^0 = E_{\text{rest}} + \text{K.E.} \tag{4.141}$$

where $K.E. = \frac{1}{2}mv^2$ is the usual kinetic energy and the rest energy $E_{\text{rest}}$ is Einstein's famous[21]

$$E_{\text{rest}} = mc^2. \tag{4.142}$$

In the Newtonian picture $E_{\text{rest}}$ is unheard of. So even though the above is the slow speed approximation, we have already discovered something new: A massive particle carries a certain amount of rest energy that is proportional to its mass. Alternatively, one can view (4.142) as the statement that mass is really a type of "locked" residual energy. The fact that this was not noticed experimentally before Einstein is due to the technical difficulties of extracting it; we simply didn't have the technology to detect or harness this rest energy. In this day and age we have exactly that technology; it is called a thermonuclear reactor. The fact that we can extract nuclear energy, whether for peaceful purposes or otherwise, is a powerful confirmation of the theory. In nature, the conversion of mass to energy is commonplace, the Sun and other stars being prime examples.

Now, although we have discovered this rest energy using a slow speed approximation, it still holds quite well in the full relativistic picture, as we will see. Based on all this, we can now write the 4-momentum as it truly is without approximation:

$$P^\mu = \left( \frac{E}{c}, \mathbf{p} \right), \tag{4.143}$$

where the total energy (of a free particle) is

$$E = m\gamma c^2 \tag{4.144}$$

and the 3-momentum is defined by (4.134). It is also possible to define a relativistic expression for kinetic energy based on (4.140) by arguing that if non-relativistically

---

[20]Which is why we didn't recognize it when we naively set $\gamma = 1$.

[21]Although this famous formula is generally attributed to Einstein, it is perhaps worthwhile to note that he was not the first to postulate a relationship between mass and energy. As mentioned briefly, the special theory of relativity is the combined work of many scientists, but Einstein is most deservedly credited with making sense of what was generally an unexplained mess. The interested reader may consult any book on the history of physics at the beginning of the twentieth century.

the kinetic energy is the total energy minus the rest energy, then relativistically it can also be defined in the same way; hence

$$\begin{aligned} \text{K.E.} &= E - E_{\text{rest}} \\ &= m\gamma c^2 - mc^2 \\ &= (\gamma - 1)\, mc^2. \end{aligned} \tag{4.145}$$

**Exercise 4.30**

1. How fast in terms of $c$ must a particle travel for its kinetic energy to be equal to its rest energy?
2. How fast in meters per second must a 2kg bowling ball travel to have the same kinetic energy as a cosmic ray proton of mass $1.6726219 \times 10^{-27}$kg moving with $\gamma = 10^{10}$?

We are now in a position where we can reexamine the question: Can a particle with mass $m$ be accelerated to exactly $v = c$? Both eqns (4.144) and (4.145) demonstrate that this is impossible. As a massive particle gains speed, the Lorentz factor increases, slowly at first, then faster and faster, as shown in Fig. [4.11]. As the speed $v$ climbs up, it becomes harder and harder to accelerate the object as it requires more and more energy. At *exactly* the speed of light $\gamma$ becomes infinite, and the energy of the particle becomes infinite as well.[22] In other words, one must "pump" an infinite amount of energy into any massive particle to accelerate it to $v = c$.

**Exercise 4.31** Consider a particle of rest mass $E_{\text{rest}}$ traveling at a speed of $v = 0.9c$. How much energy in multiples of $E_{\text{rest}}$ must be added to the particle to accelerate it from its initial velocity to $0.99c$? To $0.999c$? To $0.9999c$? Do you see the point of this exercise?

Another useful expression can be found if we consider the invariant norm of the 4-momentum; starting with (4.133),

$$\begin{aligned} P^\mu P_\mu &= -m^2\gamma^2 c^2 + m^2\gamma^2 v^2 \\ &= m^2\gamma^2 c^2 \left( \frac{v^2}{c^2} - 1 \right) = -m^2\gamma^2 c^2 \left( 1 - \frac{v^2}{c^2} \right), \end{aligned} \tag{4.146}$$

and recalling the definition of $\gamma$ we end up with

$$P^\mu P_\mu = -m^2 c^2, \tag{4.147}$$

[22]It used to be assumed that there are two types of mass: the **rest mass** $m_0$ and the **relativistic mass** $m = \gamma m_0$. Using this interpretation, the momentum and total energy equations become $\mathbf{p} = m\mathbf{v}$ and $E = mc^2$ respectively, since the Lorentz factor $\gamma$ is encoded inside $m$. In some (old) references one reads statements like "the mass of a particle increases with its speed." Experimentally there is no way (based on relativity alone) to differentiate between the two interpretations. However, it has been accepted in recent times that it is better to stick to the interpretation that mass is a constant quantity and that it is the momentum and energy that increase with speed.

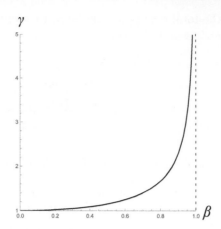

Fig. 4.11: The growth of the Lorentz factor from 1 ($v = \beta = 0$) to $\infty$ ($v = c; \beta = 1$). The same curve, multiplied by the constant $mc^2$, can represent the growth of the total energy.

which also follows from taking (4.103) and (4.126) together. Now using (4.147) together with (4.143) we get

$$P^\mu P_\mu = -\frac{E^2}{c^2} + \mathbf{p} \cdot \mathbf{p} = -\frac{E^2}{c^2} + p^2 = -m^2 c^2. \tag{4.148}$$

Rearranging we end up with the very important relation

$$E^2 = m^2 c^4 + p^2 c^2. \tag{4.149}$$

Equation (4.149) says that total energy squared of a free particle (i.e. no potential energy of any kind) is equal to the sum of the square of its rest energy and the square of an energy term that is found from the momentum of the particle. Once again, these equations are true for *only* massive particles; for massless particles (4.106) implies

$$P^\mu P_\mu = 0, \tag{4.150}$$

which leads to

$$E = pc, \tag{4.151}$$

as a special case of (4.149) when $m = 0$. This is interesting! Apparently the special theory of relativity allows for the existence of massless particles that carry both momentum and energy! We alluded earlier to the fact that such massless objects must necessarily travel at the speed of light. Let's show that this follows from (4.151). If we first substitute the definitions (4.134) and (4.144) in (4.151):

$$m\gamma c^2 = m\gamma v c, \tag{4.152}$$

we realize that in the limit $m \to 0$ this relation can only be true if $v = c$, $\gamma \to \infty$, *and* $m\gamma$ stays finite! This is then the inevitable conclusion that massless particles must *necessarily* travel at the speed of light! It also implies that the magnitude $p$ of the 3-momentum can no longer be defined by the usual $m\gamma v$, but must be found by other means. In particular, quantum mechanics teaches us that the momentum of the photon is $p_{ph} = h/\lambda$, where $h = 6.626 \times 10^{-34}$J·s is **Planck's constant**[23] and $\lambda$ is the photon's wavelength. Using (4.151), the photon's total energy is then $E_{ph} = hc/\lambda$. These expressions can be used to analyze the relativistic motion of photons and other massless particles.

Around 1923, the American physicist Arthur Holly Compton[24] demonstrated that light behaves as a particle in scattering experiments and consequently won the Nobel Prize in Physics four years later. He found that certain experimental results can be explained if one assumes that light quanta, i.e. photons, carry momentum despite the fact that they are massless, and that their momentum must be exactly as predicted by quantum mechanics.[25] It is instructive to reproduce his result using the language of 4-vectors.

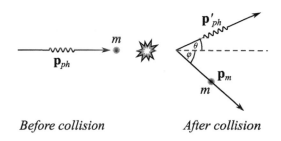

*Before collision*          *After collision*

Fig. 4.12: Schematic of the Compton scattering.

Known as the **Compton scattering** experiment, the situation may be described as shown in Fig. [4.12]. A photon with 3-momentum $p_{ph} = h/\lambda$ approaches a particle with mass $m$ (an electron in the original experiment) initially at rest. After the collision, the photon and the particle scatter with 3-momenta $p'_{ph}$ and $p_m$. Working in Cartesian coordinates and taking the positive $x$ axis to be the original direction of the photon's motion, we can write the 4-momenta of both particles before and after the collision as follows:

Before:

$$P^\mu_{ph} = \left(\frac{E_{ph}}{c}, p_{ph}, 0, 0\right) = p_{ph}(1,1,0,0) = \frac{h}{\lambda}(1,1,0,0),$$
$$P^\mu_m = mc(1,0,0,0). \tag{4.153}$$

---

[23]Max Karl Ernst Ludwig Planck: German theoretical physicist (1858–1947). Considered the father of quantum mechanics.

[24]1892–1962.

[25]Compton's original paper did not directly build on Einstein's work; however, today we identify its main result as a consequence of special relativity, as we will now show.

After:

$$P'^{\mu}_{ph} = \left( \frac{E_{ph}}{c}, p_{ph} \cos\theta, p \sin\theta, 0 \right) = p_{ph} \left( 1, \cos\theta, \sin\theta, 0 \right) = \frac{h}{\lambda'} \left( 1, \cos\theta, \sin\theta, 0 \right),$$

$$P'^{\mu}_{m} = \left( \frac{E_m}{c}, p_m \cos\varphi, -p_m \sin\varphi, 0 \right). \tag{4.154}$$

The objective is to find how the wavelength of light after the collision $\lambda'$ differs from its counterpart before the collision $\lambda$. In other words, we are looking for the change of wavelength $\Delta\lambda = \lambda' - \lambda$, and will hopefully relate it to the observable quantities in the problem; namely $m$, $\theta$, and $\phi$. Our starting point is 4-momentum conservation (4.129):

$$P^{\mu}_{ph} + P^{\mu}_{m} = P'^{\mu}_{ph} + P'^{\mu}_{m}, \tag{4.155}$$

from which we rewrite

$$P'^{\mu}_{m} = P^{\mu}_{ph} + P^{\mu}_{m} - P'^{\mu}_{ph}. \tag{4.156}$$

Squaring both sides of (4.156), or more precisely taking the scalar product of either side with itself, gives:

$$P'^{\mu}_{m} P'_{m\mu} = (P^{\mu}_{ph} + P^{\mu}_{m} - P'^{\mu}_{ph})(P_{ph\mu} + P_{m\mu} - P'_{ph\mu})$$

$$= P^{\mu}_{ph} P_{ph\mu} + P^{\mu}_{m} P_{m\mu} + P'^{\mu}_{ph} P'_{ph\mu} + 2 P^{\mu}_{ph} P_{m\mu} - 2 P^{\mu}_{ph} P'_{ph\mu} - 2 P^{\mu}_{m} P'_{ph\mu}. \tag{4.157}$$

Noting that

$$P'^{\mu}_{m} P'_{m\mu} = P^{\mu}_{m} P_{m\mu} = -m^2 c^2$$

$$P^{\mu}_{ph} P_{ph\mu} = P'^{\mu}_{ph} P'_{ph\mu} = 0, \tag{4.158}$$

we are left with

$$P^{\mu}_{ph} P'_{ph\mu} = P^{\mu}_{ph} P_{m\mu} - P^{\mu}_{m} P'_{ph\mu}. \tag{4.159}$$

Evaluating

$$P^{\mu}_{ph} P'_{ph\mu} = \frac{h^2}{\lambda^2} (-1 + \cos\theta)$$

$$P^{\mu}_{ph} P_{m\mu} = -\frac{h}{\lambda} mc$$

$$P^{\mu}_{m} P'_{ph\mu} = -\frac{h}{\lambda'} mc, \tag{4.160}$$

eqn (4.159) ends up being (after a little cleaning)

$$\lambda' - \lambda = \frac{h}{mc} (1 - \cos\theta), \tag{4.161}$$

which is the desired result (sometimes referred to as **Compton's formula**). Hence the change of the wavelength of incident light $\Delta\lambda = \lambda' - \lambda$ is directly related to its scattering angle and the mass of the other particle. This constitutes a relatively easy experimental check of relativistic mechanics that has no Newtonian counterpart.

Calculations such as this can be divided into two separate parts. The first is straightforwardly based on the physics; namely just setting up the argument leading to eqn (4.155). The remainder is figuring out what algebraic tricks one needs to do to isolate the required quantities and relate them to what is given. In this calculation, we have exploited our knowledge of the scalar product of 4-momenta, but that may not always be the way to go. Unfortunately there is no rule of thumb that one may follow to solve these kinds of problems; however, the most common techniques usually depend on some of the equations in this section, such as (4.133), (4.143), (4.147), (4.149), and (4.150), as well as our knowledge of how to take scalar products, etc.

**Exercise 4.32** Two objects, both of mass $m$, are moving toward each other at the same speed, $0.75c$, with respect to an observer in the lab. After the collision they stick together and stop. Non-relativistically, the mass of the combined object is simply $2m$. Show that this is not true relativistically and derive the combined object's new mass $M$ in terms of $m$. You will find that $M > 2m$. Where did the "extra" mass come from?

**Exercise 4.33** A subatomic particle of mass $M$, at rest with respect to the lab, suddenly decays into two particles: one of mass $m$ and the other massless. Find the total energy $E$ of the resulting massive particle in terms of $M$ and $m$ and its speed in terms of $c$.

**Exercise 4.34** A particle of mass $M$ at rest decays into two daughter particles of mass $m$ each, where $M > 2m$. Show that the daughter particles move off in opposite directions and find their (equal) speed $v$ in terms of $c$, $m$, and $M$.

**Exercise 4.35** A particle of mass $m$ and kinetic energy equal to twice its rest energy strikes a stationary particle with initial rest mass $2m$ and sticks to it. Find the mass of the composite and its velocity.

**Exercise 4.36** A particle of mass $m$ traveling with some constant velocity $\mathbf{v}$ and associated Lorentz factor $\gamma$ collides with a stationary particle, also of mass $m$. The two particles fly off with $(\mathbf{v}_1, \gamma_1)$ and $(\mathbf{v}_2, \gamma_2)$ respectively. Using conservation of 4-momentum show that
$$\gamma = \gamma_1 \gamma_2 \left( 1 - \vec{\beta}_1 \cdot \vec{\beta}_2 \right),$$
where $\vec{\beta}_1 = \mathbf{v}_1/c$ and $\vec{\beta}_2 = \mathbf{v}_2/c$ as usual. Also show that in the low-velocity limit $v \ll c$ the two particles fly off perpendicular to each other.

**Exercise 4.37** (This exercise is from [9]) A particle with mass $m_1$ moving with velocity $u_1$ along the $x$ axis collides elastically with a stationary particle of mass $m_2$, and as a result $m_1$ and $m_2$ are deflected through angles $\theta$ and $\phi$ (measured from above and below the $x$ axis respectively). If $E_B$ and $E_A$ are the total energy of $m_1$ before and after the collision respectively, show that

$$\cos \theta = \frac{\left( E_B + m_2 c^2 \right) E_A - m_2 c^2 E_B - m_1 c^4}{\sqrt{(E_B^2 - m_1^2 c^4)(E_A^2 - m_1^2 c^4)}}.$$

What is the non-relativistic limit of this result?

**Exercise 4.38** (This exercise is from [9]) In a collision process a particle of mass $m_2$, at rest in the laboratory, is struck by a particle of mass $m_1$ carrying relativistic 3-momentum $\mathbf{p}_{lab}$ and total energy $E_{lab}$, as shown in Fig. [4.13]. In the collision the two particles are transformed into two other particles (a normal occurrence in quantum-mechanical collisions) of masses $m_3$ and $m_4$. The configuration of momentum vectors in the laboratory frame as well as the center of mass frame of the system is shown in the figure. Show that the total energy of the system in the center of mass frame is given by:

$$E_{com} = \sqrt{m_1^2 c^4 + m_2^2 c^4 + 2 m_2 c^2 E_{lab}}. \tag{4.162}$$

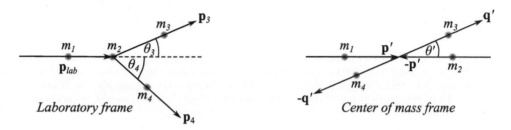

Fig. 4.13: Sketch of the situation in Exercise 4.38.

## 4.7 A Note on Units

Most of the equations in this chapter contain the speed of light constant $c$. One argues, then, that $c$, in a sense, is the constant of nature that *defines* the special theory of relativity. Now the numerical value of $c$ depends on what system of units one has chosen to work with. For instance, $c$ is equal to $299,792$ kilometers per second in standard units, but is equal to $186,282$ miles per second in British units. Note that these systems of units, which we are so used and rarely question, are *very* arbitrarily defined. Nature does not require us to define the kilometer, the mile, or even the second in the way that they are.[26] They are choices made by people for historical reasons. What's really happening here is that we began with these arbitrary definitions of units, *then* we found the values of the natural constants of the theory; in this case, just $c$. It is possible, however, and perhaps more "natural," to do it the other way around: start with setting a "natural" value for $c$, then proceed *backward* to define your units. The choice, then, can be made to take $c$ equal to unity; $c = 1$, defining the so-called **natural system of units**. Most textbooks on relativity do this right from the start. This choice has the advantage of unifying the units of a variety of quantities. For example, distances have the same units as time periods! It also tends to considerably

---

[26]The "kilometer" was originally defined in 1793 as one $1/10{,}000$ of the distance from the equator of Earth to its North Pole along a great circle, which is a very arbitrary choice indeed. The mile is even more arbitrary than that.

simplify the formulae, since $c$'s no longer appear in the equations. Here are some of the formulae we have already learned, written in natural units $(c=1)$:

$$ds^2 = -dt^2 + dx^2 + dy^2 + dz^2,$$
$$\vec{\beta} = \mathbf{v},$$
$$x^\mu = (t, x, y, z),$$
$$U^\mu = \gamma(1, \mathbf{v}),$$
$$U^\mu U_\mu = -1$$
$$E = m\gamma$$
$$E^2 = m^2 + p^2. \tag{4.163}$$

The use of such a system emphasizes the physical relationships over the numerical ones. For instance the, Minkowski metric implies that in natural units space and time measurements have exactly the same units. It is as if it is emphasized that there is no "real" difference between space and time. Another example is eqn (4.142), which becomes just $E_{\text{rest}} = m$, where mass and energy have the same units; further emphasizing their equivalence.

There are other systems of units where other fundamental constants of nature are used as a basis for the units themselves. For example, when we discuss general covariance we will need to use Newton's gravitational constant $G$ extensively. It is, along with $c$, the *defining* constant of the general theory of relativity. And yes, you guessed it, the so-called **geometrized system of units** defines $G = 1$ *as well as* $c = 1$. As far as this book is concerned, however, we will continue to explicitly show $c$ and $G$ in the equations. One exception will be in Chapter 10, as well as in some calculational examples and exercises (just for the sake of variety).

Now the most general system of units that can be defined in this way is the so-called **Planck system of units**. This is achieved by choosing the so-called **reduced Planck constant** $\hbar$ to be equal to unity in *addition* to both $c$ and $G$.[27] Since all the forces of nature, including gravity, are assumed to be quantum mechanical on a fundamental scale, this system is most commonly used in such models as superstring theory and loop quantum gravity. The first appears to be a full theory of unification of all matter and forces in the universe, while the second aspires to that but is more focused on the quantization of spacetime (i.e. gravity) itself. The choice $\hbar = G = c = 1$ sets a natural scale for these theories. For example, the so-called **Planck length**,

$$l_P = \sqrt{\frac{G\hbar}{c^3}}, \tag{4.164}$$

---

[27]The reduced Planck constant is related to the original Planck constant by $\hbar = h/2\pi$. It is conventionally defined in this way as a matter of convenience, since it is exactly this combination that appears everywhere in quantum mechanics. One could just as easily have defined the Planck system of units using $h = 1$. The reader is warned that sometimes $\hbar$ itself is referred to as Planck's constant in many sources, which may be a source of confusion to the beginner.

is the only length quantity that can be constructed using the three fundamental constants. In geometrized units it is just equal to 1, but in standard units it is of order $10^{-35}$m. This is clearly a very small distance. It is taken to be the distance scale where both general relativity and quantum mechanics act with "equal" strength; the scale of quantum gravity. Some argue that it is the smallest distance where any physical process can occur; it may simply make no sense to ask if anything can happen at shorter distances. We can also construct a **Planck time**, defined as the time it takes a photon to cross a Planck length:

$$t_P = \frac{l_P}{c} = \sqrt{\frac{G\hbar}{c^5}}. \tag{4.165}$$

This has a value of order $10^{-44}$ seconds; a very small time period indeed. Again this is sometimes taken to be the smallest time where any physical process can occur; a "quantum" of time, so to speak. Last, but most definitely not least, we can define a **Planck mass**:

$$m_P = \sqrt{\frac{\hbar c}{G}}. \tag{4.166}$$

In standard units this is of order $10^{-8}$kg. This is not a terribly small number like in the case of length or time; it is roughly the mass of, say, a flea's egg. But compared to typical subatomic particles, this is super massive! For comparison, consider that the mass of the electron is of order $10^{-31}$kg. So a subatomic particle that has mass equal to $m_P$ is about $10^{22}$ times more massive than an electron! This is a highly dense particle indeed, considering that it is of subatomic scale. To further understand what this means, we define the **Planck energy** by simply applying (4.142):

$$E_P = m_P c^2 = \sqrt{\frac{\hbar c^5}{G}}, \tag{4.167}$$

which gives an order of magnitude of $10^9$ Joules. Although this is not a huge amount of energy (it is roughly equivalent to the energy stored in a car's full gas tank), it is extremely large if one considers it as energy that has to be concentrated in a small scale. In fact, it is about $10^{15}$ times larger than the highest energy ever concentrated in subatomic particles in our human-made particle accelerators. Such energies at such small scales are where all the forces of nature are assumed to become one, the scale of complete unification; it is also the scale where a quantum theory of gravity manifests itself. Consider that such energies, if converted to heat, yield temperatures of order of magnitude $10^{32}$ degrees Kelvin[28] (the **Planck temperature**). The Big Bang is a case where such high temperatures are assumed to have existed.

One hopes that the reader has by now grasped the importance of these systems of units to fundamental physics. As mentioned earlier, we will rarely use them in this

---

[28]Or Fahrenheit or Celsius. The difference between temperature scales is very small at such large scales.

book; however, the reader interested in studying more advanced texts should prepare themselves to eventually get comfortable with them.

**Exercise 4.39** Using dimensional analysis, derive eqns (4.164) and (4.166).


## 4.8 Specially Covariant Electrodynamics

As noted earlier, Maxwell's theory of electricity and magnetism is already relativistic to begin with, so no changes need to be made to the basic theory itself. For example, the Lorentz factor $\gamma$ need not be inserted in any of its fundamental equations as we had to do with Newtonian mechanics; it can be shown that it is already *encoded* in Maxwell's equations. In this section we will give a quick review of electrodynamics and rewrite its equations in the tensor language to emphasize its covariance. The *in vacuo* electric and magnetic fields $\mathbf{E}\,(t,\mathbf{r})$ and $\mathbf{B}\,(t,\mathbf{r})$ are defined by **Maxwell's equations**,[29] giving their divergence and curl as follows:

**Gauss' law**

$$\vec{\nabla} \cdot \mathbf{E} = \frac{\rho}{\epsilon_0} \tag{4.168}$$

**The no-monopoles law**

$$\vec{\nabla} \cdot \mathbf{B} = 0 \tag{4.169}$$

**Faraday's law**

$$\vec{\nabla} \times \mathbf{E} + \frac{\partial \mathbf{B}}{\partial t} = 0 \tag{4.170}$$

**The Ampère–Maxwell law**

$$\vec{\nabla} \times \mathbf{B} - \mu_0 \epsilon_0 \frac{\partial \mathbf{E}}{\partial t} = \mu_0 \mathbf{J}, \tag{4.171}$$

where, as defined earlier in §3.4, the constants $\epsilon_0$ and $\mu_0$ are the permittivity of free space and the permeability of free space respectively.[30] They are related to the speed of light in vacuum via

$$c = \frac{1}{\sqrt{\mu_0 \epsilon_0}}. \tag{4.172}$$

---

[29]Due to the works of the German mathematician Carl Friedrich Gauss (1777–1855), the English physicist Michael Faraday (1791–1867), the French physicist André-Marie Ampère (1775–1836), and of course the Scottish mathematical physicist James Clerk Maxwell (1831–1879).

[30]We are working in the MKS system of units. In other systems the Maxwell equations have slightly different forms.

It is in fact this very relationship, combined with eqns (4.176) below, that led Maxwell to theorize that light is nothing more than an electromagnetic wave propagating with a constant speed in vacuum—a hypothesis that was demonstrated experimentally by Hertz[31] a few years after Maxwell's death. The scalar function $\rho(t, \mathbf{r})$ and the vector function $\mathbf{J}(t, \mathbf{r})$ are the *sources* of the fields; they are the electric charge density (in units of Coulomb per cubic meter) and electric current density (in units of Ampère per square meter) respectively. They are further related via the so-called **continuity equation**, which is just a statement of the conservation of electric charge:

$$\vec{\nabla} \cdot \mathbf{J} + \frac{\partial \rho}{\partial t} = 0. \tag{4.173}$$

The electric and magnetic fields are defined via

$$\mathbf{E} = -\vec{\nabla} V + \frac{\partial \mathbf{A}}{\partial t}, \qquad \mathbf{B} = \vec{\nabla} \times \mathbf{A}, \tag{4.174}$$

where $V(t, \mathbf{r})$ is the **electric scalar potential** and $\mathbf{A}(t, \mathbf{r})$ is the **magnetic vector potential**. Finally, the force that a given electric charge $q$ experiences as it moves with a velocity $\mathbf{v}$ in an electromagnetic field is given by the so-called **Lorentz force law**,

$$\mathbf{F} = q\mathbf{E} + q(\mathbf{v} \times \mathbf{B}). \tag{4.175}$$

Maxwell's equations are first-order differential equations in $\mathbf{E}(t, \mathbf{r})$ and $\mathbf{B}(t, \mathbf{r})$. They can be reduced to two second-order equations in $V(t, \mathbf{r})$ and $\mathbf{A}(t, \mathbf{r})$ by combining them with (4.174)[32] to get

$$\nabla^2 V - \frac{1}{c^2} \frac{\partial^2 V}{\partial t^2} = -\frac{\rho}{\varepsilon_0}, \qquad \nabla^2 \mathbf{A} - \frac{1}{c^2} \frac{\partial^2 \mathbf{A}}{\partial t^2} = -\mu_0 \mathbf{J}. \tag{4.176}$$

These equations are wave equations. As discussed earlier around eqn (4.31), they can be written in terms of the d'Alembertian operator:

$$\Box^2 V = -\frac{\rho}{\varepsilon_0}, \qquad \Box^2 \mathbf{A} = -\mu_0 \mathbf{J}. \tag{4.177}$$

In the special case of static fields, i.e. time-independent $V$ and $\mathbf{A}$ (and consequently $\mathbf{E}$ and $\mathbf{B}$), eqns (4.176) reduce to

$$\nabla^2 V = -\frac{\rho}{\varepsilon_0} \tag{4.178}$$

$$\nabla^2 \mathbf{A} = -\mu_0 \mathbf{J}. \tag{4.179}$$

Any equation, and there are many similar equations in physics, that has the form $\nabla^2 f(\mathbf{r}) = g(\mathbf{r})$, where $f$ and $g$ are some scalar functions, is known as a **Poisson**

---

[31] Heinrich Rudolf Hertz: German physicist (1857–1894).

[32] This is not straightforward; it requires knowledge of the so-called gauge freedom of electrodynamics. The curious reader may consult any advanced text, such as [9].

equation.[33] A special case would be for the vanishing of the right-hand side "sources" reducing the Poisson equations to the **Laplace equations**:[34]

$$\nabla^2 V = 0 \tag{4.180}$$

$$\nabla^2 \mathbf{A} = 0. \tag{4.181}$$

The question now becomes "Which of these quantities are components of what 4-vectors and/or 4-tensors?" It turns out that the potentials $V$ and $\mathbf{A}$ can be viewed as components of the **potential 4-vector**

$$A^\alpha = \left( \frac{V}{c}, \mathbf{A} \right) \tag{4.182}$$

that transform covariantly. In addition, the charge and current densities also define components of a **current 4-vector**:

$$J^\alpha = (c\rho, \mathbf{J}). \tag{4.183}$$

The factors of $c$ in (4.182) and (4.183) are essential to make the units of the components of the vector equal throughout, as the reader can check. Now the electric and magnetic fields themselves do not constitute 4-vectors; rather, they turn out to be components of an antisymmetric second-rank tensor $F^{\alpha\beta}$, usually referred to as the **electromagnetic field strength tensor** or the **Faraday tensor**. In matrix form in Minkowski-Cartesian coordinates this is:

$$[F^{\alpha\beta}] = \begin{pmatrix} 0 & -E_x/c & -E_y/c & -E_z/c \\ E_x/c & 0 & -B_z & B_y \\ E_y/c & B_z & 0 & -B_x \\ E_z/c & -B_y & B_x & 0 \end{pmatrix}. \tag{4.184}$$

The relations (4.174) can *both* be written as follows:

$$F^{\alpha\beta} = \partial^\alpha A^\beta - \partial^\beta A^\alpha, \tag{4.185}$$

and the four Maxwell equations become

$$\partial_\alpha F^{\alpha\beta} = \mu_0 J^\beta \tag{4.186}$$

$$\partial_\alpha F_{\beta\gamma} + \partial_\gamma F_{\alpha\beta} + \partial_\beta F_{\gamma\alpha} = 0, \tag{4.187}$$

where (4.186) gives both the Gauss and Ampère–Maxwell laws and (4.187) gives the no-monopoles and Faraday laws. Furthermore, the minimal substitution $\partial \rightarrow \nabla$ allows us to write the Maxwell equations in *any* arbitrary coordinate system:

[33]Baron Siméon Denis Poisson: French mathematician, engineer, and physicist (1781–1840).

[34]As discussed back in §2.5, the Laplacian and the d'Alembertian act on the *components* of vectors; in other words, the second equation of (4.179) is really the three equations $\nabla^2 A^i = -\mu_0 J^i$.

$$\nabla_\alpha F^{\alpha\beta} = \mu_0 J^\beta \tag{4.188}$$

$$\partial_\alpha F_{\beta\gamma} + \partial_\gamma F_{\alpha\beta} + \partial_\beta F_{\gamma\alpha} = 0. \tag{4.189}$$

Note that both (4.185) and (4.187) are already fully covariant, since if we substitute $\partial \to \nabla$ the terms involving the Christoffel symbols vanish, as the reader is encouraged to verify. The continuity eqn (4.173) turns out to be particularly simple:

$$\partial_\alpha J^\alpha = 0, \tag{4.190}$$

or, in covariant form,

$$\nabla_\alpha J^\alpha = 0. \tag{4.191}$$

Equation (4.191) is a conservation law. It states that since the 4-divergence of the 4-current is zero, then the 4-current itself must be a conserved quantity. This emphasizes the universal conservation of electric charge. In physics, conservation laws can always be reduced to highly symmetric and very simple expressions such as (4.191). The simplicity is almost always a sign of a deeper structure. It is as if nature is telling us that we have stumbled upon something truly beautiful. The interested reader may further explore this concept in textbooks on classical field theory under the heading **Noether's theorem.**[35]

The Lorentz force law (4.175) becomes

$$f^\mu = qF^\mu{}_\nu \frac{dx^\nu}{d\tau}, \tag{4.192}$$

where $f^\mu$ is the 4-force related to mechanics by eqn (4.127). Finally, the d'Alembertian wave eqns (4.177) are both encoded in

$$\Box A^\mu = \nabla^\alpha \nabla_\alpha A^\mu = g^{\alpha\beta} \nabla_\alpha \nabla_\beta A^\mu = \mu_0 J^\mu. \tag{4.193}$$

**Exercise 4.40**

1. Explicitly verify that (4.185) leads to (4.174).
2. Explicitly verify that (4.186) and (4.187) lead to (4.168), (4.169), (4.170), and (4.171) in Cartesian coordinates.
3. Explicitly verify that (4.190) leads to (4.173) in Cartesian coordinates.
4. Explicitly verify that (4.192) leads to (4.175).
5. Explicitly verify that (4.193) leads to (4.177).

**Exercise 4.41**

1. Starting from (4.188) and (4.189), write the Maxwell equations in cylindrical polar coordinates as well as the coordinate systems of Exercise 1.21.

[35] Amalie Emmy Noether: German mathematician (1882–1935).

2. Starting from (4.191), write the continuity equation in cylindrical polar coordinates as well as the coordinate systems of Exercise 1.21.
3. Starting from any of the forms (4.193), write the wave equation in cylindrical polar coordinates as well as the coordinate systems of Exercise 1.21.

**Exercise 4.42**

1. Explicitly show that eqn (4.189) is in fact a direct consequence of the definition (4.185). This means that the full content of the Maxwell equations is in either (4.188) and (4.189) *or* (4.188) and (4.185).
2. Explicitly show that the Cartesian continuity eqn (4.190) is a consequence of (4.186). *Hint*: Note that the field strength tensor $F^{\alpha\beta}$ is totally antisymmetric.
3. Explicitly show that the covariant continuity eqn (4.191) is a consequence of (4.188).
4. Explicitly show that the wave eqn (4.193) is also a consequence of (4.188) if we make the choice $\nabla_\mu A^\mu = 0$. This last is the so-called "gauge" choice discussed briefly in footnote 32.

**Exercise 4.43** So far no one has been able to find magnetic monopoles, i.e. point magnetic charges that generate a radial magnetic field, just as point electric charges generate a radial electric field. The situation could change any second, however, since there is no theoretical reason why magnetic monopoles shouldn't exist. If you woke up tomorrow morning and learnt that they had been found, how would the Maxwell equations change? In other words, rewrite eqns (4.168), (4.169), (4.170), and (4.171) to accommodate the presence of magnetic charges $\rho_m$ and magnetic currents $\mathbf{J}_m$ (be careful to figure out the units ahead of time and see if quantities of $\varepsilon_0$ and $\mu_0$ need to be included to make the new equations dimensionally correct). The issue of whether or not magnetic monopoles exist has always been interesting from a theoretical perspective. However, it can be shown that even if they do exist, Maxwell's equations need *not* be changed if and only if all subatomic particles have the same ratio of electric to magnetic charge! Furthermore, their existence has been shown (theoretically) to answer a variety of open questions in physics, such as the quantization of charge. The interested reader may consult any advanced text on the subject; for example [10].

**Exercise 4.44** Consider the rest frame $O'$ of a positive charge $q$. Its electromagnetic field tensor (4.184) in this frame contains only the electric field, since it is not moving and hence has no magnetic field:

$$\left[F^{\alpha\beta}\right] = \begin{pmatrix} 0 & -E_x/c & -E_y/c & -E_z/c \\ E_x/c & 0 & 0 & 0 \\ E_y/c & 0 & 0 & 0 \\ E_z/c & 0 & 0 & 0 \end{pmatrix}. \tag{4.194}$$

Now if $O'$ is moving with speed $v$ along the $x$ axis relative to an observer $O$, what electromagnetic field tensor will $O$ observe? Note that the boosted field tensor will contain components in the slots reserved to the magnetic field, as of course it must, since a moving charge *is* a current, and currents generate magnetic fields. We conclude, then, that magnetic fields are only "just" a relativistic effect.

# 5

# General Covariance

*Since the mathematicians have invaded the theory of relativity, I do not understand it myself anymore.*

Albert Einstein

## Introduction

General covariance is, as the name implies, more general than the other types of covariance we have considered so far. Recall that we have slowly expanded our understanding of space and time from ordinary Euclidean translations and rotations, through Galilean covariance, and ended with the special covariance of four-dimensional spacetime. In doing so we have always been able to specifically pinpoint the metric that defines the covariance. In general relativity, one is confronted with the fact that spacetime may no longer be flat, in other words no longer described by the Minkowski metric in any of its incarnations (4.10) and (4.11), and may not even be known in advance. Spacetime will now be allowed to *curve*, in the sense that the shortest distance between two points is no longer necessarily a straight line. Furthermore, the degree of this curvature is not *a priori* known, but rather varies depending on the physics of the situation in question. This was discovered by Albert Einstein in 1915 based on his attempts to generalize special covariance from inertial reference frames (constant speeds) to non-inertial or accelerating reference frames, which by a stroke of sheer brilliance he also found to be directly related to the phenomenon that we call "gravity." Undoubtedly, this incredible connection, more than special relativity, is the most amazing of all single-handed discoveries in the history of science, and it gave Einstein the fame and place in history that he truly deserved.

## 5.1   What is Spacetime Curvature?

Using everything that we have learned so far, if one wishes to define general relativity in the shortest possible terms, then one can write something like this: "The connection between space and time may be described by a metric of the form

*Covariant Physics: From Classical Mechanics to General Relativity and Beyond*. Moataz H. Emam, Oxford University Press (2021). © Moataz H. Emam.
DOI: 10.1093/oso/9780198864899.003.0005

$$ds^2 = g_{\mu\nu}dx^\mu dx^\nu \qquad \mu, \nu = 0, 1, 2, 3, \tag{5.1}$$

where the metric tensor $g_{\mu\nu}$ no longer necessarily describes flat spacetime, in the sense that the shortest distance between two points (in space*time*) may be something other than a straight line. The exact values of the components of $g_{\mu\nu}$ are defined by the mass and energy content of the region of space in question and can be found by solving a set of differential equations known as the Einstein field equations." To clarify, if a specific region of space is completely devoid of, and sufficiently far away from, mass and/or energy, then $g_{\mu\nu}$ is the Minkowski metric, whether Cartesian (4.10) or in any non-Cartesian coordinate system (4.11); straight lines will describe the shortest distance between two points and special covariance is in effect. If, however, there is some distribution of mass and/or energy nearby (since energy and mass are equivalent), then this necessarily bends and warps the spacetime background *itself*, causing free particles to follow non-straight trajectories, "tricking" us into thinking that the particle is being "pulled" by the invisible force of gravity. In other words, in general covariance there really is no such thing as a force of gravity; the concept *itself* has been replaced.

Another way of defining general relativity is based on the concept of covariance under transformations between reference frames. We have explored this in various contexts. The laws of physics are the same between two reference frames that are rotated with respect to each other as well as moving with a constant speed with respect to each other (low speeds: Galilean, high speeds: Lorentzian). But what about accelerating reference frames? Two reference frames accelerating with respect to each other will exhibit different physics; non-inertial pseudo-forces will appear. But if we require the invariance of the laws of physics between two accelerating reference frames, then we must absorb these pseudo-forces into something that we already observe. Surprisingly, this turns out to be gravity! In other words, gravity is a pseudo-force that appears as a consequence of general covariance.

Let's take all this one step at a time. First, we set the stage by asking the question: If curved spacetimes are defined by the fact that the shortest distance between two points is no longer a straight line but is described by some other curve, how does one find this curve? We already have a recipe for that: the so-called geodesic eqn (3.26), whose only solution in the three-dimensional Newtonian case was the straight line.[1] We must, however, upgrade (3.26) to four dimensions. We do so by first changing its indices from $i, j = 1, 2, 3$ to $\mu, \nu = 0, 1, 2, 3$:

$$\ddot{x}^\mu + \Gamma^\mu_{\alpha\beta}\dot{x}^\alpha\dot{x}^\beta = 0, \tag{5.2}$$

where the Christoffel symbols are found from the metric (5.1) via (4.33). Next, we note that the derivatives in (5.2) were originally with respect to Newtonian absolute time, i.e. $\dot{x} = dx/dt$, but in the relativistic world time no longer holds the special position it had in the Newtonian/Galilean world view; we must then change the derivatives to

---

[1]Although we *did* get a quick taste of the problem of the motion of a particle in a curved space via the sphere problem of §3.1.1.

any arbitrary parameter that the $x^\mu$'s are dependent on. The most natural choice is the proper time $\tau$:

$$\frac{d^2 x^\mu}{d\tau^2} + \Gamma^\mu_{\alpha\beta} \frac{dx^\alpha}{d\tau} \frac{dx^\beta}{d\tau} = 0. \tag{5.3}$$

However, this is not cast in stone. One can choose almost any parameter $\lambda(\tau)$ such that the geodesic curve is described by $x^\mu(\lambda)$ and still leaves eqn (5.3) in the same form, i.e. invariant. This "freedom" is particularly important if the geodesic in question describes "lightlike" curves (where $ds^2 = 0$),[2] i.e. trajectories of beams of light (no longer straight), since we have seen that light has no rest frame and thus cannot be parameterized using time, proper or otherwise. So we generalize (5.2) to

$$\frac{d^2 x^\mu}{d\lambda^2} + \Gamma^\mu_{\alpha\beta} \frac{dx^\alpha}{d\lambda} \frac{dx^\beta}{d\lambda} = 0. \tag{5.4}$$

It can be shown that reparameterizing the geodesic equation in terms of $\lambda$ leaves the geodesic equation invariant if and only if it is linearly related to $\tau$. In other words,

$$\lambda(\tau) = a\tau + b, \tag{5.5}$$

where $a$ and $b$ are arbitrary integration constants. The requirement (5.5) is known as the **affine transformation** and guarantees that any two parameters related to each other in this way will leave the geodesic equation invariant. Henceforth the parameter $\lambda$ will be referred to as the **affine parameter** since it satisfies this requirement.

---

**Exercise 5.1** Show that the affine parameter $\lambda(\tau)$ leaves the geodesic equation invariant if and only if

$$\frac{d^2 \lambda}{d\tau^2} = 0, \tag{5.6}$$

whose solution is (5.5).

---

One particularly useful choice for the affine transformation is

$$\lambda \equiv \frac{\tau}{m}; \tag{5.7}$$

i.e. setting $a = 1/m$ and $b = 0$ in (5.5). For a particle with $m = 0$, say a photon of light, there exists no proper time; however, $\lambda$ as defined by (5.7) can *still* exist, since we can require that in the limit of vanishing $\tau$ and $m$ the affine parameter remains finite, i.e.

$$\lambda = \lim_{\substack{\tau \to 0 \\ m \to 0}} \frac{\tau}{m} \equiv \text{finite for massless particles.} \tag{5.8}$$

The affine parameter can also be further understood as a measure of distance along null geodesics. When you look at a scene with your eyes, you are looking *along* null

---

[2]The definitions (4.7), (4.8), and (4.9) are still applicable.

geodesics, and the natural measure of distance to objects that you see is the affine distance.

Hence the stage is set: Given a specific metric $g_{\mu\nu}$, one calculates the Christoffel symbols, plugs them into (5.4), and solves the resulting differential equations. If they give straight lines *everywhere*, then the spacetime in question is flat and special covariance is in effect; if they don't, then (5.1) describes a curved spacetime background and general covariance is active. Defining the curvature of any space or spacetime is generally a difficult thing to do, and has historically gone through several incarnations. The definition we are using here (temporarily) is intuitive and easily visualizable. Based on the work of Riemann, however, there is a much more rigorous way of defining curved spaces/spacetimes based on the concept of parallel transporting vectors, but for our immediate purposes we can postpone this discussion to Chapter 7.

At this point the reader is reminded that the curves described by (5.4) are *spacetime* curves in four dimensions. While parts of them are spatial curves describing the motions of particles in the usual sense, the general picture involves time as well as space, which makes it slightly more difficult to visualize. What does a four-dimensional curve in four-dimensional spacetime look like? For simplicity, let's temporarily discuss pure two-dimensional surfaces without time,[3] while keeping in mind that this is just a method of providing a visual of what should essentially be four-dimensional surfaces *including* time.[4] If the image of an infinite flat plane is invoked to model flat spacetime, then when we try and visualize a curved space, images of spheres, cylinders, paraboloids, and other familiar surfaces may be used. It is important, however, to keep in mind that some of these may not actually be "truly" curved! At least not based on the definition of curvature we are using here. Consider, for example, a cylinder with a given radius $a$ whose circular section is parallel to the $x$–$y$ plane and extends to infinity in the $z$ direction like the ones in Fig. [1.4]. The parametrization of such a cylinder in Cartesian coordinates is

$$\begin{aligned} x^1 &= a\cos\varphi \\ x^2 &= a\sin\varphi \\ x^3 &= z, \end{aligned} \tag{5.9}$$

leading to the metric *on* the surface of the cylinder

$$dl^2 = a^2 d\varphi^2 + dz^2, \tag{5.10}$$

as the reader can easily verify.[5] The non-vanishing components of $g_{ij}$ are then the constants $g_{\varphi\varphi} = a^2$ and $g_{zz} = 1$. Since the Christoffel symbols are derivatives of the components of the metric tensor, they will all vanish and the geodesic equation on the

---

[3]Think of the sphere problem of §3.1.1.

[4]Sometimes these "surfaces" are dubbed **hypersurfaces**, since they are four-dimensional, while the term "surface" implies two dimensions.

[5]Equivalently we can begin with the cylindrical polar metric (1.26) and set $\rho = a$.

cylinder will give straight lines as their only solution![6] Another way of seeing this is if we make the coordinate redefinition

$$x = a\varphi \tag{5.11}$$

and plug back into (5.10) to find

$$dl^2 = dx^2 + dz^2, \tag{5.12}$$

which most certainly *is* flat! It is not surprising, then, that the geodesic equations will imply straight lines in every direction, although some of them will be "rolled" on circles of radius $a$. This implies yet another way of understanding this; cylinders can be thought of as originally flat planes that have been rolled over a circle, and what were originally "straight" lines become helical upon this rolling.[7] In other words, if we begin with the flat $x$–$z$ plane, then require the identification $x = x + 2\pi$, we get a cylinder. In that sense cylinders are considered "flat." Furthermore, parallel geodesics on a cylinder stay parallel no matter how far one extends them. They will never intersect or drift away from each other. This is another sign of flat surfaces, even if they don't look the part. A counter-example would be spheres, which are true curved two-dimensional surfaces. We have already seen in §3.1 how solutions of the geodesic equation on a sphere are great circles. They may be parallel to each other momentarily, but taken further around the sphere they are bound to intersect. Also, there is no way of "building" a sphere from a flat plane, as was possible with cylinders. Starting with a plane, try rolling it perfectly into a spherical shape without cutting or creasing it; you will find that it is impossible. The reader is encouraged to try and find a coordinate redefinition that will reduce the metric of the sphere (3.34) to the flat Euclidean metric, like we did for the cylinder. I think that pretty quickly you will be convinced that such redefinitions do not, and cannot, exist.

**Exercise 5.2** Using the definition of curvature discussed here, would you say that a circle is a curved space? Give at least two arguments to support your answer. Does your answer apply to other closed loops such as ellipses?

**Exercise 5.3** Start with the flat $x$–$y$ plane, then require $x = x + 2\pi$ *and* $y = y + 2\pi$. What geometric shape do you get? Do you think it is as flat as a cylinder or as curved as a sphere? Why?

Now, we also need to study curved spacetimes *without* reference to any higher-dimensional space. For instance, consider when we discuss the sphere; we usually think of it as a two-dimensional surface "embedded" in three dimensions. But our true interest is four-dimensional "surfaces" that are *not* embedded in anything! Think about it! *Why* would they be? These spaces/spacetimes represent the very "fabric" of our universe, and to our knowledge our universe is *not* embedded in anything. And even if

---

[6]We have already done this. See the discussion following eqn (3.26).

[7]Recall the results of Exercise 3.6.

it were embedded in some higher-dimensional *hyperspace*, what is *that* embedded into? There has got to be an end to the "embedding," or at least a point where we simply stop trying. Hence we need to develop a theory of spacetime that is *independent* of embedding. The metric approach provides a basis for this; it defines distances intrinsic to the surface in question. We will also need a definition of curvature that is *intrinsic* to the surface and makes no reference to any points "outside" of it. This is what Riemann achieved, as we will see. But how can one visualize such stuff? The answer is we don't. The visual is simply lost. This is one of the main difficulties of general relativity. We are forced to just deal with all this in purely mathematical terms. Still, visual "aids" known as **embedding diagrams** are possible. These are spacetime "grids" where time and one of the spatial dimensions are removed. As an example, Fig. [5.1a] shows an embedding diagram of flat spacetime simply denoted by an infinite two-dimensional surface. In comparison, a curved spacetime is represented in Fig. [5.1b]. This one represents the curved spacetime surrounding a massive spherical object, such as a planet.

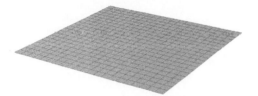

(a) Embedding diagram of flat spacetime. The grid, being made of squares, can be thought of as a visual representation of the Minkowski-Cartesian metric. Each straight line represents an infinite spatial plane in real space.

(b) Embedding diagram of the curved spacetime surrounding a massive spherical object (not shown). This grid implies spherical coordinates. The circles represent spatial concentric spheres in real space.

Fig. 5.1: Examples of embedding diagrams.

## 5.2  Gravity as Curvature

It is a credit to Einstein's genius that he reached these incredible conclusions concerning the "true" nature of spacetime just by doing thought experiments, also known in the physics literature by their German name: "gedanken" experiments. Imagine! Just by *thinking*, in a true Aristotelian fashion, he managed to connect gravity with spacetime curvature! Let us consider some of these thought scenarios in an attempt to reproduce his thinking process.

Consider two objects of different masses that are dropped from the roof of a building vertically and at the same time, $t = 0$, in an experiment very familiar to students in a freshman physics class. Ignoring friction and assuming that the only force acting on the two masses is gravity, we detect that both objects will accelerate with the same constant and uniform acceleration, $g = 9.81\text{m/s}^2$, irrespective of how different their masses are. At any instant of time during their trip, they will always be at the same height, traveling together until they eventually hit the ground at the same time. The fact that their accelerations will be the same even though their masses are different was known to Galileo, who proved it experimentally over 500 years ago.[8] Mathematically what this means is that the quantity $m$ as it appears in Newton's second law $F = ma$ and the one that appears in the weight formula $F = mg$ are one and the same. Note that there is no theoretical reason to assume this to be true, since the two definitions of mass come from completely different assumptions: Newton's second law says that any force is proportional to acceleration, and that the constant of proportionality may be called the **inertial mass** $m_I$. On the other hand, the weight formula says that the force generated by gravity on an object is proportional to $g$ and[9] that the constant of proportionality is the **gravitational mass** $m_G$. For a falling object, Newton's second law requires

$$F = m_I a$$
$$m_G g = m_I a. \tag{5.13}$$

Solving this for the acceleration gives

$$a = \frac{m_G}{m_I} g. \tag{5.14}$$

If $m_G$ and $m_I$ are not the same, then the acceleration will depend on their ratio, making it possible for different objects to fall at different accelerations depending on their masses. This is not observed, and accepting the experimentally verified ratio

$$\frac{m_G}{m_I} = 1 \tag{5.15}$$

is known today as the **weak principle of equivalence**, or the **Galilean principle of equivalence**. It was well known to both Galileo and Newton.

**Exercise 5.4** Using the usual methods of classical mechanics, find the dependence of the period $T$ of a simple pendulum on the ratio $m_G/m_I$. This dependence lends itself to an experimental way of measuring this ratio in the lab. A simple pendulum is shown in Fig. [8.3].

A few centuries later, Einstein comes along and starts thinking about the problem. He performs his gedanken experiments that further expand on (5.15). The first

---

[8]There is a famous tale that Galileo demonstrated this by dropping two objects from the top of the Leaning Tower of Pisa. Most historians today accept that this probably didn't happen. What we do know is that Galileo did perform dropping experiments in his own home, most likely by sliding or rolling objects down ramps, to make it easier to observe and document their accelerations. The actual "vertical" drop experiment was later performed by many scientists and is currently done in freshman labs all over the world.

[9]Whether $g$ is a constant close to the surface of a planet or varying as in eqn (3.75).

such experiment involves an observer sitting in an elevator. Assume that the elevator cables suddenly break and both the elevator and the observer accelerate freely downward toward their inevitable doom.[10] Because of the principle of equivalence, both the elevator and the observer will accelerate downward at the same rate, and hence they will be motionless with respect to each other. The observer will "float" in what she will perceive as a state of weightlessness. Now imagine that another observer inside a similar elevator is taken aboard a spaceship far away from Earth and the solar system and then left floating in a region of space where the effects of gravity are as close to zero as possible. This observer will also float weightless inside the elevator. Einstein argued that, short of looking outside the elevator, there is *absolutely* no way either observer will be able to tell which of the two situations they are in! Floating inside a boxed elevator can either be because both observer and elevator are freely falling under the influence of gravity, or because they are motionless in gravity-less space. The situation is sketched in Fig. [5.2].

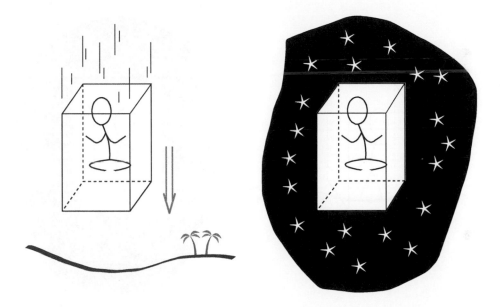

Fig. 5.2: Einstein's gedanken experiment number 1: No observer or physics experiment inside a closed box can tell the difference between uniform free fall (left) and floating weightlessly in space (right).

Now if the elevator in space happens to have rockets attached to it, as shown in Fig. [5.3], and they are firing such that the elevator accelerates in a given direction at a rate of $9.81 \text{m/s}^2$, the floor of the elevator will "catch up" with the observer's feet and the observer will feel as if they are standing with their full and normal weight on the surface of the Earth. There is absolutely no way that the observer can tell in

[10]Let's assume, just for the sake of non-morbidness, that eventually the emergency brakes kick in and the observer is saved. But our gedanken experiment occurs before that happens.

which state they are in: Are they motionless on Earth being pulled down by gravity, or are they accelerating in space? There is no experiment that can be performed that will clue the person into the answer. To Einstein, if there is no way to tell between two different phenomena, then the two phenomena must be one and the same! *However*, an important condition for all of this to work is that the elevator in the various scenarios is small enough compared to the source of gravity. Consider the situation in Fig. [5.4]. If the elevator is too large in the second scenario, then the person inside may notice that balls falling at opposite ends of the elevator are *not* doing so parallel to each other[11] and conclude that they are not accelerating uniformly but rather standing in the radial gravitational field of a planet, hence becoming able to tell the difference. We call this the condition of *locality*: that the indistinguishability between gravity and acceleration is a local phenomenon. Taken globally, i.e. on a large scale, the argument fails to work. Note that this connects to the condition that curved spaces are locally flat, as we will see in a moment.

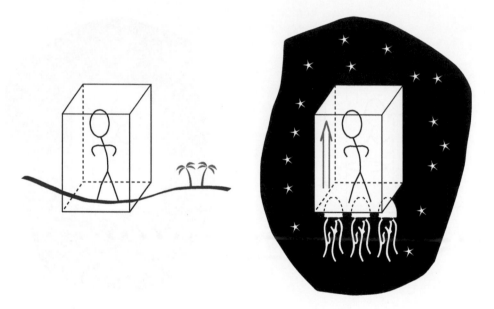

Fig. 5.3: Einstein's gedanken experiment number 2: No observer or physics experiment inside a closed box can tell the difference between a uniform gravitational field (left) and acceleration through space (right).

We can now formulate Einstein's version of the principle of equivalence; it is known as either the **strong principle of equivalence** or the **Einstein principle of equivalence** and goes something like this: Gravity and acceleration are locally equivalent, completely indistinguishable from each other. What we perceive as the invisible force of gravity is in "reality" an *accelerating reference frame*.

---

[11] As was discussed in §3.3 and Exercise 3.17.

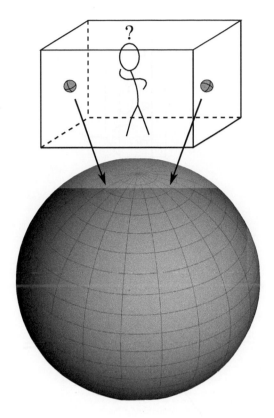

Fig. 5.4: Caveat: In the scenarios discussed, the elevators must be small compared to the source of gravity.

Okay, so what does all of this have to do with spacetime being curved? It turns out that the mathematics that describes accelerating reference frames is *exactly* identical to, and hence the same as, that which describes curved spacetimes.[12] One can demonstrate this with a simple example. Consider the situation in Fig. [5.5]. An observer $O'$ inside an elevator is falling freely in a uniform gravitational field. Since $O'$ perceives no acceleration, his is the world of special relativity, and as such his spacetime is described by the usual flat Minkowski metric. In other words, every experiment that $O'$ can possibly perform inside the elevator follows the rules of special covariance. As discussed, this is true only locally in the immediate neighborhood of $O'$. Now, let another observer $O$ be standing far enough away to see what's happening on a more global scale. He sees an *accelerating* reference frame: the elevator. As such, his spacetime cannot be described by the metric of special relativity, which deals only with non-accelerating

---

[12]See the Einstein quote at the beginning of Chapter 7 on how this was pointed out to him by a mathematician friend.

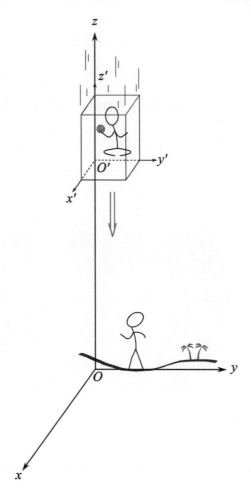

Fig. 5.5: Comparing the metrics as seen by the local freely falling observer $O'$ and the global/coordinate observer $O$.

reference frames. Observer $O$ then sees the global metric, i.e. a metric that covers all of spacetime, not just the one that covers the small local volume of the elevator.[13] Now here is where curvature comes in. The global metric as observed by $O$ *cannot* possibly be flat, for the following reason: Assume the contrary; that it *is* possible for $O$ to see a global flat metric. Then this metric is necessarily the *same* as the one observed by $O'$ because all flat metrics can be transformed into each other. But $O'$ sees his frame as not-accelerating, while $O$ sees an accelerating frame; ergo $O$'s metric must be one that cannot be transformed to a flat metric, and as such it *must be curved*. This is the crucial connection between accelerating reference frames and curved spacetimes;

---

[13] An implicit assumption here is that nothing else exists in the universe except the described situation. Or, in other words, any other phenomenon is too far away to change the metric that $O$ observes.

a stroke of true genius on the part of Einstein. One would then expect that $O$ reports that $O'$ is following a *geodesic* that can be calculated using the curved metric! We call $O'$ the **local observer**, while we call $O$ the **global** or **coordinate observer**. Both observers may disagree on time and spatial measurements, just like the inertial observers did in special relativity; however, they *will* agree on the covariant quantities. For instance, both will agree on the interval $ds^2$,

$$ds^2\big|_{\text{local}} = ds^2\big|_{\text{global}} = \eta_{\mu'\nu'}dx^{\mu'}dx^{\nu'}\big|_{\text{flat}} = g_{\mu\nu}dx^\mu dx^\nu\big|_{\text{curved}}, \qquad (5.16)$$

as well as any other covariant object, such as all scalar products of vectors. Let us explore this a bit more mathematically following Fig. [5.5]. The freely falling local observer $O'$ is now holding a ball in her hand. To her the ball is at rest, located at constant $x'$, $y'$, and $z'$. The spacetime around her in the elevator is nice and Minkowskian, hence the ball's 4-position is just

$$\mathbf{r}'_{\text{4d}} = x^{\mu'}\hat{\mathbf{e}}_{\mu'} = c\tau\hat{\mathbf{e}}_{0'} + x'\hat{\mathbf{e}}_{1'} + y'\hat{\mathbf{e}}_{2'} + z'\hat{\mathbf{e}}_{3'}, \qquad (5.17)$$

where $\hat{\mathbf{e}}_{\mu'}$ are the unit basis vectors in Minkowski-Cartesian spacetime. The metric according to $O'$ is just the familiar

$$ds^2 = d\mathbf{r}_{\text{4d}} \cdot d\mathbf{r}'_{\text{4d}} = \hat{\mathbf{e}}_{\mu'} \cdot \hat{\mathbf{e}}_{\nu'}dx^{\mu'}dx^{\nu'} = \eta_{\mu'\nu'}dx^{\mu'}dx^{\nu'}. \qquad (5.18)$$

Now if asked, observer $O$ will report that the ball is not at rest, but is falling with a constant acceleration $g = 9.81\text{m/s}^2$ (if we only consider the very approximate gravitational field close to Earth's surface). Then the height of the ball $z$ drops following the freshman physics formula:

$$z = z' + H - \frac{1}{2}gt^2, \qquad (5.19)$$

where $H$ is the initial separation between the frames, and we have also assumed the ball to have started its drop from rest. Note that the vertical measurement is the only one the two observers will disagree on. Hence $x = x'$ and $y = y'$. In this weak gravity and slow speed environment we may also take $t = \tau$, as one would not expect any time dilation effects. Hence the ball's position vector with respect to $O$ is

$$\mathbf{r}_{\text{4d}} = ct\hat{\mathbf{e}}_0 + x\hat{\mathbf{e}}_1 + y\hat{\mathbf{e}}_2 + \left(z - H + \frac{1}{2}gt^2\right)\hat{\mathbf{e}}_3. \qquad (5.20)$$

To find the metric with respect to $O$, we first compute the new basis vectors $\hat{\mathbf{g}}_\mu$ using the four-dimensional equivalent of (1.129):

$$\hat{\mathbf{g}}_\mu = \frac{\partial \mathbf{r}_{\text{4d}}}{\partial x^\mu}, \qquad (5.21)$$

which gives

$$\hat{\mathbf{g}}_0 = \hat{\mathbf{e}}_0 + \frac{gt}{c}\hat{\mathbf{e}}_3$$

$$\hat{\mathbf{g}}_1 = \hat{\mathbf{e}}_1$$

$$\hat{\mathbf{g}}_2 = \hat{\mathbf{e}}_2$$

$$\hat{\mathbf{g}}_3 = \hat{\mathbf{e}}_3. \tag{5.22}$$

Now, the metric tensor is defined via the four-dimensional equivalent of (1.130):

$$g_{\mu\nu} = \hat{\mathbf{g}}_\mu \cdot \hat{\mathbf{g}}_\nu, \tag{5.23}$$

which gives

$$[g_{\mu\nu}] = \begin{pmatrix} -1 + \frac{g^2}{c^2}t^2 & 0 & 0 & \frac{g}{c}t \\ 0 & 1 & 0 & 0 \\ 0 & 0 & 1 & 0 \\ \frac{g}{c}t & 0 & 0 & 1 \end{pmatrix}. \tag{5.24}$$

The full metric is then

$$ds^2 = -c^2\left(1 - \frac{g^2 t^2}{c^2}\right)dt^2 + 2\frac{gt}{c}dzdt + dx^2 + dy^2 + dz^2. \tag{5.25}$$

We may call this the **Galilean metric**. Its level of approximation ($g \equiv$ constant) does not quite reach Newton's full theory of gravity, which we discuss in the next section. The Galilean metric is clearly something new; it is time-dependent, and it even has cross terms $dzdt$! Using the strict definition of curvature, which we are yet to discuss, this metric is actually still flat. It is sometimes described as the flat pseudo-Minkowskian metric of an *extremely* weak uniform gravitational field, as well as for speeds much less than $c$. In other words, it represents the ordinary everyday world of falling objects that we live in. However, one would not expect the geodesics of this metric to be straight lines in spacetime. This is because if they were perfectly straight then there would be no difference between accelerating and non-accelerating motion. Let's check this and see what we get. Using (4.33) we find that the only non-vanishing Christoffel symbol is

$$\Gamma^z_{tt} = \frac{g}{c}. \tag{5.26}$$

Substituting in (5.4) with $\lambda = \tau = t$ we get the geodesic equations

$$\ddot{t} = 0$$

$$\ddot{x} = 0$$

$$\ddot{y} = 0$$

$$\ddot{z} = -\frac{g}{c}\dot{t}^2. \tag{5.27}$$

Straight spacetime geodesics would be described by the first three equations of (5.27), *as well as* $\ddot{z} = 0$. Hence the geodesics of the Galilean metric are not perfectly straight lines in *spacetime*. Now to understand what $\dot{t}$ is, let us integrate the first equation once. This gives

$$\dot{t} = \frac{dt}{d\tau} = \text{constant}. \qquad (5.28)$$

Since we are already making the approximation $t = \tau$, this constant is just unity. Using this in the fourth geodesic equation leads to

$$\ddot{z} = -g, \qquad (5.29)$$

which is *exactly* what one gets using ordinary classical mechanics. It is very important to note that we reached this result without *ever* assuming that gravity is a force acting on the ball! On the other hand, $F = ma$ also leads to (5.29) *by assuming* that gravity is a force! In a sense, then, we have just demonstrated the equivalence principle as well as making the connection between acceleration and curvature in one stroke.[14]

Let us summarize our major results:

1. A "falling" observer $O'$ lives in flat Minkowski spacetime. In other words, because her world is confined to a small scale (the elevator), her spacetime is *locally* flat, just as a small piece on the surface of a sphere can be thought of as approximately flat.
2. The *global* observer $O$, looking at things from a larger perspective, perceives a curvature in spacetime, and, when asked, she can conclude that the elevator is not "really" falling, but rather traveling uniformly along geodesics of that spacetime.
3. Gravitational forces can be thought of as the existence of curved spacetime geometry, corresponding to accelerating frames of reference.
4. The curvature of spacetime is due to the presence of a large mass nearby (for example, Earth). We will learn later that energy can generate spacetime curvature as well; not surprising, since we have already concluded that mass and energy are equivalent.

These conclusions are at the core of the general theory of relativity, and together they constitute Einstein's reformulation of the equivalence principle. From this point onward, we no longer need to talk about "gravity" as a force; all we need to do is write the metric and calculate its consequences. The "falling" observer $O'$ is not doing so because he is being pulled by an invisible force, he is simply following the shortest possible distance in a curved background; in other words, he is a freely moving object. The reader may be surprised to realize that by doing this we are really switching the roles of the two observers from the usual "common sense" ones. Recall that from a Newtonian perspective, observer $O$ in Fig. [5.5] is at rest, meaning that the forces

---

[14]I would like to point out that this calculation is terribly approximate; it is the approximation of an approximation, since $g \equiv$ constant is already an approximation of the full Newtonian theory, while the latter is itself an approximation of general relativity. However it is a useful calculation to begin with since it provides a familiar visual.

on him exactly cancel. On the other hand, $O'$ is accelerating because he is under the influence of a non-vanishing force; namely gravity. *However*, from the perspective of general covariance, $O'$ is "really" not accelerating at all. *He* is the one at rest; feeling absolutely no force acting on him. The elevator is just traveling *freely* along the "straightest" line possible, i.e. a geodesic. In other words, $O'$ is the one *not* under the influence of any forces! It is the fixed observer $O$ that is "truly" accelerating, since he is constrained (by the ground he is standing on) *not* to follow geodesics; in other words, the ground exerts a force on him stopping him from freely moving! Now, I suggest that the reader goes back and rereads this paragraph. It might take them a while to get their thinking around this strange swap of the roles of objects. It is almost as if we have redefined Newton's first law of motion to read: "An object stays at geodesic motion (which may or may not be straight) unless influenced by a force, while an object at rest does so *because* it is pushed by a force!"

Based on all of this, we now require that the laws of physics be invariant under this new transformation between accelerating and non-accelerating coordinates; i.e. we demand their general covariance. All this entails is that they are written in a form that would be invariant under change of observers, whether accelerating or not. This is simply achieved by rewriting these laws in tensor form, which at this point we are used to doing anyway. The metrics will be different from what we are used to and the transformation tensors will also vary from case to case, but the basic rules of tensor calculus won't change. In the remaining sections of this chapter, we consider specific cases of curved spacetimes, denoted by the metric tensor $g_{\mu\nu}$. All of these examples satisfy Einstein's gravitational field equations, which we will discuss in Chapter 7.

---

**Exercise 5.5**

1. Find the inverse Galilean metric tensor $g^{\mu\nu}$ then verify orthogonality: $g^{\mu\rho}g_{\rho\nu} = \delta^\mu_\nu$.
2. One way of demonstrating that (5.25) is essentially flat even though its geodesics are not precisely straight is by showing that there exists a transformation tensor $\lambda^{\mu'}_\nu$ that takes (5.25) to the Minkowski-Cartesian metric. If (5.25) was truly curved, then such a tensor couldn't exist. Find it.
3. Explicitly verify (5.26) and (5.27).

**Exercise 5.6**

1. Write down the divergence $\nabla_\mu A^\mu$ of an arbitrary 4-vector field using the Galilean metric.
2. Write down the Laplacian $\nabla^\mu \nabla_\mu f(x^\alpha)$ of an arbitrary scalar function using the Galilean metric.
3. Is (5.25) invariant under the Galilean transformations?

## 5.3 The Newtonian Limit: The Metric of Weak Gravity

Imagine four-dimensional spacetime, as we often do, as the two-dimensional surface of an elastic material such as a trampoline.[15] If there is nothing on it, the trampoline is nice and flat, such as in Fig. [5.1a]. If some massive object, say a bowling ball, is placed in the center of the trampoline, the elastic surface will curve and warp accordingly, as in Fig. [5.1b]. Similarly for spacetime, the presence of matter or energy will curve that "surface." Technically, such spaces are better described by the term **manifold**: a "surface" in any number of dimensions, in this case four, with a metric imposed on it as a grid.[16] A manifold with an associated metric is like a trampoline's surface with a coordinate grid drawn on it. One can visualize such manifolds with the embedding diagrams referred to earlier. Note that Cartesian coordinates cannot generally be used to describe curved surfaces; non-Cartesian non-Minkowskian coordinates become inevitable. Also observe that it is possible to choose different types of metrics on the same manifold; this simply amounts to different choices of coordinates. The spacetime manifold, along with any possible metrics on it, is analytically found by solving the Einstein field equations. But before we get to that, I am certain that the reader is eager to learn about specific examples of known manifolds/metrics and learn about their properties and physical implications.

Our first example of a curved spacetime is an approximate one.[17] One assumes that if the general theory of relativity is a proper theory of gravity, then it should yield Newton's theory of gravity as a special case. As a quick review, recall that Newtonian gravity requires the gravitational potential outside a uniform spherical mass $M$ to be (recall §3.3)

$$\Phi = -\frac{GM}{r}, \tag{5.30}$$

where $r$ is the spherical coordinates radial, $G = 6.67408 \times 10^{-11} \mathrm{m}^3/\mathrm{kg\ s}^2$ is Newton's constant, and the mass $M$ is centered at the origin. The Newtonian gravitation field is

$$\vec{\mathfrak{g}} = -\vec{\nabla}\Phi = -\frac{GM}{r^2}\hat{\mathbf{r}}, \tag{5.31}$$

where $\hat{\mathbf{r}}$ is a unit vector in the radial direction as shown in Fig. [5.6]. Now if another mass $m$ is placed in the gravitational field, it experiences a force

$$\mathbf{F} = m\vec{\mathfrak{g}} = -\frac{GmM}{r^2}\hat{\mathbf{r}}, \tag{5.32}$$

which is the more familiar form of Newton's law of universal gravitation. The equation of motion of the mass $m$ is of course

$$\mathbf{F} = m\ddot{\mathbf{r}}, \tag{5.33}$$

---

[15]If you go online you will find many freely available video demonstrations of general relativity using elastic surfaces.

[16]The terms "chart," "map," and "lattice" are also sometimes used.

[17]The even more drastically approximate Galilean metric of the previous section doesn't count.

where **r** is the position vector of $m$ relative to the center of $M$. Hence[18]

$$\ddot{\mathbf{r}} = \vec{\mathfrak{g}} = -\vec{\nabla}\Phi = -\vec{\nabla}\left(-\frac{GM}{r}\right). \tag{5.34}$$

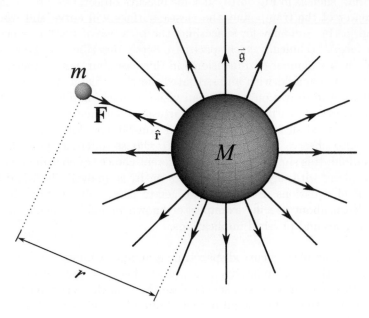

Fig. 5.6: The gravitational field $\vec{\mathfrak{g}}$ of a uniform spherical mass $M$ and the force **F** that another mass $m$ feels due to it.

Now, returning to the metric, we are trying to describe the weak gravitational field of a spherical object; in other words, weak curvature. The required metric tensor would then be a small deviation from flat space; in other words, one can say that the spacetime "trampoline" is so lightly curved around the spherical mass that it is almost flat. The components of the metric tensor would have the form

$$g_{\mu\nu} = \eta_{\mu\nu} + h_{\mu\nu}, \tag{5.35}$$

---

[18]A more detailed discussion of this particular formulation of Newtonian gravity may be found in any upper-level mechanics textbook. The reader unfamiliar with it may find it useful to note that this formulation in terms of a gravitational potential is mathematically identical to writing Coulomb's law of electrostatics in electric potential ($V$ for voltage) form. So, for example, the gravitational field $\vec{\mathfrak{g}}$ here plays the role the electric field $\mathbf{E} = -\vec{\nabla}V$ plays in Coulomb theory. We also note that it is possible to write a Gauss law (4.168) for the gravitational field, $\vec{\nabla} \cdot \vec{\mathfrak{g}} = -4\pi G\rho$, where $\rho$ is the mass density as well as $\vec{\nabla} \times \vec{\mathfrak{g}} = 0$, both of which describe the radial behavior of the gravitational field $\vec{\mathfrak{g}}$ just as their analogues do for the electric field. The reader fluent in the language of electrostatics will also realize that these imply a Poisson equation for the gravitational potential: $\nabla^2\Phi = -4\pi G\rho$. In fact, the Einstein field equations can be thought of as a spacetime generalization of the Poisson equation, as we will see. In a region of space where $\rho = 0$, i.e. a vacuum, the Poisson equation reduces to the Laplace equation $\nabla^2\Phi = 0$.

where $|h_{\mu\nu}| \ll 1$. As such, all higher order $O\left(h^2\right)$ terms appearing anywhere from now on will be even smaller (with quantities such as $h^{\mu\nu}h_{\mu\nu}$ and $h^{\mu\rho}h_{\rho\nu}$) and can be ignored. Orthogonality requires $g^{\mu\alpha}g_{\alpha\nu} = \delta^{\mu}_{\nu}$, which is satisfied by the inverse metric tensor:

$$g^{\mu\nu} = \eta^{\mu\nu} - h^{\mu\nu}, \tag{5.36}$$

as can be explicitly checked:

$$\begin{aligned} g^{\mu\alpha}g_{\alpha\nu} &= \left(\eta^{\mu\alpha} - h^{\mu\alpha}\right)\left(\eta_{\alpha\nu} + h_{\alpha\nu}\right) \\ &= \delta^{\mu}_{\nu} - h^{\mu\alpha}\eta_{\alpha\nu} + \eta^{\mu\alpha}h_{\alpha\nu} - h^{\mu\alpha}h_{\alpha\nu} \approx \delta^{\mu}_{\nu}. \end{aligned} \tag{5.37}$$

We also note that $|h_{\mu\nu}| \ll 1$ implies that indices are raised and lowered by the flat metric $\eta$. The full interval in Cartesian coordinates is then

$$ds^2 = \left(\eta_{\mu\nu} + h_{\mu\nu}\right)dx^{\mu}dx^{\nu} = -c^2\left(1 - h_{00}\right)dt^2 + \left(\delta_{ij} + h_{ij}\right)dx^i dx^j. \tag{5.38}$$

Since we are looking for a spherically symmetric solution to describe the situation in Fig. [5.6], we rewrite the metric in spherical coordinates by using the minimal substitution $\delta_{ij} \rightarrow g_{ij}$:

$$ds^2 = -c^2\left(1 - h_{00}\right)dt^2 + \left(g_{ij} + h_{ij}\right)dx^i dx^j, \tag{5.39}$$

where $g_{ij}$ is the metric tensor of spherical coordinates found from (1.27) and the coordinates $x^i$ are now the spherical coordinates. Spherical symmetry also implies a purely radial dependence of the unknown $h$ functions; i.e. $h_{00}\left(r\right)$ and $h_{ij} = f\left(r\right)g_{ij}$. Hence

$$ds^2 = -c^2\left(1 - h_{00}\right)dt^2 + \left(1 + f\right)\left(dr^2 + r^2 d\Omega^2\right), \tag{5.40}$$

where $d\Omega^2 = d\theta^2 + \sin^2\theta d\varphi^2$ is the metric of the unit sphere. Next, we construct the geodesic equations, keeping in mind that in addition to the assumption of weak gravity we should also consider slow speeds, since Newton's theory of gravity is not specially covariant. Hence if we choose the affine parameter $\lambda$ to be the proper time $\tau$, then all derivatives of $x^i$ with respect to $\tau$ appearing in the geodesic equations vanish approximately. The geodesic eqns (5.4) then reduce to

$$\frac{d^2x^{\mu}}{d\tau^2} + c^2\Gamma^{\mu}_{00}\left(\frac{dt}{d\tau}\right)^2 \approx 0. \tag{5.41}$$

The reader is encouraged to verify that

$$\Gamma^{\mu}_{00} \approx -\frac{1}{2}\left(\partial^{\mu}h_{00}\right). \tag{5.42}$$

The geodesic equation for $\mu = 0$ vanishes term by term, and only the three spatial equations survive

$$\ddot{x}^i = \frac{c^2}{2} \partial^i h_{00},$$ (5.43)

where we also set $\dot{t} = dt/d\tau \approx 1$, as one would not expect any time dilation effect for a weakly curved spacetime. In standard vector format this is

$$\ddot{\vec{r}} = \frac{c^2}{2} \vec{\nabla} h_{00}.$$ (5.44)

Comparing this result with (5.34), we conclude that[19]

$$h_{00} = -\frac{2}{c^2} \Phi = \frac{2GM}{c^2 r}$$

$$g_{00} = -1 + h_{00} = -1 - \frac{2}{c^2} \Phi = -1 + \frac{2GM}{c^2 r}.$$ (5.45)

In addition to figuring out what $g_{00}$ is in this case, note that this result implies a physical interpretation for the components of the metric tensor in general: It seems that $g_{\mu\nu}$ are functions that play the same role that the gravitational potential $\Phi$ played in the Newtonian theory. We will refer to this interpretation several times in subsequent calculations. Now it turns out that finding the remaining unknown function $f(r)$ requires using the Einstein field equations, which we will do later. It suffices here to just note that it gives $f(r) = h_{00}$. Hence the full Newtonian metric is

$$ds^2 = -c^2 \left(1 + \frac{2}{c^2}\Phi\right) dt^2 + \left(1 - \frac{2}{c^2}\Phi\right)\left(dr^2 + r^2 d\Omega^2\right)$$

$$= -c^2 \left(1 - \frac{2GM}{c^2 r}\right) dt^2 + \left(1 + \frac{2GM}{c^2 r}\right)\left(dr^2 + r^2 d\Omega^2\right).$$ (5.46)

**Exercise 5.7** Starting with (5.40), detail the calculation that leads to (5.43).

**Exercise 5.8** Check that dimensional analysis supports the interpretation that the components of the metric tensor play the role of gravitational potential. In other words, show that $g_{\mu\nu} \to c^2 \Phi$ is dimensionally consistent.

**Exercise 5.9**

1. Find the components of the inverse Newtonian metric tensor using the second form in (5.46) via (5.36). Verify orthogonality.
2. Calculate all the Christoffel symbols of the Newtonian metric using the second form in (5.46) and (4.33).

**Exercise 5.10**

1. Write down the divergence $\nabla_\mu A^\mu$ of an arbitrary 4-vector field using the Newtonian metric.
2. Write down the Laplacian $\nabla^\mu \nabla_\mu f(x^\alpha)$ of an arbitrary scalar function using the Newtonian metric.

---

[19] Up to an arbitrary constant which we can freely set to zero, i.e. assume the gravitational potential to vanish at radial infinity.

## 5.4   The Metric Outside a Spherical Mass

Next let's consider the spacetime surrounding a spherical mass $M$ *without* making any low-curvature or slow speed approximations. In other words, the trampoline is now allowed to curve as *much* as it wants. As such, we can no longer compare directly to the Newtonian case. We do expect, however, that the general case would reduce to the Newtonian metric (5.46) in the low-curvature limit. The metric in question was discovered by Karl Schwarzschild[20] shortly after Einstein published his field equations. He did so by assuming a perfectly general spherically symmetric form of the metric, plugging it into the field equations, and finding the forms of unknown functions. We will reproduce his calculation in Chapter 7. In spherical coordinate-like form, the **Schwarzschild metric** is

$$ds^2 = -c^2 \left( 1 - \frac{2GM}{c^2 r} \right) dt^2 + \frac{dr^2}{\left( 1 - \frac{2GM}{c^2 r} \right)} + r^2 d\Omega^2, \tag{5.47}$$

where, just as in the Newtonian case, $M$ is the total mass of the spherical object under study and $d\Omega^2 = d\theta^2 + \sin^2\theta d\varphi^2$ is the metric on a unit sphere.

Now let's stare at the Schwarzschild metric a bit and see if we can use all our knowledge of what a metric is to understand its physical meaning. First note that if the mass $M$ is removed, i.e. $M \to 0$, then (5.47) immediately becomes

$$ds^2 = -c^2 dt^2 + dr^2 + r^2 d\theta^2 + r^2 \sin^2\theta d\varphi^2, \tag{5.48}$$

which we recognize as the Minkowski metric in spherical coordinates (1.27). In other words, in the absence of mass the curvature vanishes and we retrieve the flat spacetime of special relativity, as one would expect; the curved trampoline of Fig. [5.1b] bounces back to the flat one of Fig. [5.1a]. Next, we also note that far away from the object, i.e. taking the limit $r \to \infty$, the Schwarzschild metric also reduces to (5.48). This is called **asymptotic flatness** and is certainly a desirable property, since the effects of curvature (a.k.a. gravity) should get weaker and weaker as one gets further and further away from the source and vanishes only at infinite distances.[21] This is depicted in Fig. [5.1b] as well. These properties are also shared with the Newtonian metric (5.46), as one would hope.

Speaking of the Newtonian metric, one should also expect that (5.47) yields the second form of (5.46) for low curvature. This can be shown to be true if we expand $\left( 1 - \frac{2GM}{c^2 r} \right)^{-1}$ binomially (4.136). Since we are now in a low-curvature (small $M$) regime we can drop the higher-order terms:

---

[20]German physicist and astronomer (1873–1916). Schwarzschild discovered his famous metric while serving in World War I. He died the following year.

[21]Notice we are assuming here that all of spacetime has no other matter or energy; it is a universe with a single non-rotating spherical mass sitting at the origin. This is a reasonable approximation if we are considerably far away from other massive objects.

$$\left(1 - \frac{2GM}{c^2 r}\right)^{-1} \approx 1 + \frac{2GM}{c^2 r}, \tag{5.49}$$

such that (5.47) becomes

$$ds^2 = -c^2 \left(1 - \frac{2GM}{c^2 r}\right) dt^2 + \left(1 + \frac{2GM}{c^2 r}\right) dr^2 + r^2 d\Omega^2. \tag{5.50}$$

This is not quite the same coordinate system used in (5.46). A coordinate redefinition of $r$ can be made such that the term $\left(1 + \frac{2GM}{c^2 r}\right)$ "appears" next to the $r^2 d\Omega^2$ term as well, finally giving the full Newtonian metric (5.46). This last is left as an exercise to the reader, noting that something very similar to it will be performed in §7.4.

**Exercise 5.11** Taking inspiration from the calculation that transforms the metric (7.41) to (7.46), show that a coordinate redefinition of $r$ transforms (5.50) to (5.46). This completes the proof that the Schwarzschild metric yields the Newtonian metric as a weak field limit.

It is common in the literature to use the shorthand

$$H(r) = \left(1 - \frac{2GM}{c^2 r}\right) \tag{5.51}$$

and rewrite (5.47) as

$$ds^2 = -c^2 H(r) \, dt^2 + H(r)^{-1} dr^2 + r^2 d\Omega^2. \tag{5.52}$$

For future reference, let's point out that the function $H(r)$ is **harmonic** in 3-space, meaning that it is a solution of Laplace's equation $\nabla^2 H(r) = 0$. This should come as no surprise, since we already know that $g_{00} \sim H(r)$ plays the role of the gravitational potential, which does satisfy the Laplace equation.[22] This can be easily seen as follows: Using (2.112) with $g = r^4 \sin^2 \theta$ being the determinant of the metric of spherical coordinates $\nabla^2 H(r) = 0$ gives

$$\partial_r \left[r^2 \left(\partial_r H\right)\right] = 0$$
$$r^2 \left(\partial_r H\right) = c_1$$
$$dH = c_1 \frac{dr}{r^2}$$
$$H(r) = c_2 - \frac{c_1}{r}, \tag{5.53}$$

where $c_1$ and $c_2$ are arbitrary constants of integration. Since we require the metric to be flat at radial infinity, we take $c_2 = 1$, and the constant $c_1$ is chosen to be $2GM/c^2$ in order to give the Newtonian potential.

---

[22]See footnote 18 in §5.3.

The Schwarzschild metric describes spacetime *outside* the surface of the sphere $M$; in other words, if the sphere (star/planet) has a specific given radius $a$, then the range of the radius $r$ in (5.47) is $a \leq r \leq \infty$. Inside the sphere there should be a different metric that depends on the distribution of matter content of the sphere; we will discuss one such metric in the next section. One would expect that both metrics *smoothly* connect at $r = a$. The situation is analogous to the Coulomb problem in electrostatics that the reader may be more familiar with from their study of the subject; namely the electrostatic potential of a uniform spherical charge distribution

$$V_{r \geq a} = \frac{Q}{4\pi\varepsilon_0 r}, \tag{5.54}$$

where $Q$ is the total charge of the sphere. The fact that all four problems (Newtonian metric, Schwarzschild metric, Newtonian gravity, and Coulomb electrostatics) are mathematically similar is no coincidence, but a direct consequence of the spherical symmetry common to all four. It can also be shown that the fact that all of these spherical potentials, including the metric components, are $1/r^n$ dependent, where $n = 1$, is related to the fact that we live in a three-dimensional space. Higher dimensions $d > 3$ would yield potentials with $n = d - 2$.

Other important similarities exist. For example, note what happens if one makes the charged or massive sphere shrink down to a point particle of zero size. All four potentials will have an infinite value at the point $r = 0$. In physics terminology, one says that we have a **singularity** at the origin. In the Newtonian/Coulomb cases it is a potential or field singularity. In the metric's case it is a **spacetime singularity**, where spacetime itself "rips" apart, to coin a phrase. As you can gather, singularities are not unusual in physics. Their appearance is usually an indication that we are missing something important. If our massive sphere is of finite radius $a \neq 0$, then we have no problem; we simply point out that the potential/metric works only outside the sphere, and the singularity is avoided. But if our sphere is a point particle $a = 0$, the appearance of the singularity there tells us that there is "missing" physics here. Such small scales, $r \to 0$, are the realm of quantum mechanics, and its rules must be brought into the mix. In the case of electricity and magnetism this has been done, and the behavior of $r \to 0$ singularities has been studied and dealt with.[23] In the gravitational case, however, a unification of quantum mechanics and spacetime is needed. Such a unified theory, usually dubbed the theory of "quantum gravity," is unfortunately still unknown,[24] so we don't have a picture, even an approximate one, as to what happens to spacetime on a zero radius point particle.

The reader may assume that a reasonable solution to the problem of $r = 0$ singularities is to simply give up the concept of massive point particles altogether. Maybe

---

[23]The quantum-mechanical theory of electricity and magnetism is known as Quantum Electrodynamics, or QED for short. Its development started in the 1920s and was completed (somewhat) in the 1960s. While it solves the $r \to 0$ singularity problem from a practical perspective, it is assumed to this day that the solution is incomplete and there is still deeper physics to be discovered.

[24]There are candidates, however, such as superstring theory and the theory of loop quantum gravity.

this is nature's way of telling us that it doesn't work with "points." Perhaps we should just postulate a new rule of nature: A massive sphere can have any radius except zero. This would certainly be a reasonable thing to consider, if it weren't for the fact that the Schwarzschild metric has another anomalous behavior that changes everything. Notice that the case $r \to 0$ is *not* the only problem with (5.47). There is a special distance from the origin, known as the **Schwarzschild radius** $r_s$, where something strange happens; this is

$$r_s = \frac{2GM}{c^2}. \tag{5.55}$$

If we substitute $r = r_s$ in (5.47), the time component of the metric vanishes and the radial component blows up to infinity! It's another singularity, but this time one that has no analogue in non-relativistic physics. Now if the radius of the spherical mass $M$ is greater than the Schwarzschild radius, i.e. $a > r_s$, then once again the problem goes away, since inside the sphere another metric (presumably a finite one) takes over. But what if our spherical mass happens to have a radius that is *less* than $r_s$? Not only is the singularity "exposed," but also note that for radii less than $r_s$ the timelike part of the metric becomes spacelike with a positive sign, while the radial part becomes timelike with a minus sign! What does all this mean? Notice that we are not even requiring $a$ to become zero. All it has to do is have a value that is less than $r_s$. This is where the problem suddenly becomes interesting. A spherically symmetric object with a radius that is less than $r_s$ is what is known as a **black hole**, specifically a **Schwarzschild black hole**, and the radius $r_s$ takes on an extremely interesting physical interpretation that we will discuss in a bit. For now let us just assume that $a > r_s$.

**Exercise 5.12** To demonstrate that not all singularities are what they seem, consider the metric

$$ds^2 = -\frac{c^2}{t^4} dt^2 + dx^2 + dy^2 + dz^2. \tag{5.56}$$

It appears to have a singularity at $t \to 0$. Demonstrate that this is not a true spacetime singularity by finding a coordinate transformation that gets rid of it. Once you have performed the transformation, you should get a (very) familiar metric.

**Exercise 5.13** The coordinate reparameterization

$$r \to r \left(1 + \frac{GM}{2c^2 r}\right)^2 \tag{5.57}$$

leads to the so-called "isotropic" form of the Schwarzschild metric:

$$.ds^2 = -c^2 A(r) dt^2 + B(r) \left(dr^2 + r^2 d\theta^2 + r^2 \sin^2 \theta d\varphi^2\right). \tag{5.58}$$

Find the unknown functions $A(r)$ and $B(r)$ in their simplest forms.

**Exercise 5.14** Consider the metric:

$$ds^2 = -c^2 dt^2 + \frac{4}{9} \left[ \frac{9GM}{2c^2 (r - ct)} \right]^{\frac{2}{3}} dr^2 + \left[ \frac{9GM}{2c^2} (r - ct)^2 \right]^{\frac{2}{3}} d\Omega^2. \tag{5.59}$$

It appears to describe a dynamic spacetime because of its dependence on $t$. Show that this is not true by finding a transformation that takes it to the static Schwarzschild solution (5.47).

## 5.5 The Metric Inside a Spherical Mass

If the Schwarzschild metric describes spacetime outside a spherical mass of radius $a > r_s$, what metric describes spacetime inside? This is dependent on the distribution of mass inside the sphere. The simplest case is a constant uniform mass distribution $M$, or in other words a sphere with density

$$\rho = \frac{M}{\frac{4}{3}\pi a^3} \equiv \text{constant}. \tag{5.60}$$

Schwarzschild himself found this metric; it is

$$ds^2 = -\frac{c^2}{4} \left( 3\sqrt{1 - \frac{2GM}{ac^2 r}} - \sqrt{1 - \frac{2GM}{a^3 c^2} r} \right)^2 dt^2 + \frac{dr^2}{\left(1 - \frac{2GM}{a^3 c^2} r\right)} + r^2 d\Omega^2. \tag{5.61}$$

This is sometimes referred to as the **interior Schwarzschild metric**, while in this context (5.47) is renamed the **exterior Schwarzschild metric**.[25] Now both solutions *must* be continuous and smooth at the surface of the sphere $r = a$. This is in analogy with the Newtonian potential of a uniform and constant spherical mass distribution. In introductory physics class we learn that the potential for the inside of such a sphere is

$$\Phi = -\frac{GM}{2a} \left( 3 - \frac{r^2}{a^2} \right), \tag{5.62}$$

which continuously connects to the outside solution (5.30) via the **continuity condition**:

$$\Phi_{r \geq a}|_{r=a} = \Phi_{r \leq a}|_{r=a}. \tag{5.63}$$

It also *smoothly* connects to (5.30) at $r = a$, meaning that there are no "kinks" or sharp turns in the potential at the surface of the sphere. This is satisfied by the **smoothness condition**:

---

[25]Be aware, though, that in the rest of this book, as well as every *other* reference, when one says the "Schwarzschild metric," or the "Schwarzschild solution," one *always* means the exterior one. The term also refers to the entire scope of $r > 0$ for the black hole case.

$$\frac{d}{dr}\Phi_{r\geq a}\bigg|_{r=a} = \frac{d}{dr}\Phi_{r\leq a}\bigg|_{r=a}.$$  (5.64)

If these conditions are not satisfied, then we will have a problem defining the gravitational field at the surface $r = a$. Continuity guarantees that the gravitational potential is single valued at $r = a$, while smoothness guarantees that the gravitational field $\vec{\mathfrak{g}}$ is single valued at the boundary. The same exact argument is made for Coulomb electrostatics in the similar problem of finding the potential inside and outside a uniform distribution of electric charge.

**Exercise 5.15** Study the Newtonian case of the gravitational potential inside a uniform distribution of mass $M$:

1. Derive (5.62). The easiest way to do this is by applying the gravitational version of Gauss' law, which states that the flux of the gravitational field $\vec{\mathfrak{g}}$ on a closed Gaussian surface $S$ is proportional to the mass content of the volume $V$ surrounded by $S$:

$$\int_S \vec{\mathfrak{g}} \cdot d\mathbf{a} = -4\pi G \int_V \rho dV.$$  (5.65)

If it is not immediately obvious what you need to do, I suggest reviewing the topic of Gauss' law for electric fields. A similar problem to ours is in every introductory physics textbook (such as [1]), except applied on a uniform distribution of electric charge.

2. Explicitly verify that the solution (5.62) is continuously and smoothly connected to (5.30).
3. Demonstrate that (5.62) is the Newtonian limit of Schwarzschild's interior solution by following an argument similar to the one in §5.3.

Since we have already deduced that the components of any metric tensor play the role of the potential in general relativity, we expect the inside and outside metrics of our massive sphere to satisfy both continuity *and* smoothness. Continuity requires

$$g_{\mu\nu}^{\text{Interior}}\bigg|_{r=a} = g_{\mu\nu}^{\text{Exterior}}\bigg|_{r=a}.$$  (5.66)

If we rewrite both the interior and exterior solutions in terms of the Schwarzschild radius $r_s$

$$ds_{\text{Interior}}^2 = -\frac{c^2}{4}\left(3\sqrt{1-\frac{r_s}{a}} - \sqrt{1-\frac{r_s}{a^3}r^2}\right)^2 dt^2 + \frac{dr^2}{\left(1-\frac{r_s}{a^3}r^2\right)} + r^2 d\Omega^2$$

$$ds_{\text{Exterior}}^2 = -c^2\left(1-\frac{r_s}{r}\right)dt^2 + \frac{dr^2}{\left(1-\frac{r_s}{r}\right)} + r^2 d\Omega^2,$$  (5.67)

and set $r = a$, then (5.66) gives

$$g_{tt}^{\text{Interior}}\Big|_{r=a} = g_{tt}^{\text{Exterior}}\Big|_{r=a} = -\left(1 - \frac{r_s}{a}\right)$$

$$g_{rr}^{\text{Interior}}\Big|_{r=a} = g_{rr}^{\text{Exterior}}\Big|_{r=a} = \frac{1}{\left(1 - \frac{r_s}{a}\right)}$$

$$g_{\theta\theta}^{\text{Interior}}\Big|_{r=a} = g_{\theta\theta}^{\text{Exterior}}\Big|_{r=a} = a^2$$

$$g_{\varphi\varphi}^{\text{Interior}}\Big|_{r=a} = g_{\varphi\varphi}^{\text{Exterior}}\Big|_{r=a} = a^2 \sin\theta. \tag{5.68}$$

The smoothness condition is satisfied if the derivatives of the metric components match at the boundary, i.e.

$$\frac{d}{dr} g_{\mu\nu}^{\text{Interior}}\Big|_{r=a} = \frac{d}{dr} g_{\mu\nu}^{\text{Exterior}}\Big|_{r=a}, \tag{5.69}$$

which the reader can check is indeed satisfied for our case. The complete spacetime describing this uniform sphere inside and out is shown schematically in Fig. [5.7].

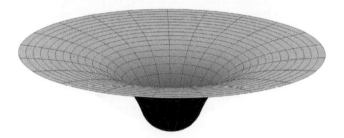

Fig. 5.7: Embedding diagram of the spacetime inside and outside a uniform spherical mass. The interior and exterior Schwarzschild metrics match smoothly at the boundary of the sphere $r = a$, provided that $a > r_s$. For this plot, we set $r_s = 3$ and $a = 4$.

**Exercise 5.16** Explicitly verify (5.69).

**Exercise 5.17**

1. Find the inverse metric tensor $g^{\mu\nu}$ for the interior solution then verify orthogonality: $g^{\mu\rho} g_{\rho\nu} = \delta^\mu_\nu$.
2. Find the Christoffel symbols for the interior solution.
3. Write down the divergence $\nabla_\mu A^\mu$ of an arbitrary 4-vector field using the interior solution metric.
4. Write down the Laplacian $\nabla^\mu \nabla_\mu f(x^\alpha)$ of an arbitrary scalar function using the interior solution metric.

## 5.6    Areas and Volumes in Curved Spaces

Now that we have seen some examples of curved spacetimes, let's make note of something the astute reader may have inferred already: This is the fact that the usual Euclidean *definitions* of "length," "area," and "volume" are *not* quite satisfied in a curved space. To see what we mean by this, consider an ordinary two-dimensional plane and draw on it a square of sides $b$. The area of the square is just $A_E = b^2$ ($E$ for Euclidean). Now consider the two-dimensional surface of a sphere of radius $R$, and draw on *that* a square of sides $b$, as shown in Fig. [5.8].

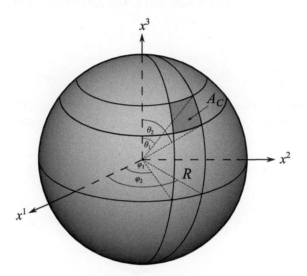

Fig. 5.8: Area of a "square" on a sphere of radius $R$. Note that the shape of the area changes as one changes the angle $\theta$. So much so that it becomes a triangle if $\theta_1 = 0$.

Now the area of the "square" on the sphere is clearly different from that on the plane; it is no longer found using the Euclidean formula. To construct a formula for the area of a square on the sphere, let's begin with the two-dimensional element of area in Cartesian coordinates

$$da = dxdy. \tag{5.70}$$

The metric on the sphere is

$$dl^2 = R^2 \left(d\theta^2 + \sin^2\theta d\varphi^2\right). \tag{5.71}$$

Using (2.126), (2.136), and $\sqrt{|g|} = R^2 \sin\theta$, we have

$$da = R^2 \sin\theta d\theta d\phi. \tag{5.72}$$

The area of a "square" on the sphere defined between angles $(\theta_1, \theta_2)$ and $(\varphi_1, \varphi_2)$ is then

$$A_C = R^2 \int_{\varphi_1}^{\varphi_2} \int_{\theta_1}^{\theta_2} \sin\theta d\theta d\varphi = R^2 \int_{\varphi_1}^{\varphi_2} d\varphi \int_{\theta_1}^{\theta_2} \sin\theta d\theta = R^2 (\varphi_2 - \varphi_1)(\cos\theta_1 - \cos\theta_2),$$

(5.73)

where $C$ stands for "Curved." Note that this is not even a square anymore, particularly if the integration is done starting with $\theta_1 = 0$, in which case the area is that of a (spherical) triangle. In order to make it as close to a square as possible, let's assume the area in question is on the equator, such that it has equal top and bottom sides, as shown in Fig. [5.9]. Each side would then be equal to $R$ times the difference of angle such that[26]

$$R(\theta_2 - \theta_1) = R(\varphi_2 - \varphi_1) = b.$$

(5.74)

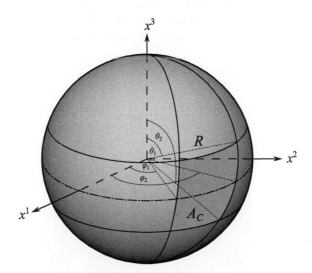

Fig. 5.9: Area of a "square" on a sphere of radius $R$. In this case we chose to move the square around the equator to make sure its top side is equal to its bottom side.

Our interest is to just demonstrate that the area defined by (5.73) is necessarily different from a similar area in the plane, and I think this is now obvious. However, just to finish the calculation let's put in some numbers. Choose $b = 1\text{m}$ such that the area in the plane is just $A_E = 1^2 = 1\text{m}^2$. If we choose the radius of the sphere to be $R = 1\text{m}$ as well, then from (5.74):

$$(\theta_2 - \theta_1) = (\varphi_2 - \varphi_1) = 1.$$

(5.75)

---

[26] Recall from elementary geometry that the length of an arc of a circle radius $R$ subtending an angle $\alpha$ (in radians) is just $R\alpha$.

Using our choice for the location of the "square," it is clear that $\theta_1 = \frac{1}{2}(\pi - 1)$ and $\theta_2 = \frac{1}{2}(\pi + 1)$. If we choose $\varphi = 0$, then $\varphi_2 = 1$; all in radians. Hence the area of this spherical "square" is

$$A_C = 1^2 (1 - 0) \left[ \cos\left( \frac{\pi}{2} - \frac{1}{2} \right) - \cos\left( \frac{\pi}{2} + \frac{1}{2} \right) \right] = 0.958851 < A_E. \qquad (5.76)$$

It is interesting that even the *definition* of what a "square" is has caused us some headache. One then must tread carefully when questions like these arise in a curved space. We do have all the tools for it, though; in fact, most of what one needs to do similar calculations was discussed as early as §2.6. A simpler problem than the "square" would be calculating the surface area of a circle of latitude $\theta$ on the sphere. Notice that we do not mean the surface area of the *disk* surrounded by the latitude circle. What we mean is the area *on* the sphere *bounded* by the circle of latitude, or in other words the area of the "cap" surrounded by the circle, as shown in Fig. [5.10]. This is easier than the square problem because a circle has a high degree of symmetry. A circle on a sphere will *remain* a circle no matter where you place it. Not so with the square, as we saw.

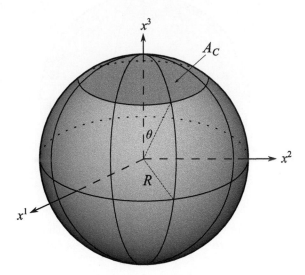

Fig. 5.10: Calculating the area bounded by a circle *on* a sphere vs the area of the *disk* bounded by the same circle.

**Exercise 5.18** On a sphere of radius $R$, find the area $A_C$ of the "circle" depicted in Fig. [5.10]; more precisely, find the area of the cap between $\theta = 0$ to the circumference of the shown circle. The coordinate distance, i.e. the "actual" length on the sphere, is certainly larger than the "straight radius" $R \sin\theta$. Hence the area will be larger than the naive Euclidean formula $\pi r^2 = \pi (R \sin\theta)^2$.

All of this is not something that is just of theoretical interest. Consider the massive sphere of radius $a$ in the previous section, and recall that it can be used to approximate *real* planets and stars. Suppose that for astrophysical purposes we wish to calculate the volume of such an object. As we will see, the answer is significantly larger than the ordinary Euclidean formula for the volume of a sphere:

$$V_E = \frac{4}{3}\pi a^3. \tag{5.77}$$

An astrophysicist interested in, say, estimating the life of a star would need to know how much nuclear fuel the star contains. Which means she will need to know the "real" volume of said star. Hence the importance of understanding such relationships. Once again, recall from §2.6 that the volume element in arbitrary, but spherically symmetric, coordinates is

$$dV = \sqrt{|g_{\text{space}}|}\,dr d\theta d\phi, \tag{5.78}$$

where $g_{\text{space}}$ is the determinant of the spatial part of the metric only, since we are not integrating over time. The volume of a sphere of radius $a$ and concentric with the origin is then

$$V_C = \int_0^{2\pi}\int_0^{\pi}\int_0^{a} \sqrt{|g_{\text{space}}|}\,dr d\theta d\varphi. \tag{5.79}$$

For the Schwarzschild massive sphere $M$ with radius $a$ we must use the interior Schwarzschild metric (5.61), for which

$$\sqrt{|g_{\text{space}}|} = \frac{r^2 \sin\theta}{\sqrt{1 - \frac{r_s}{a^3}r^2}}. \tag{5.80}$$

Hence

$$V_C = \int_0^{2\pi} d\varphi \int_0^{\pi} \sin\theta d\theta \int_0^a \frac{r^2}{\sqrt{1 - \frac{r_s}{a^3}r^2}}dr = 4\pi \int_0^a \frac{r^2 dr}{\sqrt{1 - \frac{r_s}{a^3}r^2}}$$
$$= 2\pi \left[ \frac{a^{\frac{9}{2}}}{r_s^{\frac{3}{2}}} \sin^{-1}\left( \sqrt{\frac{r_s}{a}} \right) - \frac{a^4}{r_s}\sqrt{1 - \frac{r_s}{a}} \right], \tag{5.81}$$

which is most certainly a far cry from (5.77). This can be shown to be larger than $V_E$. We can get a feel of how much larger by plugging in some numbers, say $r_s = 1$, $a = 2$ (recall that $a$ must be greater than $r_s$, otherwise we fall into a black hole, *literally*), then we get

$$V_E = 33.5103$$
$$V_C = 40.5757. \tag{5.82}$$

Clearly the real volume of this sphere (or star) is close to 25 percent larger than our naive Euclidean estimate. That is a significant difference and should be taken into account in any real-life astrophysical calculations. But remember that this is only an estimate that assumes a uniform static sphere. You can find better and better approximations by considering other metrics that take into consideration the spin of the star, as well as any variation in its mass density.

**Exercise 5.19** For a sphere of uniform mass $M$ and radius $a$ whose interior is described by (5.61), find the true distance between $r = 0$, and $r = a$.

**Exercise 5.20** For the spacetime described by the metric

$$ds^2 = \left(1 - br^2\right)^2 \left(-c^2 dt^2 + dr^2\right) + r^2 d\Omega^2, \qquad (5.83)$$

where $b$ is some arbitrary positive constant, find:

1. The true distance from $r = 0$ to $r = a$.
2. The surface area of a sphere of coordinate radius $r = a$.
3. The volume of a sphere of coordinate radius $r = a$.

## 5.7 Non-Rotating Black Holes

A scenario where a spherical mass $M$ has a radius that is less than the Schwarzschild radius is not impossible. In fact, it has long been known that such objects exist in our universe and that the most common way they form is by the collapse of stars. The idea goes like this: Imagine a non-rotating star with a given mass $M$ (once again, yes—this is an idealization since stars do rotate; there is another metric (8.169) that describes the more realistic rotating case). Outside the star the spacetime background is described by (5.47), while inside it is described by something else. The radius of the star is initially much larger than $r_s$, and there are no singularities of any kind anywhere. Now the star is so massive that its constituent particles are pulling on each other gravitationally, tending the mass of the star to collapse inward on itself. However, the star remains stable for billions of years because the pressure of its nuclear reactions pushes outward countering the inward gravitational pull exactly. One day, however, the star runs out of nuclear "fuel" and the furnace no longer burns. This is the moment when the gravitational pull takes over and catastrophically collapses the star inward. If the pull is of such intensity that the star's size collapses to *less* than the Schwarzschild radius, then the star becomes a black hole. As a side note, let's point out that a black hole is not the only possible end state of a star. If the collapse is not strong enough to pull the star inside the Schwarzschild radius, other objects can form: either **white dwarves** or **neutron stars**. While interesting from an astrophysical perspective, these are of less interest to us, since their radii are still greater than $r_s$ and no singularities exist. Our own star, the Sun, is expected to form a white dwarf at the end of its life, because it is not massive enough to collapse to a black hole. It has been calculated that a star that is less than about 1.4 times the mass of our Sun will become a white dwarf at the end of its life. This value is known as the **Chandrasekhar limit**.[27] On the other hand, any star that is larger than 1.4 but less than about 2.17 solar masses would most likely become a neutron star. The "2.17" figure is known as the **Tolman–Oppenheimer–Volkoff limit**.[28]

Now if a given star collapses strongly enough, not only does its radius drop below $r_s$, exposing the singularity at that distance, but the constituent mass of the star continues to pull in on itself, having no outward force to balance it anymore until it becomes almost zero in size. In other words, the entire mass of the star collapses into a point located at the origin of coordinates $r = 0$, reaching the *other* singularity there. Thus, a Schwarzschild black hole has two singularities: one at $r = r_s$ and the other at $r = 0$. The mathematics used in these calculations made it clear that a spherically symmetric object, whose spacetime is described by the Schwarzschild solution and whose mass is greater than the Tolman–Oppenheimer–Volkoff limit, must necessarily collapse into a singularity and consequently form a black hole. But stars, or any massive object such as a sufficiently large cloud of dust, do not usually satisfy perfect spherical symmetry. The question, then, was whether black hole solutions are simply artifacts of the symmetry assumed and are not realistic objects that can form in nature.[29] This issue was settled in 1965 by Sir Roger Penrose[30] who mathematically showed that any collapsing object with sufficient mass would *necessarily* form a black hole, irrespective of its initial symmetry. He did this by showing that the geodesics that individual particles in the object follow during collapse would always end on a single point; the singularity. For his work, later referred to as a **singularity theorem**, Penrose was awarded half the 2020 Nobel Prize in Physics. The Nobel committee described his achievement as "the discovery that black hole formation is a robust prediction of the general theory of relativity."

We have previously discussed the implications of the $r = 0$ singularity. To understand what really happens there, we need a quantum-mechanical theory of spacetime that we do not yet have at our disposal. So as far as general relativity is concerned, the entire mass of the original star is now cramped in an infinitely small volume that consequently has an infinitely high density. What about the singularity at the Schwarzschild radius? This one turns out to be not a true spacetime singularity; in other words, nothing really blows up there and no true spacetime infinities appear. The fact that something strange does happen to (5.47) at $r = r_s$ is due to the collapse of the coordinate system *itself* at this distance. To understand this, recall our very early discussions of coordinate systems in Chapter 1. We noted that sometimes coordi-

---

[27]Named after its discoverer the American-Indian astrophysicist Subrahmanyan Chandrasekhar (1910–1995).

[28]Richard Chace Tolman: an American mathematical physicist and physical chemist (1881–1948); Julius Robert Oppenheimer: an American theoretical physicist (1904–1967); and George Michael Volkoff: a Russian-Canadian physicist (1914–2000).

[29]Even if one considers that other black hole solutions, like (8.169), are not spherically symmetric, they still have other symmetries that beg the same question.

[30]English mathematical physicist (b. 1931).

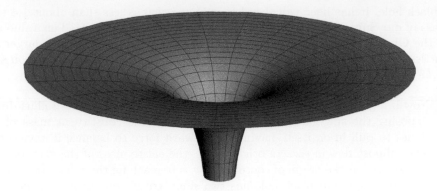

Fig. 5.11: Embedding diagram of a Schwarzschild black hole. The "throat" extends indefinitely.

nate systems do not cover all of space. In polar coordinates in the plane, for example, the origin $O$ cannot be described by a unique set of numbers $(\rho, \varphi)$. We pointed out that this is an inherent defect in the polar system of coordinates, and if one needs to describe $O$, then the only way is either to use two coordinate systems or to switch to another coordinate system that would cover $O$, such as Cartesian coordinates. The problem at the Schwarzschild radius is exactly of that sort. The sphere described by $r = r_s$ is not a true spacetime singularity like $r = 0$; rather it is what is known as a **coordinate singularity**, a defect in the coordinate system that can be remedied by choosing a different coordinate system. Notice, however, that this does *not* mean that nothing of importance happens at the sphere $r = r_s$. Something does indeed happen there. This distance happens to be the point of no return. You have heard that nothing can escape a black hole; well $r = r_s$ is the sphere where nothing inside of it can ever escape to the outside, as we will show. It is called the **event horizon**.

To prove the claim that $r = r_s$ is not a real physical singularity, the only thing we need to do is find at least one other system of coordinates that covers all of spacetime (including $r = r_s$ but excluding the real spacetime singularity at $r = 0$). One such possibility is known as the **Lemaître coordinate system**,[31] defined by the set $\tau$, $\rho$, $\theta$, and $\varphi$ as follows:

$$cd\tau = cdt + \sqrt{\frac{r_s}{r}}\frac{dr}{H} \tag{5.84}$$

$$d\rho = cdt + \sqrt{\frac{r}{r_s}}\frac{dr}{H} \tag{5.85}$$

$$\theta = \theta, \quad \varphi = \varphi. \tag{5.86}$$

This transforms (5.47) to the **Lemaître metric**:

---

[31] Georges Lemaître: Belgian priest, astronomer and physicist (1894–1966).

$$ds^2 = -c^2 d\tau^2 + \frac{r_s}{r} d\rho^2 + r^2 d\Omega^2, \quad \text{where} \quad r = \left[ \frac{3}{2} \sqrt{r_s} \left( \rho - c\tau \right) \right]^{\frac{2}{3}}. \tag{5.87}$$

The metric (5.87) is "designed" such that particle trajectories with constant $\rho$ are radial geodesics, with $\tau$ being the proper time along them. Clearly in this metric nothing special happens at $r = r_s$, which corresponds to $r_s = \frac{2}{3}(\rho - c\tau)$, meaning that the "grid" described by (5.87) covers all of space (minus $r = 0$), in contrast with Schwarzschild's original coordinates.

**Exercise 5.21** Derive the Lemaître metric using the transformations (5.84), (5.85), and (5.86).

Another popular system of coordinates goes by the title the **Kruskal–Szekeres coordinate system**.[32] They are defined by the parameters $(T, R, \theta, \varphi)$ related to the standard Schwarzschild coordinates as follows:

$$R = e^{\frac{c^2 r}{4GM}} \sqrt{\frac{c^2 r}{2GM} - 1} \cosh\left( \frac{c^3 t}{4GM} \right)$$

$$T = e^{\frac{c^2 r}{4GM}} \sqrt{\frac{c^2 r}{2GM} - 1} \sinh\left( \frac{c^3 t}{4GM} \right)$$

$$\theta = \theta, \varphi = \varphi, \tag{5.88}$$

where we recall that the hyperbolic sine "sinh" and the hyperbolic cosine "cosh" are defined by (4.72). This leads to the **Kruskal–Szekeres metric**

$$ds^2 = \frac{32 G^3 M^3}{c^6 r} e^{-\frac{c^2 r}{2GM}} \left( -dT^2 + dR^2 \right) + r^2 d\Omega^2, \tag{5.89}$$

or, in terms of $r_s$,

$$ds^2 = \frac{2 r_s^3}{r} e^{-\frac{r}{r_s}} \left( -dT^2 + dR^2 \right) + r^2 d\Omega^2. \tag{5.90}$$

Also, we have not quite removed all of the old coordinates, as $r$ still appears in (5.90). It is, however, related to the new coordinates $T$ and $R$ as follows:[33]

---

[32] Found by the American physicist Martin David Kruskal (1925–2006) and the Hungarian-Australian mathematician George Szekeres (1911–2005).

[33] It is possible to invert (5.92) to $r(T, R)$ as follows:

$$r(T, R) = r_s \left[ 1 + W \left( \frac{-T^2 + R^2}{e} \right) \right], \tag{5.91}$$

where the so-called Lambert $W$ function is defined as the solution of

$$z = W(z) e^{W(z)}.$$

$$-T^2 + R^2 = \left(\frac{r}{r_s}\right) e^{\frac{r}{r_s}}. \tag{5.92}$$

Clearly, just like in (5.87), if we substitute $r = r_s$ in (5.90), no anomalous behavior happens, while $r = 0$ still blows up—once again confirming that the Schwarzschild radius is a coordinate singularity: a defect of the original coordinates, and not a true physical spacetime singularity like the one at $r = 0$. We will develop a more rigorous method of checking singularities and differentiating between true singularities and coordinate ones in Chapter 7.

**Exercise 5.22**  Derive the Kruskal–Szekeres metric using the transformations (5.88).

**Exercise 5.23**

1. Using the Lemaître relations (5.84), (5.85), and (5.86), find the transformation tensor $\lambda_\nu^{\mu'}$ that transforms (5.87) to the original Schwarzschild metric (5.52).
2. Find the inverse tensor $\lambda_{\nu'}^{\mu}$. You may wish to review your work in Exercise 2.6.

If you found this difficult, it is even more tedious to find $\lambda_\nu^{\mu'}$ for the transformation between (5.89) and (5.47). You may still attempt to do so and see how far you get.

We still have not answered the question of what physical phenomenon the so-called event horizon represents? The claim is that it is the surface of no return. Anything falling into it can never escape. In terms of the old Newtonian picture, we know that to escape the gravitational pull of any object, one has to achieve escape velocity, whose value depends on the size and the mass of the object. From this point of view the escape velocity from the surface, or inside, of an event horizon is *greater* than the speed of light, and since we know that nothing can travel faster than light, nothing can escape the event horizon. We will continue this discussion in the next chapter.

Although literally black and impossible to see optically, there are several observational techniques for confirming the existence of black holes. The most common such method is based on observing the speeds of stars that happen to be in orbit around the black hole. The faster they are, the more massive the black hole is. Via this indirect technique, it is now well known that a supermassive black hole exists at the center of every observable galaxy, including our own Milky Way. This method was pioneered by two scientific groups in the early years of the twenty-first century; one based in the United States and the other in Germany. The leaders of these groups, Andrea M. Ghez[34] and Reinhard Genzel,[35] shared half the 2020 Nobel Prize in Physics for this achievement. The other half was awarded to Sir Roger Penrose for his 1965 singularity theorem as discussed earlier.

[34] American astrophysicist (b. 1965).
[35] German astrophysicist (b. 1952).

A more direct method of observation is based on detecting the electromagnetic radiation, mostly radio waves, emitted from a large amount of matter falling *into* a given black hole. Whether this is from a large cloud of debris or a star that has wandered too close to the black hole, usually this matter gets trapped rotating about the black hole in the form of the so-called **accretion disk** as it spirals toward the event horizon. Friction in the disk (just high-speed particles rubbing against each other) generates energy strong enough to be detected far away from the event itself. The detection of these energies can be converted into what amounts to a photographic image of the accretion disk, with the black hole's event horizon showing as a dark circle in the center. An international collaboration of scientists was put together in the second decade of the twenty-first century to achieve exactly such a feat: to take "photographs" of black holes. The project was collectively named the "Event Horizon Telescope" collaboration. During the writing of this book, specifically on April 10, 2019, the collaboration announced the release of the first photograph of a supermassive back hole at the center of a distant galaxy by the name of M87: Fig. [5.12].

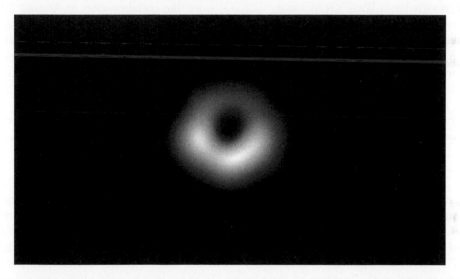

Fig. 5.12: The first ever "photograph" of a black hole, or, more precisely, the accretion disk surrounding the event horizon (dark circle) of a black hole. To understand the scale of this supermassive black hole at the center of the M87 galaxy, suffice it to note that the width of the event horizon is approximately 1.5 light days. This is the distance light covers in an Earth day and a half, approximately equal to 23.6 billion miles; more than *ten* times the diameter of Neptune's orbit around the Sun! This black hole's mass is equivalent to about 6.5 *billion* Suns (our Sun has a mass of roughly $1.989 \times 10^{30}$kg)! By all standards this is a gigantic black hole. Image credit: "Event Horizon Telescope Collaboration" https://eventhorizontelescope.org.

At the time of writing, plans are underway to take photographs of other supermassive black holes, such as the one at the center of our own Milky Way galaxy, known as "Sagittarius A*" (pronounced "Sagittarius A-Star" and abbreviated "Sgr A*"). The discovery that supermassive black holes exist at the centers of galaxies came as a surprise to the astrophysical community. It cannot be a coincidence, and there are studies that seem to indicate that the presence of such objects has in fact contributed to the early formation of galaxies. A growing number of astrophysicists believe that black holes have played, and continue to play, an important role in the evolution of the universe as we observe it, rather than being rare accidental objects.

A third method of black hole detection is based on observing them gravitationally. It has long been known that general relativity predicts the existence of **gravitational waves**.[36] These are "waves" of "contracting" and "expanding" spacetime that emanate from catastrophic events. In theory, any accelerating mass generates gravitational waves; however, they are so weak that it takes events of a massive magnitude to generate detectable ones, and even then they are still so weak that they require very precise measurement techniques. About a century after they were first predicted by Einstein himself, gravitational waves were finally detected by the Laser Interferometer Gravitational-Wave Observatory (LIGO) scientific collaboration in August 2017; the feat has been replicated several times since then. That first detection was of such a nature that the only event that could explain it was the collision of two black holes in the distant past. This discovery confirmed the existence of gravitational waves and provided further confirmation for the existence of black holes. It is thought that with the development of these gravitational wave detectors we are now at the beginning of a new era of observational astronomy, one that is comparable to the era that followed the invention of the telescope during Galileo's time.[37] Astronomy and cosmology were never the same after that, and one would think that they will never be the same after LIGO.

One last comment before we temporarily leave this topic: Contrary to popular belief, black holes do not "suck." They are *not* some cosmic vacuum cleaner that would uncontrollably pull you toward them. This should be obvious from our discussion of the Schwarzschild metric. Assume, for example, that a massive star with some number of planets in orbit around it suddenly collapses into a black hole of the same mass. The fact that its radius has dropped below the Schwarzschild radius does *not* change the metric at the location of the planets. In other words, said planets would continue in their orbits unhindered by the sudden collapse of their star. Realistically, such a collapse would be a catastrophic supernova explosion that would probably wipe out the star's entire planetary system, but assuming a scenario where this doesn't happen, the planetary orbits would not be affected. In other words, if one begins with a spherical

---

[36] A solution of the theory that we will not discuss in this book. However, see Exercise 7.28.

[37] It is interesting that there is no consensus on who in fact first invented the optical telescope. The first person to apply for a patent for it was a Dutch eyeglass maker named Hans Lipperhey, who did so in 1608, but there are many historical indications that using two lenses to enlarge distant images was known before Lipperhey, whether by Europeans or other civilizations such as the Arabs, the Indians, or the Chinese. It is interesting to note, however, that Galileo himself, having heard of the telescope and without having seen one, immediately understood the concept and not only built his own, but also improved on the original design. It is accepted today that Galileo was the first to realize the telescope's importance for astronomy, and was the first to direct it toward the heavens. The rest, of course, is history.

Schwarzschild object of radius $a > r_s$ and smoothly transforms it to $a < r_s$, effectively generating a black hole, any object whose location was originally at any distance $r > a$ will feel no change as far as gravitational pull is concerned.


## 5.8  Cosmological Spacetimes

Our next example is a spacetime metric that describes the universe as a whole.[38] To construct such a metric, we rely on the observation that the entire universe, as observed on a very large scale, seems to be homogeneous and isotropic. "Homogeneity" means that physical conditions are the same at every point in space, while "isotropy" means that these conditions are the same in every *direction* in space. Another way of saying this is that the universe has no special points, for instance no center, as well as no preferred directions. To understand what this means, consider the fact that our solar system is most certainly not homogeneous or isotropic. Conditions are not the same everywhere; for example, the mass distribution is not uniform, and neither is the local temperature. One can also define a preferred direction to our solar system; namely the plane in which the planets orbit (the **ecliptic plane**, as it is called), or more precisely the axis perpendicular to that plane around which the planets orbit. Expanding further, we note that even an entire galaxy is neither homogeneous nor isotropic. Local clusters of galaxies aren't either. However, once we get to distance scales roughly larger than 100Mpc,[39] the mass and energy distribution becomes approximately uniform in such a way that homogeneity and isotropy become a very reasonable approximation. In fact, the entire universe starts behaving like a perfect fluid with uniform density; each "molecule" in such a fluid is an *entire* cluster of galaxies. We will say more about this perfect fluid approximation when we return to the subject in Chapter 7. Now, the universe continues to be homogeneous and isotropic until we reach ultra-large distances: roughly larger than 3,000Mpc. At these scales some inhomogeneous structures appear in the way clusters of galaxies are distributed. This so-called **large scale structure of the universe** is assumed to have arisen from inhomogeneities that formed very early on in the history of the universe. Despite the presence of these structures, one can still use a homogeneous and isotropic metric to approximate the behavior of the universe, even at such large scales.

Returning to our topic, mathematicians tell us that there are exactly three types of spaces that are simultaneously homogeneous and isotropic: flat space, hyperspheres of constant positive curvature, and hyperbolic spaces of constant negative curvature. Constant curvature means that the way the surface curves is exactly the same at every point in such a space. For example, in two dimensions a flat space is just an infinite plane, as in Fig. [5.1a]. Its curvature is zero everywhere (homogeneity), and

---

[38]Some of the discussion in this section is based on arguments in [11].

[39]**Mega parsecs**: a unit of distance commonly used in cosmology. A parsec is equivalent to 3.26 light years (ly), where a light year (the distance light travels in a year) is about 5.79 trillion miles. Hence one Mpc is about $1.917 \times 10^{19}$ miles.

there is obviously no preferred direction in which to define an axis (isotropy). A two-dimensional surface of constant positive curvature is simply the surface of an ordinary sphere (see Fig. [3.2]) where one may define any point, for instance the north pole, anywhere, arbitrarily (homogeneity). Similarly, any great circle may be taken as an axis, say a zero meridian (isotropy). Finally, a two-dimensional surface of constant negative curvature is the so-called hyperbolic paraboloid; Fig. [5.13]. Homogeneity and isotropy may not be as immediately obvious for that one, but they are satisfied nevertheless. One may further describe the notion of constant curvature by considering parallel lines: in flat space, two lines that are initially parallel to each other will continue indefinitely without ever meeting. On a sphere, however, lines that are initially parallel will eventually meet if extended long enough, while on a hyperbolic paraboloid they will continue to move away from each other indefinitely.

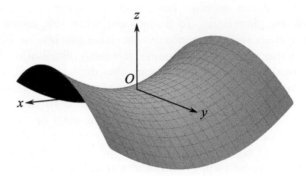

Fig. 5.13: The hyperbolic paraboloid is an infinite surface with constant negative curvature. Shown here is just the part of it centered around the origin. All negative curvature surfaces have the so-called "saddle points," named so for the obvious reason.

Our interest is in three-dimensional "surfaces" that may describe the homogeneous and isotropic space of our universe. In this case a flat "surface" is just the three-dimensional space of Euclidean geometry, while three-dimensional hyperspheres or hyperbolic hypersurfaces no longer lend themselves to visualization and have to be dealt with mathematically. One way of constructing metrics for such spaces is to begin by finding two-dimensional metrics for the two-dimensional versions of these spaces, then generalizing to the required three dimensions. For example, a two-dimensional plane can be described by the Cartesian metric $dl^2 = dx^2 + dy^2$, and one easily just adds a $dz^2$ to generalize to a three-dimensional flat space. If we choose to start with polar coordinates, the metric is $dl^2 = d\rho^2 + \rho^2 d\varphi^2$; this can be generalized to three-dimensional flat space by the substitution $d\varphi^2 \to d\Omega^2$, leading to the three-dimensional Euclidean spherical coordinate metric (1.27). It is possible to do this sort of calculation for all three surfaces simultaneously. Begin with the equation of a two-dimensional sphere with radius $a$:

$$x^2 + y^2 + z^2 = a^2. \tag{5.93}$$

This equation describes the sphere, a *two* dimensional surface, using the *three* coordinates $x$, $y$, $z$. In other words, this is a sphere *embedded* in three-dimensional space. If one wishes to describe the surface of the sphere using only two coordinates and without reference to the third dimension, in other words remove the embedding, then one needs to eliminate one of the $x$, $y$, and $z$ coordinates from (5.93). We can do this as follows: First take the differential of (5.93), leading to

$$dz = -\frac{xdx + ydy}{z} = \pm\frac{xdx + ydy}{\sqrt{a^2 - x^2 - y^2}}, \tag{5.94}$$

then substitute this in the ordinary three-dimensional Euclidean metric

$$dl^2 = dx^2 + dy^2 + dz^2 = dx^2 + dy^2 + \frac{(xdx + ydy)^2}{a^2 - x^2 - y^2}. \tag{5.95}$$

As such, we have successfully described the surface of the sphere without reference to the third dimension; in other words, the sphere is now described only by $x$ and $y$, provided that these are related by $x^2 + y^2 \le a^2$. Switching to "polar" coordinates using the usual

$$x = \rho\cos\varphi, \qquad y = \rho\sin\varphi \tag{5.96}$$

leads to

$$dl^2 = \frac{d\rho^2}{1 - \frac{\rho^2}{a^2}} + \rho^2 d\varphi^2. \tag{5.97}$$

For finite positive values of $a^2$, interpreted as the radius squared, this equation describes a metric on a sphere. However, if we allow $a \to \infty$, then we see that (5.97) reduces to the polar coordinate metric of a flat plane (1.23). Furthermore, if one takes $a^2$ to be *negative*, it can be shown that (5.97) now describes a metric on the hyperbolic paraboloid of Fig. [5.13].[40] In other words, (5.97) includes *all* the three cases of constant curvature that describe homogeneity and isotropy. Now, rather than dealing with weird values of the constant $a$, it is convenient to redefine the radial coordinate using the following transformation,

$$\rho \to \rho\sqrt{|a^2|}, \tag{5.98}$$

which leads to

$$dl^2 = |a^2|\left(\frac{d\rho^2}{1 - k\rho^2} + \rho^2 d\varphi^2\right), \tag{5.99}$$

where now the three cases correspond to

1. $k = 0$: Flat space.
2. $k = +1$: Positive curvature sphere.
3. $k = -1$: Negative curvature hyperbolic space.

---

[40]The quantity $a^2$ may *formally* be chosen to be negative, but in this case its root $a$ loses its physical meaning as a radius.

**Exercise 5.24** Show that (5.99) is equivalent to (5.97) via the transformation (5.98).

Once again, it is important to emphasize that (5.99) describes two-dimensional surfaces using two parameters only, $\rho$ and $\varphi$, both living *on* the surface, making no reference to a third dimension. In other words, the embedding has been removed. Also note that $|a^2|$ in the case of the plane ($k = 0$) is just a scaling number that can take any value and in fact may be absorbed in the definition of the coordinates. We now generalize (5.99) to three-dimensional surfaces by simply replacing $d\varphi^2$ with $d\Omega^2$:

$$dl^2 = a^2 \left( \frac{dr^2}{1 - kr^2} + r^2 d\Omega^2 \right) = a^2 \left[ \frac{dr^2}{1 - kr^2} + r^2 \left( d\theta^2 + \sin^2 \theta d\varphi^2 \right) \right], \qquad (5.100)$$

where we have also replaced $\rho$ with $r$ to make it more spherical coordinate-like. In (5.100) the quantity $a^2$ is understood to be a positive scaling factor.

**Exercise 5.25** In cylindrical coordinates, the equation of a given paraboloid embedded in three-dimensional Euclidean space is

$$z^2 = 2\rho - 1. \qquad (5.101)$$

1. Remove the embedding. In other words, write a two-dimensional metric mapping the surface of the paraboloid without reference to the third dimension.
2. Transform the metric you found to a reasonable choice of coordinates that simplifies it as much as possible.
3. Generalize the metric you found to a three-dimensional paraboloid.

There are probably many correct answers to this exercise. It all depends on what coordinate choice you made in part 2. Would the paraboloidal coordinates (1.9) be the best choice?

Next, we turn to upgrading the three-dimensional (5.100) to a four-dimensional spacetime metric. We also have to include the possibility that the spatial part of the metric, i.e. the entire universe, is evolving in time. The only way to do so while preserving the homogeneity and isotropy of (5.100) is to make the scale factor $a^2$ time dependent. This way the space in question changes in size evenly over all points. As such, we end up with:[41]

$$ds^2 = -c^2 dt^2 + a^2 \left( t \right) \left( \frac{dr^2}{1 - kr^2} + r^2 d\Omega^2 \right). \qquad (5.102)$$

This is the **Friedmann–Lemaître–Robertson–Walker metric**, also known by various combinations of these names (Friedmann–Robertson–Walker, Robertson–Walker,

---

[41]There are two conventions for the units of the constant $k$. The first is that it is allowed to have units of inverse length squared, making $a$ dimensionless (despite the fact that it started out as the "radius" of a sphere). This guarantees that $r$ has units of length. The second is defining both $k$ and $r$ as dimensionless, which ascribes units of length to $a$. As far as our purposes are concerned, either choice is valid.

or Friedmann–Lemaître).[42] The time-dependent factor $a(t)$ can be thought of as the size scale of the universe. It can allow for a static universe $a(t) = $ constant, an expanding universe $\dot{a}(t) > 0$, or a contracting universe $\dot{a}(t) < 0$, where an over dot is a derivative with respect to $t$. As mentioned, the parameter $k$ sets three possibilities for the universe:

1. For $k = 0$, the metric becomes

$$ds^2 = -c^2 dt^2 + a^2(t)\left(dr^2 + r^2 d\Omega^2\right), \qquad (5.103)$$

    which is spatially flat, except that the distance between any two points is constantly changing as the scale factor $a$ evolves.

2. The $k = +1$ case yields

$$ds^2 = -c^2 dt^2 + a^2(t)\left(\frac{dr^2}{1 - r^2} + r^2 d\Omega^2\right). \qquad (5.104)$$

    This metric has a coordinate singularity at $r = 1$ which may be dealt with using a suitable change of coordinates as in the Schwarzschild case.

3. For $k = -1$,

$$ds^2 = -c^2 dt^2 + a^2(t)\left(\frac{dr^2}{1 + r^2} + r^2 d\Omega^2\right). \qquad (5.105)$$

**Exercise 5.26** Using the coordinate transformation

$$d\chi = \frac{dr}{\sqrt{1 - kr^2}}, \qquad (5.106)$$

show that the Robertson–Walker metric (5.102) can be written in the form

$$ds^2 = -c^2 dt^2 + a^2\left[d\chi^2 + \Phi^2(\chi) d\Omega^2\right], \qquad (5.107)$$

where for a flat universe $\Phi(\chi) = \chi$, for a universe with positive curvature $\Phi(\chi) = \sin\chi$, and for a universe with negative curvature $\Phi(\chi) = \sinh\chi$. This metric, while less intuitive than the original, is convenient for a variety of calculations.

**Exercise 5.27** Another convenient redefinition of coordinates that, when used with the one in Exercise 5.26, gives a particularly simple form for the Robertson–Walker metric is

$$d\eta = \frac{dt}{a(t)}. \qquad (5.108)$$

Using this, along with (5.106), show that the metric can be written in the form

$$ds^2 = a^2(\eta)\left[-c^2 d\eta^2 + d\chi^2 + \Phi^2(\chi) d\Omega^2\right]. \qquad (5.109)$$

Here is why this is a most convenient form for a metric describing the entire universe. Consider that we, the observer, are located at the origin of coordinates. All light rays coming

---

[42]Alexander Friedmann: Russian physicist (1888–1925), Howard P. Robertson: American physicist (1903–1961), and Arthur Geoffrey Walker: French mathematician (1909–2001).

toward us from every point in the universe are then necessarily radial geodesics. In other words, they are described by $ds^2 = 0$ and $d\Omega = 0$. Taken together with (5.109), the equation that describes radial geodesics is then

$$-c^2 d\eta^2 + d\chi^2 = 0. \tag{5.110}$$

When integrated, (5.110) gives

$$\chi(\eta) = \pm c\eta + b. \tag{5.111}$$

When plotted in the $c\eta$–$\chi$ plane, the equation describes straight lines with $\pm 1$ slopes and intercept $b$. In other words, it gives light cones similar to those of special relativity just like Fig. [4.3] if $b = 0$, or more generally Fig. [6.5] for various values of $b$. In other words, the coordinate transformations (5.106) and (5.108) allow us to treat the entire universe as *if* it were the flat spacetime of special relativity, at least from a mathematical perspective. This naturally simplifies a variety of calculations. The interested reader may further explore this topic in any book on modern cosmology, e.g. [11].

Large-scale observations seem to indicate the simplest of the above possibilities: The universe is large-scale flat and is approximately described by (5.103). It is also a dynamic universe; the scale factor $a(t)$ is a function in time, implying that the universe may be constantly changing in size. The problem now boils down to the following question: What is the exact form of the function $a(t)$ that describes the dynamics of our universe? Analysis of the Einstein field equations, which we will eventually get to, indicate that the form of $a(t)$ should depend on the matter and energy content of the universe. If we define $\rho(t)$ and $p(t)$ as the density of the matter/energy content of the universe and their pressure respectively, then the following relationships (for $k = 0$),

$$\left(\frac{\dot{a}}{a}\right)^2 = \frac{8\pi G}{3}\rho(t)$$

$$2\left(\frac{\ddot{a}}{a}\right) + \left(\frac{\dot{a}}{a}\right)^2 = -\frac{8\pi G}{c^2}p(t) \tag{5.112}$$

are true. These are known as the **Friedmann equations**, and together they describe the behavior of the scale factor. It is sometimes convenient to substitute the first of (5.112) into the second to get

$$\frac{\ddot{a}}{a} = -\frac{4\pi G}{3}\left(\rho + \frac{3p}{c^2}\right). \tag{5.113}$$

One can also define the so-called **Hubble parameter**:[43]

---

[43]Edwin Hubble: American astronomer (1889–1953). Hubble is considered the father of modern cosmology. In 1924 he discovered that the universe is much larger than previously thought, and that certain cloudy patches in the sky, known as **nebulae**, are in fact other galaxies. Five years later he discovered that the universe is expanding, i.e. $\dot{a} > 0$.

$$H(t) = \frac{\dot{a}}{a}. \tag{5.114}$$

This quantity was originally defined as part of **Hubble's law**, also sometimes referred to as the **Hubble–Lemaître law**, which describes the rate with which points in the universe are receding away from us, a discovery that led to the current understanding that the universe is, and has always been, expanding. Hubble's law has the form

$$v = H(t) D, \tag{5.115}$$

where $D$ is the distance to a specific point in space, usually a distant galaxy, and $v$ is the speed with which this point is receding away from us. The current value of the Hubble parameter is known as **Hubble's constant** $H_0$ and is approximately 67.77±1.30 km/sMpc.[44]

Clearly, then, the rate of change of the scale factor of the universe is positive, i.e. $\dot{a} > 0$. The exact solutions for $a$ depend on the density $\rho$ and pressure $p$. Since we are treating the matter/energy content of the universe as an ideal fluid, thermodynamics tells us that the pressure and density are related by the so-called equation of state:

$$p = \omega c^2 \rho, \tag{5.116}$$

where the dimensionless constant $\omega$ depends on our assumptions of the type of matter/energy that pervades the universe. For example, consider the case of a "dust"-filled universe: The density $\rho$ is considered rarefied enough that the pressure vanishes, i.e. $\omega = 0$. In this case, the density $\rho$ would be inversely proportional to the cube of the scale factor. In other words,

$$\rho = \frac{\rho_0}{a^3}, \tag{5.117}$$

where $\rho_0$ is the value of $\rho$ at $a = 1$. As such, eqn (5.113) gives $a \sim t^{2/3}$.

**Exercise 5.28** Show that $a \sim t^{2/3}$ is indeed the solution for the flat universe Friedmann equations if $p = 0$ and $\rho \sim a^{-3}$.

**Exercise 5.29** Find $a(t)$ for a universe filled entirely with ultra-relativistic particles, also known as radiation. In this case, $\rho \propto \frac{1}{a^4}$ and $\omega = \frac{1}{3}$.

More detailed analysis may be found in any modern textbook on cosmology. Suffice it here to note that the current understanding of the evolution of the universe can be summarized as follows (see the sketch in Fig. [5.14]):

[44]As most recently reported (nine days prior to this writing) by the Dark Energy Survey (DES) collaboration.

1. The universe began from an extremely small size; all matter and energy observed in our current cosmos were squeezed into a very small, possibly infinitely small, and extremely hot volume. In other words, $a(t=0) \to 0$: a clear case of a singularity. As in the case of the central singularity of a black hole, at such small scales the rules of quantum mechanics kick in and a theory of quantum gravity is needed. As such, no one currently knows whether or not $t=0$ was a true spacetime singularity. It is interesting to note that, inspired by Penrose's singularity theorem for black holes, briefly discussed in §5.7, Stephen Hawking discovered another singularity theorem that mathematically proves that all timelike geodesics in the universe, i.e. the world lines of the individual particles that constitute the universe, must necessarily end on a singularity if traced back far enough in time. In other words, it is inevitable that our universe started from a singularity.[45] Today, both theorems are collectively referred to as the **Hawking–Penrose singularity theorems**.

2. For unknown reasons, this primordial singularity started to expand. This event is popularly known as the **Big Bang**.[46]

3. Almost immediately afterwards (best estimates $t \sim 10^{-36}$ second), the universe went through an extremely rapid period of expansion known as **inflation**. During an extremely short period of time (ending at possibly $t \sim 10^{-32}$ second or thereabout) the universe's scale factor $a(t)$ increased by at least $10^{26}$-fold, possibly even higher. The reason this happened is unknown (there are many hypotheses), but evidence seems to strongly support it.[47]

4. Immediately following the end of the inflationary epoch, the universe, cooling down considerably, settled into a long period of slow expansion. It is assumed that whatever happened in the early universe that caused such a rapid inflation had by now subsided.

5. Currently there is strong evidence that the universe is passing through a second accelerated phase. It is not only expanding, but it is doing so at a higher rate every second.

The discovery of the current accelerated rate of expansion of the universe was a considerable surprise to physicists and astronomers. The reason is that as the universe gets larger, one would assume that the initial "push" from the Big Bang begins to slow down and the gravitational attraction between the universe's constituents would take over, slowing down the rate of the expansion or even collapsing the universe in

[45] Stephen William Hawking: English theoretical physicist and cosmologist (1942–2018). He published this discovery in his 1966 PhD thesis.

[46] The term "Big Bang" is an unfortunate misnomer. It implies an "explosion," and explosions are events that happen *in* space. This is incorrect; the term describes the first instant in the expansion *of* space itself. Some would even interpret it as the very beginning of the universe, evolving from "nothing." It is hard to imagine exactly what it was, but an explosion it most definitely wasn't.

[47] The astute reader may have wondered how quickly the universe expanded during inflation, and if you guessed "faster" than the speed of light, you would be correct. This doesn't violate relativity in any way. Recall that the speed of signals or objects moving *in* space is not allowed to exceed $c$, but what is happening here is different: Space *itself* is expanding, and there are no restrictions on how fast this is allowed to be (that we know of).

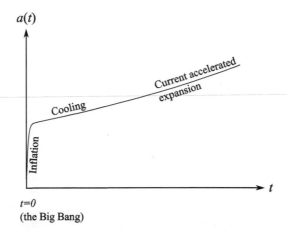

Fig. 5.14: Sketch showing, very roughly, and completely *not* to scale, the evolution of the scale factor as it is understood today.

on itself, depending on how much matter and energy there is in total. The accelerated expansion implies that "something" is "pushing" the universe, causing it to expand faster and faster. And just as there are hypotheses to explain the original inflation, there are also many theories to explain the current accelerating phase. The most popular theory is the presence of a possibly new type of energy, which was dubbed **dark energy**, that is causing a positive pressure on the universe, counteracting the negative gravitational pull and pushing the expansion faster. But no one knows for sure exactly how this works. There are also numerous proposals to modify general relativity itself,[48] effectively adding terms to the Einstein equations that may (or may not) explain the "reversal" of the direction of gravity from "pull" to "push." Such an effect, if it exists, clearly manifests itself only on large scales and is negligible on small ones. Finally, we note that the question of what the universe is "expanding into" is one that relativists and cosmologists simply ignore. The entire theory, based on the concept of metric, describes spacetime from *within itself*, making no reference to any embedding into higher dimensions. As such, most physicists just surrender to the idea that the expansion is "internal," in the sense that it needs no external space to be consistent. That is not to say that such an embedding cannot exist; there certainly may *be* higher dimensions. In fact, such an option has been heavily studied in the scientific literature, and some theories, namely superstring theory, posit as many as 11 spacetime dimensions; one of time and ten of space. There is currently no empirical evidence as to whether or not such ideas are true.[49]

---

[48]We will review some of these in Chapter 10.

[49]Strictly speaking, superstring theory itself requires 10 spacetime dimensions, but a higher-order generalization of it, known by the enigmatic name "M-theory," requires 11! There have even been studies in *12* spacetime dimensions, *two* of time and ten of space, that go by the even stranger name of "F-theory." It has been proposed, rather tongue in cheek, that the "M" in M-theory stands for "Mother of all theories." In that context the "F" in "F-theory" stands for "Father."

**Exercise 5.30** Find $a(t)$ for a flat universe filled entirely with dark energy. In this case $\rho =$ constant, and $\omega = -1$. Demonstrate explicitly that this universe's expansion is positively accelerating.

**Exercise 5.31** Derive the so-called "cosmological continuity equation,"

$$\dot{\rho} = -3H \left( \rho + \frac{p}{c^2} \right), \tag{5.118}$$

from the Friedmann equations. Generally speaking, a "continuity equation" is the term usually given to denote the conservation of something: in this case, the conservation of the matter/energy density $\rho$. In this context it is equivalent to other continuity equations in physics, such as (4.173), which represents the conservation of electric charge.

**Exercise 5.32** Using the continuity eqn (5.118), derive the dependence of $\rho$ on $a$ for the cases of:

1. Dust: $\omega = 0$.
2. Radiation: $\omega = \frac{1}{3}$.
3. Dark energy: $\omega = -1$.
4. A general $\omega$.

**Exercise 5.33** Using the result of part 4 in Exercise 5.32, find the dependence of $a$ on time for *any* value of $\omega$.

**Exercise 5.34**

1. Find the inverse metric tensor $g^{\mu\nu}$ for the Robertson–Walker metric (5.102) then verify orthogonality: $g^{\mu\rho} g_{\rho\nu} = \delta^{\mu}_{\nu}$.
2. Find the Christoffel symbols for the Robertson–Walker metric.
3. Write down the divergence $\nabla_\mu A^\mu$ of an arbitrary 4-vector field using the Robertson–Walker metric.
4. Write down the Laplacian $\nabla^\mu \nabla_\mu f(x^\alpha)$ of an arbitrary scalar function using the Robertson–Walker metric.

# 6

# Physics in Curved Spacetime

*Physics is really figuring out how to discover new things that are counterintuitive.*

Elon Musk

## Introduction

In this chapter we present and discuss the ultimate problem in covariance: How do objects behave in a curved spacetime background? We will develop a formalism for the generally covariant mechanics of point particles and discuss major issues and problems, some of which may have no clear answers in today's physics, while others are clearly defined and well structured, although they may seem very counterintuitive. The reader may take this chapter as the ultimate test for their understanding of the topic. Following the first section, which is dedicated to formulating generally covariant mechanics, we will focus on the very important problem of calculating the motion of freely falling particles. There is no new physics in doing so, just mathematical techniques that some readers, especially those who have taken an advanced course in classical mechanics, may recognize. Following that, we will briefly give an overview of generally covariant electrodynamics.

## 6.1 Generally Covariant Mechanics

If we go back and reread Chapter 4, we will notice that everything we did there almost always started with the metric. This is our approach throughout the whole book: Start at the metric and derive everything you need from it. The same thing can be done in general relativity. The main difference here is that the metric varies depending on the physical situation, as we have seen in the previous chapter. In this section, we will demonstrate how to define mechanics using an arbitrary metric. However, we do have to be extra careful. As discussed in more detail below, some concepts in curved backgrounds are difficult to pin down exactly. Let us also remember that, based on the principle of equivalence, when considering the rest frame of an event the local spacetime is necessarily flat and the Minkowski metric is in effect. Curved spacetimes

*Covariant Physics: From Classical Mechanics to General Relativity and Beyond*. Moataz H. Emam, Oxford University Press (2021). © Moataz H. Emam.
DOI: 10.1093/oso/9780198864899.003.0006

will appear on a global scale, meaning that the curved interval will only be apparent to an observer far away from the event, not local with respect to it (remember the global observer $O$ as she observed the falling elevator in §5.2).

Let us begin. First, we list the scalar products that all observers will agree on in general relativity just like they agreed on in special relativity. These are

$$dx^\mu dx_\mu = ds^2 = -c^2 d\tau^2 \tag{6.1}$$
$$U^\mu U_\mu = -c^2 \tag{6.2}$$
$$P^\mu P_\mu = -m^2 c^2, \tag{6.3}$$

for massive timelike particles. Their lightlike counterparts are:

$$dx^\mu dx_\mu = ds^2 = 0$$
$$U^\mu U_\mu = 0$$
$$P^\mu P_\mu = 0, \tag{6.4}$$

in addition to

$$A^\mu U_\mu = 0 \tag{6.5}$$

for both. The indices are raised and lowered by the metric tensor $g_{\mu\nu}$, which may be one of the cases presented in the previous chapter or any other solution of the Einstein field equations. We also take the following to be true for massive particles *by definition*:

$$x^\mu = (ct, \mathbf{r}) \tag{6.6}$$
$$U^\mu = \frac{dx^\mu}{d\tau} = \left( c\frac{dt}{d\tau}, \frac{d\mathbf{r}}{d\tau} \right) \tag{6.7}$$
$$A^\mu = \frac{dU^\mu}{d\tau} \tag{6.8}$$
$$P^\mu = mU^\mu. \tag{6.9}$$

For massless lightlike particles we can always use the affine parameter $\lambda$ instead of $\tau$, as we will see in specific applications later. The rules of tensor calculus, which by now should be more or less intuitive to the reader, still apply in all of their gory details.

### 6.1.1   Time Dilation

Curved spacetimes imply a gravitational time dilation effect. Consider two observers, $O$ and $O'$, both *fixed* in space in the background of some curved spacetime manifold mapped by a metric tensor $g_{\mu\nu}$, as shown schematically in Fig. [6.1]. Assume a certain *static* event $P$ occurring locally in the $O'$ frame. The duration of time of this event is $O'$'s proper time $d\tau$, and the metric is just $ds^2 = -c^2 d\tau^2$ since the event is motionless, i.e. the $dx^i$ terms vanish. On the other hand, the coordinate time $dt$ as measured by $O$

is found from $ds^2 = c^2 g_{00} dt^2$, where again $O$'s $dx^i$ terms vanish. Since both observers will agree on $ds^2$, we can write

$$-c^2 d\tau^2 = c^2 g_{00} dt^2. \tag{6.10}$$

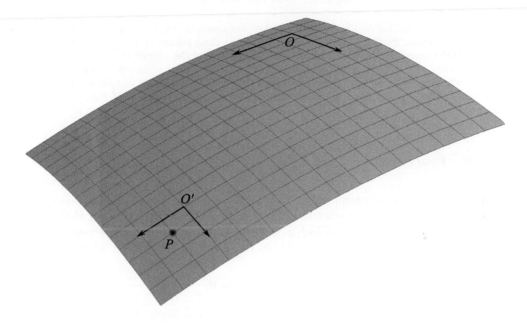

Fig. 6.1: Embedding diagram showing two observers in an arbitrary spacetime background. An event $P$ occurs locally in the $O'$ frame.

For simplicity in subsequent expressions let's define $g_{00} = -T^2$, where $T^2$ is whatever happens to be in the metric; for example, in the Schwarzschild case $T^2 = H(r) = \left(1 - \frac{2GM}{c^2 r}\right)$. We can then write the so-called **gravitational time dilation** formula:

$$dt = \frac{d\tau}{T}. \tag{6.11}$$

As an example, in the Schwarzschild case we find

$$dt_{\text{Sch}} = \frac{d\tau}{\sqrt{H(r)}} = \frac{d\tau}{\sqrt{1 - \frac{2GM}{c^2 r}}}, \tag{6.12}$$

which works only outside the event horizon, if there is one. *At* the event horizon, where $H \to 0$, the coordinate time becomes infinite; time, as measured by $O$, literally stops! *Inside* the event horizon, $H$ becomes complex and coordinate time doesn't seem to even exist, signaling that the coordinate observer cannot make time measurements at radii less than the Schwarzschild radius. We will further explore these conclusions later in this chapter.

**Exercise 6.1** Consider an observer $O'$ in a spaceship hovering at a distance $r = 2r_s$ in a Schwarzschild spacetime, e.g. a black hole. Let's say she is watching an hour-long TV show on her computer. How long would the TV show appear to be to a coordinate observer $O$ hovering at a far-away distance? How far from the event horizon should $O'$ be for the TV show to appear twice as long as her local time measurement?

**Exercise 6.2** In the movie *Interstellar* the protagonists stayed on a planet in orbit around the "Gargantua" black hole for three hours. Upon their return to Earth, the equivalent of 23 years had passed! Using this information and assuming the planet they landed on is located at $r = 2r_s$, estimate Gargantua's mass and the diameter of its event horizon. How does Gargantua compare to M87, the supermassive black hole that was "photographed" by the "Event Horizon Telescope" collaboration in 2019? See the caption of Fig. [5.12] for M87's data.

### 6.1.2   Four-Velocity

At this point, it is very tempting to use (6.11) in (6.7) to define an analogous expression to (4.100), but we must remember that (6.11) was found by assuming that the object is motionless with respect to *both* observers. This means that $dr/dt$ vanishes for $O$ as well as $O'$. As such, we can use (6.11) in the 4-velocity only if the 3-velocity vanishes. In other words,

$$U^\mu = \frac{1}{T}\left(c, \mathbf{0}\right) \tag{6.13}$$

is true for the fixed global frame $O$ (in the $O'$ local frame $T = 1$), while something like $\frac{1}{T}\left(c, \frac{d\mathbf{r}}{dt}\right)$ isn't true for moving ones. To generalize (6.11), and consequently $U^\mu$, we have to replace (6.10) with

$$ds^2 = -c^2 d\tau^2 = g_{\mu\nu} dx^\mu dx^\nu \tag{6.14}$$

in full. Consequently,

$$U^\mu U_\mu = \left(\frac{ds}{d\tau}\right)^2 = g_{\mu\nu}\frac{dx^\mu}{d\tau}\frac{dx^\nu}{d\tau} = \gamma_g^2 g_{\mu\nu}\frac{dx^\mu}{dt}\frac{dx^\nu}{dt} = -c^2, \tag{6.15}$$

where we define

$$\gamma_g = dt/d\tau, \tag{6.16}$$

named after the Lorentz $\gamma$. Note that this $\gamma_g$ must reduce to $1/T = 1/\sqrt{-g_{00}}$ for vanishing 3-velocity. It should also reduce to the Lorentz $\gamma$ (4.48) for $g_{\mu\nu} = \eta_{\mu\nu}$. We can now write

$$U^\mu = \gamma_g\left(c, \frac{d\mathbf{r}}{dt}\right) = \gamma_g\left(c, \mathbf{v}\right), \tag{6.17}$$

where, from (6.15),

$$\gamma_g = \frac{1}{\sqrt{-\frac{1}{c^2}g_{\mu\nu}\frac{dx^\mu}{dt}\frac{dx^\nu}{dt}}} = \frac{1}{\sqrt{-g_{00} - \frac{1}{c^2}g_{ij}\frac{dx^i}{dt}\frac{dx^j}{dt} - \frac{2}{c}g_{0i}\frac{dx^i}{dt}}}, \tag{6.18}$$

where we allowed for the possibility that the metric may not be diagonal, i.e. $g_{0i} \neq 0$. If it *is* diagonal, then

$$\gamma_g = \frac{1}{\sqrt{-g_{00} - \frac{1}{c^2} g_{ij} \frac{dx^i}{dt} \frac{dx^j}{dt}}} = \frac{1}{\sqrt{-g_{00} \left(1 + \frac{g_{ij}}{g_{00} c^2} \frac{dx^i}{dt} \frac{dx^j}{dt}\right)}}$$

$$= \frac{1}{\sqrt{-g_{00} \left(1 + \frac{v^2}{g_{00} c^2}\right)}} \quad \text{(for diagonal metrics)}, \tag{6.19}$$

or even more compactly

$$\gamma_g = \frac{1}{T\sqrt{\left(1 - \frac{\beta^2}{T^2}\right)}} \quad \text{(for diagonal metrics)}, \tag{6.20}$$

where $\beta$ is (4.47). The definition (6.20) may be upgraded to non-diagonal metrics if one redefines

$$\beta = \frac{1}{c} \sqrt{v^2 - 2g_{oi} \frac{dx^i}{dt}}. \tag{6.21}$$

The $\gamma_g$ factor clearly reduces to the Lorentz $\gamma$ (4.48) for $T = 1$ as required. Note that this is true not only for a flat spacetime but also for *any* metric where $g_{00} = -1$, such as (5.102).

---

**Exercise 6.3** Compute $\gamma_g$ in terms of $v$ for the metrics we have learned about in the previous chapter:

1. The Newtonian metric (5.46).
2. The exterior Schwarzschild metric (5.47).
3. The interior Schwarzschild metric (5.61).
4. The Robertson–Walker metric (5.102).

---

The velocity **v** is not a constant in the above equations as it was assumed to be in special relativity, since the particle's rest frame is "accelerating" with respect to $O$. There are, however, special cases where the magnitude of **v** is a constant even though the frame is accelerating; a particle in uniform circular motion about a curvature source is such a case. Let's consider the example of an object, or alternatively a local observer $O'$, on the surface of a slowly rotating sphere of radius $R$ and uniform mass $M$, as shown in Fig. [6.2]. For example, one may think of $O'$ as you, dear reader, as you sit in a chair reading this. Earth is, up to a good approximation, a slowly rotating sphere; hence its surrounding spacetime may be described by either the Newtonian weak field metric (5.46) or the full Schwarzschild metric (5.52). If this approximation bothers you, just consider a perfectly motionless sphere with a particle that is made to move in perfect circles close to its surface: a circle that may coincide, for instance,

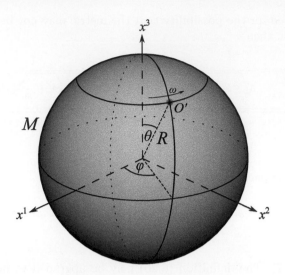

Fig. 6.2: Calculating time dilation in a rotating local reference frame $O'$ in the Schwarzschild metric.

with a circle of latitude on Earth. This can be practically performed if $O'$ happens to be onboard an airplane traveling with a constant angular speed $\omega$ along a circle of latitude $\theta$. As a side note, let's point out that the physicist's circle of latitude $\theta$ starts at zero at the north pole and varies to 180° at the south pole, while a geographer's $\theta$ (the one you see on maps) is zero at the equator and varies to 90° at the north pole and −90° at the south pole.[1] Now an observer $O$ that is far away from the sphere (say, on the Moon) observes the full curved metric as usual. Hence we can write

$$-c^2 d\tau^2 = -c^2 H\left(r\right) dt^2 + H\left(r\right)^{-1} dr^2 + r^2 \left(d\theta^2 + \sin^2\theta d\varphi^2\right), \qquad (6.22)$$

where we have chosen to use the Schwarzschild metric and $H\left(r\right)$ is defined by (5.51). Now for simplicity let's align the $z$ axis with the sphere's (slow) rotational axis, or alternatively with the particle's orbit axis of symmetry, and also choose observer $O'$ to be at the equator on the surface of the sphere as in Fig. [6.3]. Consequently $r = R$, $\theta = 90°$, $dr = d\theta = 0$, and $\sin\theta = 1$:

$$-c^2 d\tau^2 = -c^2 H\left(R\right) dt^2 + R^2 d\varphi^2,$$
$$\left(\frac{d\tau}{dt}\right)^2 = \frac{1}{\gamma_g^2} = H - \frac{R^2}{c^2}\left(\frac{d\varphi}{dt}\right)^2. \qquad (6.23)$$

In this case, the velocity of the particle's rest frame is simply the angular velocity $\mathbf{v} = \omega\hat{\mathbf{g}}_\varphi = R\omega\hat{\mathbf{v}}_\varphi$, where $\omega = \frac{d\varphi}{dt}$. Thus we end up with[2]

---

[1]The experiment we are describing here has actually been done several times. Keep reading.
[2]Compare with Exercise 4.23.

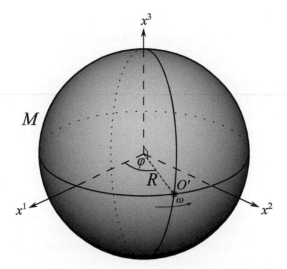

Fig. 6.3: Calculating time dilation in a rotating local reference frame $O'$ in the Schwarzschild metric. In this sketch we have chosen $\theta = 90°$.

$$\gamma_g = \frac{1}{\sqrt{H\left(1 - \frac{R^2\omega^2}{Hc^2}\right)}} = \frac{dt}{d\tau}, \tag{6.24}$$

and the particle's 4-velocity (as measured by $O$) is

$$U^\mu = \frac{1}{\sqrt{H\left(1 - \frac{R^2\omega^2}{Hc^2}\right)}}\left(c, 0, 0, R\omega\right). \tag{6.25}$$

Note that if the particle is placed at rest with respect to $O$, i.e. $\omega = 0$, then the 4-velocity becomes just (6.13). The time dilation effect (6.23) can be thought of in two components. The presence of $H$ is the gravitational piece, while the other term in $\omega$ is the special relativistic effect of the object's speed which we studied in Exercise 4.23. For Earth, one can evaluate this explicitly:

$$\begin{aligned}
\frac{1}{\gamma_g^2} &= 1 - \frac{2GM_{\text{Earth}}}{c^2 R_{\text{Earth}}} - \frac{R_{\text{Earth}}^2 \omega^2}{c^2} \\
&= 1 - \frac{2\left(6.67408 \times 10^{-11}\text{m}^3/\text{kg} \cdot \text{s}^2\right)\left(5.972 \times 10^{24}\text{kg}\right)}{\left(299,792,458\text{m/s}\right)^2 \left(6,371,000\text{m}\right)} \\
&\quad - \frac{\left(6,371,000\text{m}\right)^2 \left(7.2921159 \times 10^{-5}\text{s}^{-1}\right)}{\left(299,792,458\text{m/s}\right)^2} \\
&= 0.99999996567\ldots. \tag{6.26}
\end{aligned}$$

Hence

$$\frac{dt}{d\tau} = \gamma_g = 1.00000001716\cdots.$$   (6.27)

In other words, observer $O'$'s clock will be *slower* than observer $O$'s by *less* than two parts in a hundred million! Clearly a minuscule difference that cannot be detected within our everyday experience. In 1970 Joseph C. Hafele[3] and Richard E. Keating[4] conducted the first experiment to detect this effect on Earth. They took four cesium atomic clocks aboard commercial airplanes that flew twice around the world, first eastward, then westward. Comparing the clocks against each other and against others housed at the United States Naval Observatory, they found that they ran differently in full agreement with the predictions of relativity: a calculation similar to the one we just did. Other experiments that tested the special relativistic effect alone and the gravitational effect alone were done previously, also with excellent agreement with prediction. Similar experiments have been done countless times since then, every time with better and better instruments, and every time verifying the predicted effects of both speed *and* the curvature of spacetime.

**Exercise 6.4** Show that $U^\mu U_\mu = -c^2$ for (6.25).

**Exercise 6.5** Calculate the 4-acceleration of the particle whose 4-velocity is given by (6.25). What is your interpretation of the result?

**Exercise 6.6** Repeat the calculation in the text, starting at eqn (6.22), using the Newtonian metric (5.46) instead. How different is your result?

**Exercise 6.7**

1. Finding the time dilation using a time-dependent metric is slightly more subtle than doing so for a static metric. Explain why this is the case, using the Robertson–Walker metric (5.102) as an example.
2. Consider a distant galaxy (local observer $O'$ whose time is $\tau$) that is receding from us (the coordinate observer $O$) by an observed radial speed $v \sim 0.3c$. If we observe an event in this galaxy that seems to have a duration of $\Delta t = 1$ hour, how long did that event last with respect to a local observer? Take $a \sim t^{2/3}$ and $k = 0$ in the Robertson–Walker metric (5.102).
3. Consider a light beam that leaves the galaxy in part 2 with a specific wavelength $\lambda_s$, where $s$ stands for "source." Due to the Hubble expansion of the universe this light beam arrives to us with a different wavelength, $\lambda_o$, where $o$ stands for "observed." Since generally any wavelength is $\lambda = ct$, use your result in part 2 to show that the light will arrive into our telescopes *red-shifted*; in other words, its wavelength will shift toward the lower end of the wavelength spectrum. This is the famous **galactic red-shift** discovered by Edwin Hubble back in 1929 and used to conclude that the universe is expanding. It has since become a standard technique in observational cosmology.

---

[3]American physicist (1933–2014).
[4]American astronomer (1941–2006).

### 6.1.3 Four-Acceleration, Geodesics, and Killing Fields

Let's define the 4-acceleration of a massive particle in a curved background:

$$A^\mu = \frac{dU^\mu}{d\tau}. \tag{6.28}$$

To put this in a more familiar form, we apply the chain rule

$$\frac{dU^\mu}{d\tau} = \frac{dx^\nu}{d\tau} \frac{\partial U^\mu}{\partial x^\nu} = U^\nu \partial_\nu U^\mu. \tag{6.29}$$

Recalling that the partial derivative is not covariant, we apply the minimal substitution $\partial \to \nabla$,

$$\begin{aligned} A^\mu &= U^\nu \nabla_\nu U^\mu \\ &= U^\nu \partial_\nu U^\mu + \Gamma^\mu_{\alpha\nu} U^\alpha U^\nu, \end{aligned} \tag{6.30}$$

and, not surprisingly, we end up with

$$A^\mu = \frac{dU^\mu}{d\tau} + \Gamma^\mu_{\alpha\nu} U^\alpha U^\nu. \tag{6.31}$$

This is the definition of coordinate-invariant acceleration we have found before in Newtonian mechanics (3.18), upgraded to four dimensions. Now, if we let the particle move on a geodesic, i.e. in its own rest frame in free fall, then the acceleration vanishes; hence

$$\frac{dU^\rho}{d\lambda} + \Gamma^\rho_{\mu\nu} U^\mu U^\nu = 0, \tag{6.32}$$

or equivalently

$$\frac{d^2 x^\mu}{d\lambda^2} + \Gamma^\mu_{\alpha\beta} \frac{dx^\alpha}{d\lambda} \frac{dx^\beta}{d\lambda} = 0, \tag{6.33}$$

which is the geodesic equation after replacing $\tau$ with the more general affine parameter $\lambda$. Let us reiterate the discussion toward the end of §5.2 and point out that requiring the 4-acceleration to vanish along geodesics is at the heart of the equivalence principle. There are *no* accelerating reference frames for a free particle; they simply do not exist. Yes, we may talk about them out of habit, but motion along geodesics essentially replaces the concept of free-fall acceleration and consequently replaces gravity itself. On the other hand, a particle that is *not* allowed to follow geodesics, for example one that is made to "stop" motionless in space by some extraneous agent (or is pushed by something), now *that* particle is truly accelerating, and $A^\mu \neq 0$. Recall the discussion around Fig. [5.5].

Another useful form of the geodesic equation can be taken directly from (6.30):

$$U^\nu \nabla_\nu U^\mu = 0. \tag{6.34}$$

We will now use this version to define an important property of general covariance based on the so-called **Killing vector**[5] concept. This is a quantity used to find symmetries of the metric, which in turn can be interpreted as conservation rules. The question is: What quantities does a particle moving along a geodesic conserve? Energy? Momentum? Angular momentum? Something else? In non-relativistic gravity such conservation rules are relatively (sorry) easy to spot. In general relativity the concept of Killing vectors introduces a method of finding conservation laws for geodesic motion.

First begin with a velocity vector pointing along a geodesic $U^\mu = dx^\mu/d\lambda$. Next consider an arbitrary 4-vector $\xi^\mu$. Construct the derivative

$$U^\nu \nabla_\nu \left( \xi_\mu U^\mu \right) = U^\nu U^\mu \nabla_\nu \xi_\mu + \xi_\mu U^\nu \nabla_\nu U^\mu, \tag{6.35}$$

where we have applied the product rule. The second term on the right-hand side of (6.35) vanishes because of (6.34), leaving

$$U^\nu \nabla_\nu \left( \xi_\mu U^\mu \right) = U^\nu U^\mu \nabla_\nu \xi_\mu. \tag{6.36}$$

Now if the quantity $\xi_\mu U^\mu$ happens to be *conserved* along the particle's path, then the left-hand side of (6.36) becomes a conservation law[6]

$$U^\nu \nabla_\nu \left( \xi_\mu U^\mu \right) = 0, \tag{6.37}$$

and a necessary and sufficient condition for this to happen is that $\xi_\mu$ satisfies the vanishing of the right-hand side of (6.36). Now, because the expression $U^\nu U^\mu \nabla_\nu \xi_\mu$ is symmetric under the exchange $\mu \leftrightarrow \nu$, i.e.

$$U^\nu U^\mu \nabla_\mu \xi_\nu = U^\mu U^\nu \nabla_\nu \xi_\mu, \tag{6.38}$$

then the vanishing of the right-hand side implies that the vector $\xi$ satisfies

$$\nabla_\mu \xi_\nu + \nabla_\nu \xi_\mu = 0. \tag{6.39}$$

This is known as **Killing's equation**. Its solutions constitute a set of vectors (more precisely vector *fields*) $\xi^\mu$ that define symmetries of the metric, or in other words, quantities $\xi_\mu U^\mu$ that are conserved for *free* particles; thus

$$\xi_\mu U^\mu = \text{constant}. \tag{6.40}$$

Solving (6.39) in general for a specific metric is not an easy task. However, in the most interesting cases one can "guess" at the Killing vectors just by looking at

[5]Wilhelm Karl Joseph Killing; German mathematician (1847-1923).
[6]Similar to (4.191).

the metric. Once found, (6.40) can immediately be used to find explicitly conserved quantities. For later convenience, we define the Killing vector with upper indices as follows: $\xi^\mu \equiv (-1, 0, 0, 0)$, $\xi^\mu \equiv (0, 1, 0, 0)$, and so on. A lower-index Killing vector will necessarily contain metric components following $\xi_\mu = g_{\mu\nu}\xi^\nu$. Consider Minkowski-Cartesian spacetime as an example. Clearly, $\xi_\mu = (1, 0, 0, 0)$ is trivially a solution of Killing's equation; hence (6.40) becomes

$$\xi_\mu U^\mu = \eta_{\mu\nu}\xi^\mu U^\nu = \eta_{00}\xi^0 U^0 = U^0 = \text{constant}. \tag{6.41}$$

Since the particle is in its own rest frame, (6.41) is simply $c = \text{constant}$, which it certainly is. Furthermore, since $P^0 = mU^0 = \frac{E}{c}$, this immediately gives

$$E = \text{constant}. \tag{6.42}$$

So one says that any metric that admits a timelike Killing vector conserves a particle's total energy as it moves along geodesics. Similarly, one can show that spacelike Killing vectors such as $\xi^\mu = (0, 1, 0, 0)$, $\xi^\mu = (0, 0, 1, 0)$, and $\xi^\mu = (0, 0, 0, 1)$ yield momentum conservation in flat spacetime. The momentum here may be linear or angular, depending on the coordinate system used. For example, in Minkowski-Cartesian coordinates, $\xi^\mu = (0, 0, 0, 1)$ implies conservation of the $z$ component of momentum ($p_z = \text{constant}$) while in Minkowski-spherical coordinates the same vector implies the conservation of the angular momentum in the $\phi$ direction: $p_\varphi = \text{constant}$, since $x^3 = \varphi$ in this case. The Minkowski metric in particular is one that is maximally symmetric; it conserves energy and all momenta, as well as Lorentz boosts, each of which has its own associated Killing vector. This makes sense, of course, since a free particle in a Minkowski background is "truly" free, while a particle in a curved background is subjected to "gravity."

Mathematically speaking, Killing vector fields originate from a specific symmetry of the metric known in mathematical circles as an **isometry**. What this means is as follows: Given a metric $g_{\mu\nu}$, if there is a specific coordinate $x^\alpha$ in the metric, where $\alpha$ is a specific value from 0 to 3, that is shifted by some infinitesimal change, $x^\alpha \to x^\alpha + \epsilon \xi^\alpha$, where $\epsilon \ll 1$, and leaves the metric invariant, i.e. $g_{\mu\nu}(x^\alpha + \epsilon\xi^\alpha) = g_{\mu\nu}(x^\alpha)$, then we call this an isometry. If that is the case, then it should be clear that if all the components of the metric tensor $g_{\mu\nu}$ are independent of $x^\alpha$, then this is an isometry and $\xi^\mu$ exists with a single component along $x^\alpha$.

**Exercise 6.8**

1. Given an arbitrary metric tensor $g_{\mu\nu}(x^\alpha)$, show that the coordinate transformation

$$x^\alpha \to x^\alpha + \epsilon\xi^\alpha, \tag{6.43}$$

where $\epsilon \ll 1$, in other words, a small change in the coordinates, leads to

$$g_{\mu\nu}(x^\alpha) \to g_{\mu\nu}(x^\alpha) - \epsilon \left[ g_{\mu\rho}(\partial_\nu\xi^\rho) + g_{\rho\nu}(\partial_\mu\xi^\rho) \right] \tag{6.44}$$

if we ignore terms of order $\epsilon^2$ and higher, $\epsilon$ already being a very small quantity. *Hint*: Expand $g_{\mu\nu}(x^\alpha + \epsilon\xi^\alpha)$ in a Taylor series (3.83).

2. Show that the coordinate transformation in part 1 is an isometry, i.e. leaves the metric tensor invariant, if and only if $\xi^\rho$ satisfies the condition:

$$g_{\mu\rho}\left(\partial_\nu \xi^\rho\right) + g_{\rho\nu}\left(\partial_\mu \xi^\rho\right) + \xi^\rho\left(\partial_\rho g_{\mu\nu}\right) = 0. \tag{6.45}$$

3. Show that (6.45) leads to Killing's eqn (6.39).

As an example, the Schwarzschild metric (5.47) has no functions that depend on time ($dt^2$ doesn't count, as we are looking at the components $g_{\mu\nu}$). This immediately implies that an isometry in time exists; in other words, replacing $t \to t + \epsilon\xi^0 = t + \epsilon$ everywhere in the metric gives the same metric back. Hence a timelike Killing vector exists. In other words, the energy of a particle moving along Schwarzschild geodesics is conserved. Similarly, the lack of dependence on the angle $\varphi$ implies the existence of an angular Killing vector, implying conservation of the angular momentum $p^\varphi$. However, since the function $H$ is dependent on $r$, any shift $r \to r + \epsilon\xi^1 = r + \epsilon$ gives a different metric and an isometry doesn't exist. One concludes that no radial Killing vectors exist for Schwarzschild, and the radial momentum $p^r$ is not conserved. From a Newtonian perspective this should make sense, since a particle falling in a radial gravitational field is not force-free, hence its momentum is changing.

**Exercise 6.9** Show that the vector $\xi^i = (-y, x)$ is a Killing vector in the two-dimensional space described by the metric

$$ds^2 = \left[1 + \left(\frac{x^2 + y^2}{4a^2}\right)\right]^{-2}\left(dx^2 + dy^2\right), \tag{6.46}$$

where $a$ is an arbitrary constant. What is the conserved quantity that corresponds to this Killing vector?

**Exercise 6.10** List both the upper- and lower-index Killing vectors of the Schwarzschild solution (5.47). Explicitly show that Killing's eqn (6.39) is satisfied for all of them.

**Exercise 6.11** Find the upper and lower-index Killing vectors for the following metrics. What quantities do freely falling particles conserve as a consequence of the existence of these vectors? Two of these are from later in the book. We do not need to understand what they represent in order to find their Killing symmetries.

1. The Newtonian metric (5.46).
2. The interior Schwarzschild metric (5.61).
3. The Robertson–Walker metric (5.102).
4. The Reissner–Nordström metric (8.165).
5. The Kerr metric (8.169).

**Exercise 6.12** Show that the Killing vectors you found in Exercise 6.11 satisfy Killing's equation. Do so for:

1. The Newtonian metric (5.46).

2. The interior Schwarzschild metric (5.61).
3. The Robertson–Walker metric (5.102).

### 6.1.4   Four-Momentum

For a massive particle we can still define

$$P^\mu = mU^\mu, \tag{6.47}$$

just as we did in special relativity. We also require

$$P^\mu P_\mu = g_{\mu\nu}P^\mu P^\nu = -m^2c^2 \tag{6.48}$$

to hold for all observers, since if it is true in the particle's local Minkowskian frame, then it must be true in all reference frames.

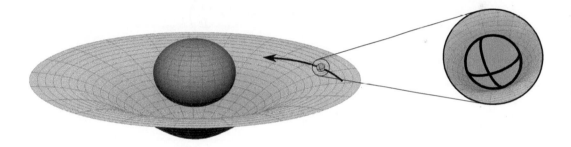

Fig. 6.4: The motion of a particle in the curved spacetime background of a massive sphere. Looking closely, the moving particle itself curves the spacetime, effectively changing the metric.

Now there is a subtlety concerning general relativistic 4-momentum. Although we have been discussing physics in curved spacetime backgrounds, we are yet to explore where that comes from; in other words, what *warps* spacetime? As noted earlier, and as we will see in more detail later, the curvature arises from the presence of mass, energy, and momentum. In other words, the components of $P^\mu$ *themselves* add to the curvature of spacetime. The problem arises when one realizes that energy is exchanged between objects and the spacetime *itself*. This can be thought of as follows: Think of a flat elastic trampoline-like surface. In the center a heavy spherical mass is placed and the "spacetime" curves following the Schwarzschild metric. Now place a small marble of mass $m$ on the surface, give it a little push, and watch it travel in that background; as sketched in Fig. [6.4]. You will see the marble following the timelike geodesic trajectories that arise as solutions of the geodesic equation, and you may be thinking that its 4-momentum (6.47), including total energy as well as ordinary momentum, must be conserved. However, look even more closely at the moving marble.

You will notice that the marble *itself* is curving the surface of the trampoline, causing a small dent underneath it that gets bigger and more pronounced if the marble is moving faster. This change in curvature implies that the spacetime is no longer described by the original metric, but is now something else that is changing all the time! In other words, the marble's energy and momentum are not, strictly speaking, conserved, but rather continuously "exchanging" contributions with the spacetime itself! Locally, in a small enough area where the spacetime is approximately flat and the particle's rest frame is described by the Minkowski metric, 4-momentum is conserved and special relativity is in effect. But "globally," on a larger scale, we run into problems and the issue of energy and momentum exchange become a lot more complicated. In Newtonian terms, the marble's own gravity affects it, in addition to the gravitational force coming from the central spherical mass. This is the sort of behavior that mathematicians call **non-linear**, and it is one of the major difficulties of general relativity. As such, the consensus is that while the 4-momentum vector can still be globally defined, the interpretation of its components as the usual energy and momentum is strictly *local*.

Now, following (6.47),

$$P^\mu = m\gamma_g \left( c, \frac{d\mathbf{r}}{dt} \right).$$ 

(6.49)

One can easily show that (6.48) leads to

$$\left( P^0 \right)^2 = \frac{1}{T} \left( m^2 c^2 + p^2 \right),$$ 

(6.50)

where $p^2 = \mathbf{p} \cdot \mathbf{p} = g_{ij} P^i P^j$ and

$$\mathbf{p} = m\gamma_g \mathbf{v} = m\gamma_g \frac{d\mathbf{r}}{dt}.$$ 

(6.51)

If we *define* (emphasis on "define") $E = P^0/c$, then (6.50) becomes

$$E^2 = \frac{1}{T^2} \left( m^2 c^4 + p^2 c^2 \right),$$ 

(6.52)

which can be thought of as the generally covariant version of (4.149). A massive particle with $p = 0$ leads to

$$E = \frac{mc^2}{T}.$$ 

(6.53)

And in the massless lightlike case

$$E = \frac{1}{T} pc.$$ 

(6.54)

Note that the choice $E = P^0/c$ that led to the last three equations is a matter of definition. A more general one arises from the following argument: If $P^\mu$ is the particle's 4-momentum and $U^\mu$ is an observer's 4-velocity, one can *define* energy as the projection of one on the other:

$$E = -P^\mu U_\mu. \qquad (6.55)$$

If the observer is in the particle's rest frame, then $U^\mu = (c, \mathbf{0})$ and we get

$$E = -cg_{00}P^0 - g_{i0}P^i. \qquad (6.56)$$

If the metric is diagonal this gives

$$E = -cg_{00}P^0, \qquad (6.57)$$

which, in Minkowski spacetime, is exactly $E = cP^0$, justifying the definition (6.55). This approach can be put in the form of isometries if we note that the rest frame 4-velocity is $U^\mu = -c\xi^\mu$, where $\xi^\mu$ is a timelike Killing vector $\xi^\mu = (-1, 0, 0, 0)$, such that

$$E = -P^\mu U_\mu = -cP^\mu \xi_\mu = cP^0. \qquad (6.58)$$

It should be obvious here that, in the absence of the appropriate Killing vectors, the total 4-momentum of a collection of particles moving and colliding freely in a curved spacetime background is *not* conserved except locally (where special relativity is in effect)! In other words, eqn (4.129) is no longer necessarily true on a global scale. Physically one can understand this if we switch briefly to the Newtonian picture of gravity: Particles in free fall are *not* force free. In other words, the forces on them do not vanish; they are "pulled" by gravity. Hence eqn (4.128) is no longer satisfied (Newton-wise), and the 4-momentum, including the total energy, is no longer conserved.

The complications in the (global) study of 4-momentum which we pointed out in this subsection make it difficult as well as unrealistic to approach the motion of particles in curved spacetime backgrounds using the same methods of Chapters 3 and 4. A more productive use of our time is in studying the free motion of particles along geodesics. We will do so in §6.2 using the (exterior) Schwarzschild solution as an example. The techniques we will develop may then be applied to any other metric.

## 6.2 The Schwarzschild Metric as a Case Study

As noted earlier, studying the free motion of a massive particle (timelike) or a massless particle (null or lightlike) in a specific spacetime background is of great importance. Not only does it provide confirmation that these spacetimes are indeed curved, but it also has major practical applications. Our case study is the Schwarzschild metric describing the spacetime around a massive static and uniform spherical object of radius $a$. This is clearly of great importance to planetary and stellar mechanics, as it can be used to model orbits. Planets can be thought of as massive particles traveling along geodesics in the approximately Schwarzschild-like spacetime of the Sun; see Fig. [6.4]. Although stars, planets, and black holes rotate and are not static like the Schwarzschild metric assumes, their rotation rates are usually not too high and can be ignored as a first approximation. In fact, just by using the Schwarzschild metric physicists have been able to find very accurate descriptions of the orbital motions of the planets of our solar system as well as the paths of light around the Sun. Of course, when the object in question is rotating at rates too high to be ignored (like a pulsar for example[7]), a different metric is needed. This would be the so-called Kerr metric, which we will discuss very briefly toward the end of §8.6.

In this section we will make no assumptions about the size of the spherical mass $M$ in question. As such, we will be considering the entire range of spacetime points appearing in the Schwarzschild solution, including the ambiguous event horizon $r = r_s$. We will discuss some strange phenomena that happen at that radius, and then move on to solve the geodesic equations for some interesting cases.

### 6.2.1 Escape Velocity

To begin, let us first fulfill our earlier promise to calculate the escape velocity from a black hole (Schwarzschild with $a < r_s$) as a precursor to asking what "really" happens at the event horizon. Recall that at the Schwarzschild radius $r = r_s$ a coordinate singularity occurs in the original Schwarzschild coordinates. Although not a true spacetime singularity, something is still special about that radius. We know this because, as noted earlier, at radii less than $r_s$ the time and radial terms of the metric switch signs, indicating some strange behavior in that region. Recall that in non-relativistic classical mechanics the escape velocity needed for an object to escape the surface of a given spherical mass (planet or star) to infinity is found by calculating energy conservation between the surface and infinity. One argues as follows: The total work required to move an object of mass $m$ from a distance $r$ from a gravitational source of mass $M$ all the way to radial infinity is equal to the negative of the Newtonian potential energy:

$$W = -V = \frac{GmM}{r}. \tag{6.59}$$

To move this object, it must be given at least that much energy in the form of kinetic energy; hence

$$\frac{1}{2}mv_{\text{escape}}^2 = \frac{GmM}{r}. \tag{6.60}$$

---

[7] A pulsar is a spinning neutron star that emits radio waves in fast succession; i.e. "pulsates." Most such objects spin only a few times per second. The fastest pulsar known to us was discovered in 2004 by astronomers in McGill University and is cataloged by the code number "PSR J1748-2446ad." It is estimated to be twice as massive as the Sun, but with a radius of only 16km, i.e. the width of a few blocks in a typical city! It spins at a rate of 716 times per second, which is so high that a point on its equator would travel at 24 percent of the speed of light, or about $70,000$km/s!

Solving for $v_{\text{escape}}$, we find

$$v_{\text{escape}} = \sqrt{\frac{2GM}{r}}. \tag{6.61}$$

Note that at infinity the mass $m$ stops completely and thus has vanishing total energy. A similar calculation within general relativity is possible. The total energy of an object starting from rest at a radius $r$ from the center of a black hole is given by (6.53):

$$E = \frac{mc^2}{\sqrt{1 - \frac{r_s}{r}}}, \tag{6.62}$$

where we have substituted the value of $T$ from the Schwarzschild metric (5.47). In this context, the energy (6.62) may be interpreted as playing the role of the potential energy required to move the object from an arbitrary radius $r$ to infinity, since at $r \to \infty$ (6.62) reduces to just the particle's rest energy (in contrast with the Newtonian case, which reduces to zero). Now, in the object's local frame, the special relativistic expression (4.144) is also valid, so equating

$$\frac{mc^2}{\sqrt{1 - \frac{r_s}{r}}} = \frac{mc^2}{\sqrt{1 - \frac{v^2}{c^2}}} \tag{6.63}$$

and solving

$$v_{\text{escape}} = c\sqrt{\frac{r_s}{r}} = \sqrt{\frac{2GM}{r}}, \tag{6.64}$$

which is the escape velocity in the Schwarzschild metric at any distance $r \geq r_s$. Note that right *on* the event horizon $r = r_s$ the escape velocity is exactly the speed of light, while inside the event horizon $r < r_s$, $v_{\text{escape}}$ is greater than $c$. We conclude, then, that nothing, not even light, can escape from $r < r_s$. It is worth noting that this is one of those cases where the Newtonian picture of gravity agrees exactly with general relativity. This result is interesting, but it is also not very rigorous. The assumption here is that the object is given an initial velocity $v_{\text{escape}}$, then left to travel under its own inertia. This does not exclude the possibility of a rocket continuously accelerating away from the event horizon. As such, a more rigorous understanding is needed.

**Exercise 6.13** One of the arguments used by the people who deny that we have ever landed on the Moon is that one needed a huge rocket ("Saturn V") with a lot of fuel to leave Earth, while to leave the Moon at the end of the trip the much smaller moon lander (the "Eagle") managed to leave the Moon with much less fuel expenditure. Show that this is not a problem, as the deniers think. First compare the escape velocity from Earth $v_E$ to that of the Moon $v_m$. Then to show that the amount of energy needed to leave the Moon is very much less than that needed to leave Earth, also compare the kinetic energy needed to leave Earth to that needed to leave the Moon. Data you may require: Earth's mass: $5.972 \times 10^{24}$kg. The Moon's mass: $7.35 \times 10^{22}$kg. Earth's radius: $6,356$km. The Moon's radius: $1,737.1$km. Mass of "Saturn V:" $2.8 \times 10^6$kg. Mass of the "Eagle:" $15,200$kg.

**Exercise 6.14** The idea of a black hole is much older than general relativity. Scientists conjectured the possible existence of what they called "dark stars" as far back as Newton's time. It was unclear at the time if light behaved as massive particles would, i.e. if it would have an escape velocity, but assuming it did, Laplace calculated the minimum diameter of a star with the same density as that of Earth where its escape velocity became larger than that of light and consequently wouldn't shine. Calculate the value he found and express your answer in multiples of the Earth's diameter. The data you will need is found in Exercise 6.13.

## 6.2.2   The Event Horizon and Light Cones

Another way to demonstrate that nothing can escape the inside of the event horizon is by invoking the light cone concept discussed back in §4.1. Recall that the light cone drawn in spacetime at a point $O$ "decides," in a sense, the future of $O$. In other words, looking at Fig. [4.2] or [4.3], an object at $O$ at time $t = 0$ has no choice but to move "up" within the light cone. Light cones can be drawn at any point in spacetime. Let's see how this is done systematically for flat spacetime first. Using the Minkowski-spherical metric (4.13), let's suppress the angular components, or equivalently choose $\theta$ and $\phi$ to be constants; e.g. $\theta = 90°$ and $\varphi = 0$. Hence (4.13) becomes

$$ds^2 = -c^2 dt^2 + dr^2 = 0 \tag{6.65}$$

for a light signal. From this, we find

$$\int_{r_0}^{r} dr = \pm c \int_{t_0}^{t} dt$$

$$r\left(t\right) = \pm c\left(t - t_0\right) + r_0, \quad \text{or}$$

$$ct\left(r\right) = \mp r \pm r_0 + ct_0. \tag{6.66}$$

The last equation describes the straight world lines of light in flat spacetime. If plotted using $x^0 = ct$, their slopes are $\pm 1$—as you can see in Fig. [6.5]. Choosing any point $(r_0, t_0)$, the future of that point lies inside the top part of its light cone. As a side note, such diagrams are also useful in demonstrating the concept of **causality**; i.e. what points in spacetime can send signals that affect other points and when. For example, consider a person standing still at $r = 10$ in Fig. [6.5]. At time $ct = 10$, i.e. point $A$, he sends a signal (for example, throws a rock) in the negative $r$ direction; the world line of the rock is shown as a solid line. Now another person located at $r = 2.5$ throws another rock at $ct = 12.5$, i.e. point $B$. The second rock's world line is shown dashed. Both rocks must of course travel more slowly than light; hence they will both plot world lines inside their respective light cones and will only collide when the two world lines meet: in this case at approximately $ct = 17.4$. The fastest signals possible between $A$ and $B$ are beams of light, whose paths are the lines making 45° angles. Hence the overlap between the two light cones; the shaded region, is the **causality region**. In other words, nothing starting at $A$ and $B$ can interact outside of that region.

**Exercise 6.15**

1. In flat spacetime two people are standing a distance $\Delta r = 10$m apart. Use Fig. [6.5] to find the shortest period of time in units of $ct$, as well as in seconds, within which either one of them can become aware of the other's presence.
2. Answer the question in part 1 using eqn (6.66).

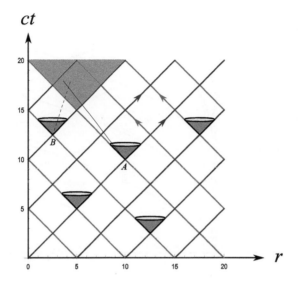

Fig. 6.5: Future light cones of points in flat spacetime. The outgoing geodesics (in blue) correspond to the positive slopes in eqn (6.66), while the ingoing geodesics (in red) correspond to the case of negative slopes.

Now let's plot a similar diagram for the Schwarzschild metric (5.47):

$$-c^2 \left(1 - \frac{r_s}{r}\right) dt^2 + \frac{dr^2}{\left(1 - \frac{r_s}{r}\right)} = 0, \tag{6.67}$$

where we have used (5.55). Upon rearranging, we end up with

$$\int_{r_0}^{r} \frac{dr}{\left(1 - \frac{r_s}{r}\right)} = \pm c \int_{t_0}^{t} dt. \tag{6.68}$$

Solving this and rearranging, we get the equation

$$ct(r) = \pm (r - r_0) \mp r_s \ln \left(\frac{r - r_s}{r_0 - r_s}\right) + ct_0, \tag{6.69}$$

which is the equation of the radial lightlike geodesics of the Schwarzschild metric. The thing about this equation is that it "flips" signs when $r < r_s$. This was clear from the metric, as we noted before. The result is that when plotted (Fig. [6.6]) the light cones inside the event horizon are "tipped" onto their sides. This means that any point inside the event horizon has a future *only* toward the central singularity; objects in $r < r_s$ not only cannot escape, but they also have no choice but to fall straight into $r = 0$.

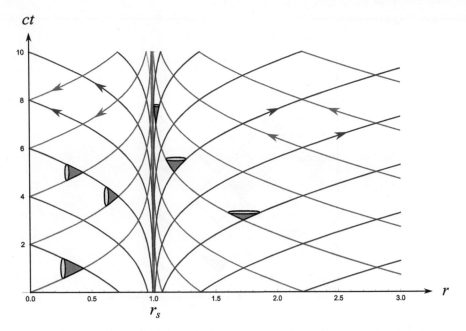

Fig. 6.6: Future light cones of points in Schwarzschild spacetime. The outgoing geodesics (in blue) have positive slopes outside the event horizon. They switch slopes inside the horizon. The ingoing geodesics (in red) follow the opposite pattern.

**Exercise 6.16** Consider two people located at $r_1 \approx 1.37$ and $r_2 \approx 2.2$ in the spacetime of Fig. [6.6].

1. Using the figure, estimate the shortest period of time, in units of $ct$, within which either one of these people can become aware of the other's presence.
2. Answer the question in part 1 using eqn (6.69).

**Exercise 6.17** Imagine that two people are located *inside* the event horizon of the black hole described by Fig. [6.6] (yes, they are both doomed to die, but let's ignore that morbid fact for a moment). Remembering that space and time flip for $r < r_s$, their locations are measured by $ct_1 = 4$ and $ct_2 = 6$.

1. Using the figure, estimate the shortest period of time, in units of $r$, within which either one of these people can become aware of the other's presence.
2. Answer the question in part 1 using eqn (6.69).

**Exercise 6.18** Write down the equations representing the light cones of the Robertson–Walker metric (5.102) for the dust-filled flat universe case: $k = 0$ and $a \sim t^{2/3}$. Sketch the spacetime diagram for these cones as seen in Fig. [6.6]. Recall from Exercise 5.27 that the time redefinition used there has the advantage of making these cones "look" like the flat spacetime ones in Fig. [6.5].

**Exercise 6.19** The so-called **Alcubierre metric**[8]

$$ds^2 = -c^2 dt^2 + [dx - v_s(t) f(r_s) dt]^2 + dy^2 + dz^2, \tag{6.70}$$

is an interesting solution of general relativity. It allows for the (hypothetical) possibility of constructing a device, popularly known as the **Alcubierre warp drive**, that would allow a spaceship to travel with *effective* faster-than-light speeds. It does, however, require a type of matter with negative energy density; i.e. negative mass, which is something that doesn't seem to exist naturally. The idea is that a spaceship with such a device would "warp" spacetime around it in such a way that the ship is effectively enclosed inside a spacetime "bubble;" which is free to move *faster* than light, even if the spaceship itself is *not* locally doing so.

The metric in the form given here refers to a specific trajectory of the superluminally traveling spaceship described by $x = x_s(t), y = 0, z = 0$ such that

$$v_s(t) = \frac{dx_s}{dt} \tag{6.71}$$

is the velocity along the trajectory, $r_s$ determines the distance of any point from the trajectory:

$$r_s^2 = (x - x_s)^2 + y^2 + z^2, \tag{6.72}$$

and $f$ is a smooth positive function with $f(0) = 1$. Write down the equations representing the light cones of the Alcubierre solution and plot some of them on a spacetime diagram.

## 6.2.3 Falling Past the Event Horizon

Consider free fall toward *or* away from the event horizon along radial geodesics.[9] First, assume the object to be lightlike, e.g. a beam of light or a single photon, for which $ds^2 = 0$, or equivalently (4.106). Since the geodesic we are looking for is radial, all angular components of the metric vanish, i.e. $d\Omega^2 = 0$, and the metric (5.52) becomes

$$ds^2 = -c^2 H dt^2 + \frac{dr^2}{H} = 0, \tag{6.73}$$

which leads to

$$\frac{dr}{dt} = cH. \tag{6.74}$$

We have already solved this in the previous section. This time we integrate between two arbitrary radii $r = r_1$ and $r = r_2$:

---

[8]Discovered by the Mexican theoretical physicist Miguel Alcubierre Moya (b. 1964).

[9]Yes, an object can "fall" away from a source of gravity. This is not just a relativistic phenomenon; even in ordinary mechanics an object thrown upward in Earth's gravitational field is said to be in "free fall," since gravity is the only force acting on it irrespective of its direction of motion.

$$\Delta t = \frac{1}{c} \int_{r_1}^{r_2} \frac{dr}{1 - \frac{r_s}{r}} = \frac{1}{c} \left[ (r_2 - r_1) + r_s \ln \left( \frac{r_2 - r_s}{r_1 - r_s} \right) \right]. \tag{6.75}$$

This expression is nice and finite for most values of $r_1$ and $r_2$, *except* if either of them is exactly equal to $r_s$, in which case the logarithm diverges! So if the light beam starts its journey right *on* the event horizon, i.e. $r_1 = r_s$, it takes an infinite amount of time for it to escape to any value $r_2 > r_s$; in other words, it simply doesn't! Furthermore, if the light beam is falling from any $r_1 > r_s$ *toward* the event horizon, i.e. $r_2 = r_s$, it *also* takes an infinite amount of time for it to arrive! That means that a coordinate observer watching the beam will *never* see it crossing the event horizon! I have used many exclamation points in the last few sentences, since these conclusions do not seem to make sense; not to our Newtonian intuition anyway. Also observe what happens if the beam starts its journey from *within* the event horizon, i.e. $r_1 < r_s$: The logarithm becomes undefined because its argument becomes negative! It is as if the definition of "time" itself collapses in this region!

---

**Exercise 6.20** Equation (6.74) can be interpreted as the speed of a photon as it moves along radial geodesics toward or away from the event horizon of a black hole. Sketch a graph of this speed versus $r$. What's the speed of the photon far away from $r_s$? What happens to it as it travels closer and closer to the event horizon? What happens to it *inside* the event horizon? Interpret your results.

---

Equation (6.75) yields the time $\Delta t$ that a coordinate observer would measure for a moving photon. It doesn't make sense, however, to ask how time passes by the photon itself.[10] This is because in order to do so we must ask what is the photon's proper time $\Delta \tau$, i.e. the time as measured in the photon's rest frame. But such a frame does *not* exist in the first place; hence the question itself is nonsensical. On the other hand, we can now discuss the motion of a massive object with rest frame $O'$ as it falls along radial geodesics toward $r_s$. Note that when asked, observer $O'$ must report that they felt absolutely nothing as they traveled toward and past the event horizon. This is because in their local frame they are essentially at rest. They are still the falling observer inside the elevator of Fig. [5.5]. The metric according to $O'$ is then just

$$ds^2 = -c^2 d\tau^2. \tag{6.76}$$

Now a coordinate observer $O$ sees the full Schwarzschild metric (5.52) minus the angular pieces. Since, as usual, both observers will agree on $ds^2$, we can write

$$-c^2 d\tau^2 = -c^2 H dt^2 + H^{-1} dr^2, \tag{6.77}$$

which leads to

---

[10]No matter how many times beginning students of relativity have asked this question, it simply has no answer!

$$H\dot{t}^2 - \frac{\dot{r}^2}{c^2 H} = 1, \tag{6.78}$$

where an over dot is a derivative with respect to the proper time $\tau$.

**Exercise 6.21** Show that (6.78) can also be found using the invariant $U^\mu U_\mu = -c^2$.

Now we would like to find an equation in $\dot{r}$ similar to (6.74), so the question arises as to what exactly one does with $\dot{t}$. This is where our discussion concerning Killing vectors becomes pertinent. One can show that $\dot{t}$ is in fact related to the conserved energy of the falling object. Since the Schwarzschild metric is independent of time, a timelike Killing vector exists; ergo

$$\xi_\mu U^\mu = \text{constant}. \tag{6.79}$$

Using $\xi^\mu = (-1, 0, 0, 0)$ and the 4-velocity of a falling object in Schwarzschild spacetime, (6.79) gives

$$cH\dot{t} = \text{constant}. \tag{6.80}$$

Since $H\dot{t}$ is dimensionless, the constant must have units of speed, and since all observers will agree on $\xi_\mu U^\mu$ and the only speed they will agree on is the speed of light, we take the constant to be proportional to $c$; hence

$$H\dot{t} = e, \tag{6.81}$$

where $e$ is some dimensionless positive number. Equation (6.81) must somehow be related to the total energy of the falling object, since it was found by considering the timelike Killing vector. One possible interpretation arises if we assume that the object starts its motion from rest at radial infinity. Then its starting energy was just its rest energy: $E_{\text{rest}} = mc^2$. Since energy is conserved, the particle's total energy is *always* equal to its rest energy. In that case $\frac{E}{mc^2} = e = 1$, where $E$ is the energy of the particle at any point in its motion. If, however, the object starts its motion with some energy that is *greater* than its rest energy, then the constant $e$ will be some other value greater than unity. In other words, $e$ is the object's total energy per its own rest energy:

$$H\dot{t} = \frac{E}{mc^2} = e. \tag{6.82}$$

We can now finally eliminate $\dot{t}$ from (6.78) and rearrange to give

$$\frac{dr}{dt} = \pm \frac{c}{e} H \sqrt{e^2 - H}. \tag{6.83}$$

The explicit value of $e$ doesn't change the conclusion we are aiming at, so for simplicity let's just set $e = 1$. We have found, then, an equation whose solution gives the radial path $r(t)$ of the falling object, similar to (6.74) for the case of the photon.

**Exercise 6.22** Show explicitly that (6.79) leads to (6.80).

**Exercise 6.23** Recall that the Lemaître metric (5.87) was constructed in such a way that curves of constant $\rho$ are *exactly* the radial geodesics of the Schwarzschild spacetime. Show that (6.83) follows directly from (5.85).

Rearranging

$$dt = -\frac{1}{c}\sqrt{\frac{r}{r_s}}\frac{dr}{H} \tag{6.84}$$

and integrating, we find the coordinate time

$$\Delta t = -\frac{2}{3cr_s^{\frac{1}{2}}}\left[\sqrt{r}\,(r + 3r_s) - 3r_s^{\frac{3}{2}}\tanh^{-1}\left(\sqrt{\frac{r}{r_s}}\right)\right]_{r_1}^{r_2}. \tag{6.85}$$

Now let's interpret what that means. The inverse tanh of any value equal to or greater than one is infinite, so this expression blows up for any $r = r_s$. If the starting point is $r > r_s$, then it would also take an infinite amount of time for the falling observer to reach $r = r_s$, as measured by the coordinate observer, just like in the case of the falling photon. In other words, the conclusions are similar to the ones we found in the case of the falling photon: No matter where $O'$ starts from, they will take an infinite amount of time to reach or cross the event horizon from either direction, as measured by $O$. In this case, however, we can ask how much time $O'$ himself measures as he falls into the black hole. In other words, what's $\Delta\tau$? One way of doing this is to solve (6.78) for $\dot{r}$ and rearrange. Or, like in Exercise 6.23, we can use (5.84). Both approaches lead to

$$\Delta\tau = -\frac{1}{c}\int_{r_1}^{r_2}\sqrt{\frac{r}{r_s}}\,dr = -\frac{2}{3c}\left(r\sqrt{\frac{r}{r_s}}\right)_{r_1}^{r_2}. \tag{6.86}$$

This result is *finite* for any value of $r$, including $r = r_s$. Just as we argued, even though the coordinate observer $O$ will never see the falling observer crossing the event horizon, $O'$ himself will experience nothing unusual. He will simply pass by the Schwarzschild radius unscathed. Weird? In a Newtonian sense, yes, absolutely.

**Exercise 6.24** Derive (6.86) using (6.78) and then again using (5.84).

**Exercise 6.25** An observer $O'$ is falling toward the event horizon of a Schwarzschild black hole. Calculate the time $\tau$ that $O'$ measures as he travels from $r = 2r_s$ to $r = r_s/2$. If a coordinate observer $O$ is timing the fall of $O'$, what approximate distance $r$ would $O'$ be at when the clock according to observer $O$ reads $t = \tau$?

**Exercise 6.26** Duplicate the calculations in this subsection for the Robertson–Walker universe. In other words, find $\Delta t$ for radial photons, then find $\Delta \tau$ and $\Delta t$ that radial local and coordinate observers would measure respectively. Approach the problem generally at first, then apply your results to the three cases of $k$. If you need to, you may assume $a \sim t^n$ where $n$ is any arbitrary real constant.

## 6.2.4 The Geodesic Equations

Let us now formally approach the problem of finding *all* of the geodesics of the Schwarzschild metric. In the previous subsections we have found the (null and timelike) radial geodesics. In those cases, the spatial part of the geodesic curve was in fact straight, just going from $r = 0$ to $r \to \infty$. The curvature of the geodesic manifested itself only in the temporal component $t(\lambda)$, which appears to us as a time dilation effect. Now of course non-radial, also known as "orbital," geodesics exist, and we will spend the remainder of this section discussing them. But first I wish to remind the reader of the discussion in the previous section concerning energy and momentum. Moving objects contribute to the deformation of spacetime. As such, their trajectories in a given background metric $g_{\mu\nu}$ are necessarily ambiguous, since their very presence *changes* $g_{\mu\nu}$. We approach this problem by taking the following stand: What we will be calculating in the rest of this section (as well as what we have already calculated in the previous subsections) are *not*, strictly speaking, the trajectories of particles! We will calculate *geodesics*; the lines defining the shortest distances between two points on a curved spacetime manifold. These come into three types: lightlike, timelike, and spacelike, depending on the classifications (4.7), (4.8), and (4.9). Geodesics are *exact*, and they exist whether or not a particle is following them. *However*, when we say that free particles follow geodesics, we are *necessarily* making an approximation. The true motion of particles requires including the new spacetime metric that their very presence induces; in other words, we have to include the non-linear effect of the particle's gravitational field on itself—a topic not very well understood to this day. It is assumed, however, that this approximation is very minor, in the sense that the deviation between the geodesics and the particles' true trajectories is extremely small. And if our particles are large enough, like planets orbiting around a star, then the difference will be virtually undetectable.

Having said this, let's begin by computing the Christoffel symbols of the Schwarzschild metric using (4.33). These are

$$\Gamma^t_{rt} = \Gamma^t_{tr} = \tfrac{1}{2}H'/H, \qquad \Gamma^r_{tt} = \tfrac{1}{2}HH', \qquad \Gamma^r_{rr} = -\tfrac{1}{2}H'/H$$

$$\Gamma^r_{\theta\theta} = -rH \qquad \Gamma^r_{\varphi\varphi} = -rH\sin^2\theta \qquad \Gamma^\theta_{\theta r} = \Gamma^\theta_{r\theta} = 1/r \qquad (6.87)$$

$$\Gamma^\theta_{\varphi\varphi} = -\sin\theta\cos\theta, \qquad \Gamma^\varphi_{r\varphi} = \Gamma^\varphi_{\varphi r} = 1/r, \qquad \Gamma^\varphi_{\theta\varphi} = \Gamma^\varphi_{\varphi\theta} = \cot\theta,$$

where a prime on $H$ is a derivative with respect to $r$. Inserting in (6.33) and simplifying leads to the Schwarzschild geodesic equations

$$\ddot{t} = -\dot{r}\dot{t}\frac{H'}{H}$$

$$\ddot{r} = \frac{1}{2}\frac{H'}{H}\dot{r}^2 + Hr\dot{\theta}^2 + Hr\sin^2\theta\dot{\varphi}^2 - \frac{c^2}{2}\dot{t}^2HH'$$

$$\ddot{\theta} = -\frac{2}{r}\dot{r}\dot{\theta} + \sin\theta\cos\theta\dot{\varphi}^2$$

$$\ddot{\varphi} = -\frac{2}{r}\dot{\varphi}\left(\dot{r} + r\cot\theta\dot{\theta}\right), \tag{6.88}$$

where an over dot is a derivative with respect to $\lambda$. Equations (6.88) are the equations that need to be solved to find the parametric functions $t(\lambda)$, $r(\lambda)$, $\theta(\lambda)$, and $\varphi(\lambda)$ describing *all* possible four-dimensional geodesic curves in the Schwarzschild spacetime.

**Exercise 6.27** Derive (6.87) and (6.88) in detail.

Mathematically speaking, (6.88) are four coupled non-linear second-order differential equations, and yes this is indeed as scary as it sounds. It is possible, however, to significantly simplify them by appealing once again to the isometries of the metric. This will allow us to solve one of them completely for $\theta$ and integrate the remaining ones once to yield first-order differential equations. We already have (6.82), which we again use to remove $\dot{t}$ from everywhere in (6.88):

$$\ddot{t} = -\dot{r}e\frac{H'}{H^2}$$

$$\ddot{r} = \frac{1}{2}\frac{H'}{H}\left(\dot{r}^2 - e^2c^2\right) + Hr\dot{\theta}^2 + Hr\sin^2\theta\dot{\varphi}^2$$

$$\ddot{\theta} = -\frac{2}{r}\dot{r}\dot{\theta} + \sin\theta\cos\theta\dot{\varphi}^2$$

$$\ddot{\varphi} = -\frac{2}{r}\dot{\varphi}\left(\dot{r} + r\cot\theta\dot{\theta}\right). \tag{6.89}$$

Next we note that the metric does not depend on $\varphi$, a sign of the existence of a Killing vector of the form $\xi^\mu = (0,0,0,1)$; hence

$$\xi_\mu U^\mu = g_{\mu\nu}\xi^\mu U^\nu = g_{\varphi\varphi}\xi^\varphi U^\varphi = r^2\sin^2\theta\dot{\varphi} = \text{constant}. \tag{6.90}$$

This quantity has units of angular momentum per unit mass; as such, we conclude that angular momentum is conserved. Since angular momentum is a vector, it *must* be conserved in direction as well as magnitude. It follows that orbital geodesic motion happens in a single plane specified by a specific value for $\theta$. As such, we set $\theta = $ constant and rewrite (6.90)

$$r^2\dot{\varphi} = \text{constant} = l \tag{6.91}$$

$$\rightarrow \quad \dot{\varphi} = \frac{l}{r^2}. \tag{6.92}$$

The reader familiar with the problem of finding orbits in the Newtonian theory of gravity[11] will recognize that using conservation laws to simplify the equations of motion of a particle happens there as well, the main difference being that the differential equations in the non-relativistic case arise from $\mathbf{F} = m\mathbf{a}$[12] rather than the geodesic equations. Now, since $\ddot{\theta} = \dot{\theta} = 0$, the third equation in (6.89) implies $\sin\theta\cos\theta = 0$, which leads to either $\theta = 0$ or $\theta = 90°$. However, the fourth equation blows up for $\theta = 0$; hence we are led to $\theta = 90°$ as the only physical possibility. In other words, the $z$ axis coincides with the direction of the conserved angular momentum vector. The first and fourth equations of (6.89) are just the time derivatives of (6.81) and (6.90) respectively, as the reader can check. Because of this, eqns (6.82) and (6.92) usually go by the name **first integrals of motion** here and in the equivalent Newtonian problem. Hence we have

$$\dot{t} = \frac{e}{H} \tag{6.93}$$

$$\ddot{r} = \frac{1}{2}\frac{H'}{H}\left(\dot{r}^2 - e^2 c^2\right) + l^2\frac{H}{r^3} \tag{6.94}$$

$$\theta = \frac{\pi}{2} \tag{6.95}$$

$$\dot{\varphi} = \frac{l}{r^2}. \tag{6.96}$$

**Exercise 6.28** Verify that taking the time derivatives of (6.93) and (6.96) leads to the first and fourth equations of (6.89) respectively.

In addition to (6.93) and (6.96), we actually have a third first integral for $r$ given by the equivalence of $ds^2$ to the two observers $O$ and $O'$. This is what we exploited in the previous section to find the radial geodesics. Hence

$$ds^2 = g_{\mu\nu}dx^\mu dx^\nu = \begin{cases} -c^2 d\tau^2 & \text{for massive particles} \\ 0 & \text{for massless particles} \end{cases} \tag{6.97}$$

which, upon dividing both sides by $-c^2 d\tau^2$, can be put in the form

$$-\frac{1}{c^2}g_{\mu\nu}U^\mu U^\nu = \epsilon, \tag{6.98}$$

where $\epsilon = 0$ for lightlike geodesics, $+1$ for timelike geodesics, and $-1$ for spacelike geodesics. This leads to

$$e^2 c^2 = \dot{r}^2 + H\left(\frac{l^2}{r^2} + \epsilon c^2\right), \tag{6.99}$$

where we set $\theta = 90°$ and use the $e$ and $l$ first integrals. Equation (6.99) represents the first integral of (6.94). Hence we have the complete set:

---

[11]The so-called "Kepler problem."
[12]Or equivalently from Lagrange's equations, if you know what that is.

$$\dot{t} = \frac{e}{H} \tag{6.100}$$

$$\dot{r}^2 + H\left(\frac{l^2}{r^2} + \epsilon c^2\right) = e^2 c^2 \tag{6.101}$$

$$\theta = \frac{\pi}{2} \tag{6.102}$$

$$\dot{\varphi} = \frac{l}{r^2}, \tag{6.103}$$

which we will proceed to solve starting in the next subsection.

---

**Exercise 6.29**  Verify that taking the derivative of (6.101) leads to (6.94), as claimed.

---

Before we proceed, let's discuss a couple of matters concerning interpretation: As we argued in the previous subsection, $e = E/mc^2$ is understood as the ratio of the total energy of the particle to its rest energy. Similarly, we can understand $l$ as the particle's angular momentum divided by its mass. More precisely, it is the angular momentum $L$ divided by the so-called "reduced mass" $\mu$:

$$l = \frac{L}{\mu}$$

$$\mu = \frac{mM}{m + M}, \tag{6.104}$$

and (6.100) may also be updated to

$$H\dot{t} = \frac{E}{\mu c^2} = e. \tag{6.105}$$

The reduced mass is constructed based on the analogous Newtonian problem, as may be found in any advanced book in classical mechanics. Physically this is understood by the fact that two masses $m$ and $M$ with a mutually attractive radial force orbit their common center of mass, rather than one mass orbiting the other. However, if the mass $M$ of the source is much larger than the mass of the particle $m$, as would be the case for, say, the Sun and a small planet like Earth, then the reduced mass becomes just the mass of the particle:

$$\lim_{M \to \infty} \mu = m. \tag{6.106}$$

In other words, the particle is taken to orbit the center of mass of $M$, which itself is taken to be stationary at the origin of coordinates. The other thing to note is that these interpretations involving $m$ are true only for massive particles. However, the reader will note that the introduction of the mass $m$ was done by hand in eqns (6.82) and (6.104). But the discussions that led to (6.100) and (6.103) make *no* assumptions about the mass of the particle! In other words, the full eqns (6.100), (6.101), (6.102), and (6.103) are true for *both* massive *and* massless particles, with only $\epsilon$ to be specified. Based on the discussion at the beginning of this subsection, this argument is equivalent to saying that the first integrals (6.100) and (6.103) are true for geodesics, but their *interpretation* as the energy and angular momentum of a particle moving along these geodesics is something that we impose on the analysis and is not to be taken as exact.

**Exercise 6.30** Duplicate the calculations in this subsection for the full (unspecified $k$ and $a$) Robertson–Walker metric (5.102). In other words, find the geodesic equations and simplify them as much as possible using the isometries of the metric. You might want to use the results of Exercise 5.34.

**Exercise 6.31** It can be shown that the metric of *flat* spacetime in a frame that is rotating with an angular speed $\omega$ about the $z$ axis of a reference frame at rest is

$$ds^2 = -c^2 \left[1 - \omega^2 \left(x^2 + y^2\right)\right] dt^2 + 2\omega \left(y dx - x dy\right) dt + dx^2 + dy^2 + dz^2. \tag{6.107}$$

Find the geodesic equations and show that in the non-relativistic limit they reduce to the Newtonian equations of motion for a free particle in a rotating frame with the usual centrifugal and Coriolis forces. You can find these in any standard book on classical mechanics such as [4] or [12]. Also compare with Exercise 3.3.

**Exercise 6.32** Construct the geodesic equations for the Alcubierre metric of Exercise 6.19 and show that the path $x_s$ defined therein is a geodesic. For a certain coordinate time difference $\Delta t$, i.e. a time period measured by a coordinate observer, how much time elapses for an observer traveling along $x_s$? In other words, find $\Delta \tau$. What is their 4-velocity?

## 6.2.5 The Way of the Potential

Our objective now is to solve the first integrals (6.100), (6.101), and (6.103) to find expressions for $t(\lambda)$, $r(\lambda)$, and $\varphi(\lambda)$ ($\theta$ already having been found to be a constant) that parameterize the Schwarzschild geodesics. We have already found one class of solutions—the radial geodesics—in previous subsections. These can be reproduced by simply setting $l = 0$ in (6.101). Hence we now look for the so-called orbital solutions; i.e. geodesics where $l$ is non-vanishing (henceforth "orbits" for short). Using lessons learned from the problem of Newtonian orbits, we can begin by classifying these solutions even before explicitly finding them, using the method of the so-called **effective potential**.

Using the explicit form for $H$ and defining the constant $\mathfrak{E} = \frac{c^2}{2}\left(e^2 - \epsilon\right)$, eqn (6.101) becomes

$$\mathfrak{E} = \frac{1}{2}\dot{r}^2 - \epsilon\frac{GM}{r} + \frac{l^2}{2r^2} - \frac{GMl^2}{c^2 r^3}. \tag{6.108}$$

Now, drawing on our Newtonian experiences, this equation should look a bit familiar. It has the form "Constant related to energy = Kinetic energy-like term (one half velocity squared) + Potential energy-like terms (dependent only on $r$)." In fact, in the equivalent non-relativistic problem, conservation of energy gives a similar formula: $E = K.E. + V_{eff}(r)$: "Constant total energy = Kinetic energy + Effective potential term," where the effective potential has two components: the true $1/r$ Newtonian gravitational potential plus a "centrifugal" $1/r^2$ term arising from the conservation of angular momentum. We have the exact same structure in (6.108) with one extra piece: the $1/r^3$ term. As such, we may interpret (6.108) as the statement of conservation of energy per unit mass, which is applicable to timelike, lightlike, and even spacelike particles pending the value of $\epsilon$.[13] For simplicity we call $\mathfrak{E}$ the "total energy," $\frac{1}{2}\dot{r}^2$ the "kinetic energy" and combine the remaining terms into an "effective potential:"

$$\mathfrak{E} = \frac{1}{2}\dot{r}^2 + V_{eff} \qquad \text{where} \qquad (6.109)$$

$$V_{eff} = -\epsilon\frac{GM}{r} + \frac{l^2}{2r^2} - \frac{GMl^2}{c^2r^3}. \qquad (6.110)$$

This last is a very powerful formula. It describes the radial behavior of a particle in a Schwarzschild background. Its analogous equation in Newtonian mechanics is[14]

$$V_{eff} = -\frac{GM}{r} + \frac{l^2}{2r^2}. \qquad (6.111)$$

**Exercise 6.33** As a refresher to the reader on how eqn (6.111) is found in the so-called Kepler problem of Newtonian mechanics, consider a particle $m$ in orbit around a source of gravity $M$. Write the conservation of energy equation $E = \frac{1}{2}mv^2 + V(r)$, where $v^2 = \mathbf{v} \cdot \mathbf{v}$ in spherical coordinates and $V(r)$ is the Newtonian gravitational potential energy defined by (3.90). Next set $\theta = 90°$ and $l = mr^2\dot{\varphi}$ as constants of motion. Finally, take all terms involving $r$ only to be the effective potential. Once finished, you should have eqn (6.111). The details we have skipped involving why $\theta = 90°$ and $l = mr^2\dot{\varphi}$ are constants of motion are found by solving the equations of motion of $\theta$ and $\phi$ that arise from $\mathbf{F} = m\mathbf{a}$. If so inclined, you may do this using the methods of Chapter 3, particularly the results of Exercise 3.2. An alternative way of reaching these conclusions is via the so-called "Lagrangian method," which we will discuss in Chapter 8.

In comparing (6.110) with (6.111), we note that they differ in one major way: the appearance of the inverse cube term in (6.110). This particular term is the one responsible for the physical differences between orbits in the Newtonian problem and those in the Schwarzschild problem. Prior to the advent of general relativity, the orbits of most of the planets of the solar system seemed to follow the effective potential (6.111), with $M$ being the Sun's mass, quite closely. There was one exception: Mercury.[15] Around

---

[13]If one does not wish to refer to particles, eqn (6.108) is still true as a constant at any point along geodesics. In this context $\dot{r}$ is not the radial speed of an object, it is just the change of $r$ on the geodesic with respect to an affine parameter. It's just easier to think of all this in the context of a moving particle.

[14]As may be found by consulting any text in classical mechanics such as [12].

[15]One must point out that this statement is not exactly true. The effective potential (6.111) assumes that there is only one planet in orbit around the Sun. The fact that in reality there are others makes for additional corrections that we are ignoring. However, these were well known at the time, and the discussion here implicitly assumes that they have been taken into consideration.

1859 it was noticed that Mercury does not follow the exact orbits predicted by (6.111) and that there is a very small deviation. It was later realized that this deviation is due to the inverse cube term, neglected in (6.111). The reason it only appears in Mercury's orbit is that Mercury is the closest planet to the Sun. Planets further away—Venus, Earth, and onwards—do not exhibit this behavior largely enough to be observed. This is true because an inverse $r$ cubed dependence dies faster than both the inverse $r$ and $r$ squared terms in (6.110); in other words, the effect of the $1/r^3$ term becomes relatively insignificant. You can see this clearly in Fig. [6.7].

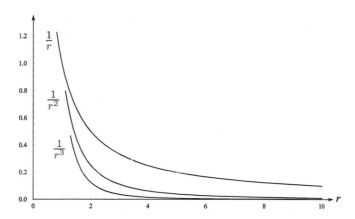

Fig. 6.7: Comparing the curves $1/r$, $1/r^2$, and $1/r^3$. Clearly the inverse cube curve dies out faster, i.e. "gets" to zero first. This explains why planets that are further away from the Sun are not as affected by the $1/r^3$ term in the effective potential as Mercury is. Imagine, for instance, that Mercury is "located" at around $r = 3$ on the scale of this graph, while Venus is at $r = 6$, where clearly the effect of the inverse cube function becomes insignificant compared to the other two.

Now, notice that although the problem of orbits is in reality a four dimensional problem, we have been able to reduce the degrees of freedom down to only one: the radial parameter $r$. The other parameters $(t, \theta, \text{and } \varphi)$ were effectively replaced by the constants of motion $\mathfrak{E}$, $90°$, and $l$ respectively. We can just solve for $r$ and then use (6.103) to find the angular behavior $\varphi(\lambda)$, and also (6.100) to find $t(\lambda)$; and of course $\theta$ has already been found in (6.102).

Before doing so, let us see if it is possible to deduce some general conclusions from the forms of (6.109) and (6.110). Both in here and in the Newtonian case one would want the kinetic energy term in the total energy to be positive, since otherwise the speed $\dot{r}$ will be imaginary, and that's not physical. On inspecting (6.109) it is easy to see that this condition is satisfied for all values $\mathfrak{E} \geq V_{eff}$: in other words, the regions where $\mathfrak{E} < V_{eff}$ are *forbidden* for any object.[16] This breaks down the possible orbits into five categories. Before looking at graphs of our case of interest, the Schwarzschild

---

[16]It is interesting to note that the case of negative kinetic energy arises in the analogous quantum-mechanical analysis and yet is completely physical! It has indeed been observed that quantum particles

effective potential (6.110), let's consider for the sake of demonstrability the arbitrary effective potential curve in Fig. [6.8], which exhibits the five possible cases.

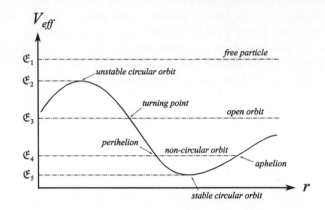

Fig. 6.8: A sample effective potential for demonstration.

1. $\mathfrak{E} = \mathfrak{E}_1 > V_{eff}$: This particle is approaching the gravitational center from radial infinity, traveling straight into $r = 0$. Note, though, that its speed changes along this path: The vertical difference between the horizontal $\mathfrak{E}_1$ line and the effective potential curve is the particle's kinetic energy. This case could be a particle that is falling radially straight into the center of gravitational attraction (i.e. has no angular momentum $l$) or it could be exhibiting a rotational motion as it approaches, in other words "spiralling" inwards.

2. $\mathfrak{E} = \mathfrak{E}_2 = V_{eff}$: This particle has a fixed radius from the gravitational center ($\dot{r} = 0$). This is a perfectly circular orbit. However, it is an *unstable* orbit in the sense that, if given a small push, known as a perturbation, such that its energy rises above $\mathfrak{E}_2$, the particle's orbital radius is no longer a constant. In other words, the particle jumps to the first case and spirals away.

3. $\mathfrak{E} = \mathfrak{E}_3$: The particle is coming in from radial infinity, approaching the potential to a minimum distance (the so-called "turning point"), orbiting around it, and shooting back to infinity. This is the case of an **open orbit**. In Newtonian gravity it corresponds to parabolic or hyperbolic curves. We will look at this in more detail by considering the case of the so-called gravitational lensing, where light is found to "bend" around a massive object such as the Sun or, even more intensely, a black hole.

4. $\mathfrak{E} = \mathfrak{E}_4$: This particle is stuck between two radii in a **bound orbit**. In the Newtonian version these are elliptical orbits. The point of closest approach is known as the **perihelion**, while the point of farthest distance is the **aphelion**. In the Schwarzschild case, the orbits *precess*; in other words, they do not close in perfect ellipses. This is the case most notable in the planet Mercury.

can "exist" inside the potential in a phenomenon known as **quantum tunneling**. Classically, such a situation does not and could not arise either in the Newtonian or in the general relativistic cases.

5. $\mathfrak{E} = \mathfrak{E}_5 = V_{eff}$: The particle here is also in a perfectly circular orbit: $r =$ constant. However, this one is stable; one can see that if the object is given a small extra amount of energy, it will just rise above the potential's minimum and become a non-circular bound orbit. Stability in this context means that the particle stays bound and does not escape like in the second case.

Now that we understand the general idea of the effective potential method, let's go back to our case of interest: the effective potential (6.110). We make two plots in Fig. [5.11]: one for the lightlike case ($\epsilon = 0$) and one for the timelike case ($\epsilon = 1$). The spacelike case is doable as well, but it does not represent the paths of real particles. For each of these cases the potential depends on the particle's angular momentum, so we make five plots for different values of $l$, just to see what happens. We also take $c = 1$ and $GM = 1$ for ease of plotting.

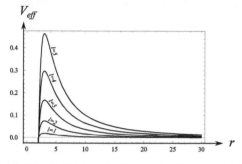

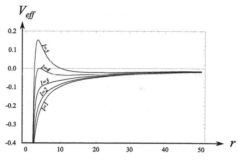

(a) The lightlike effective potential ($\epsilon = 0$).      (b) The timelike effective potential ($\epsilon = 1$).

Fig. 6.9: The effective potential of lightlike and timelike Schwarzschild geodesics for various values of $l$ ($c = 1, GM = 1$).

One can see that both cases have unstable circular orbits. There don't seem to be any stable circular orbits for the lightlike case. For the timelike case it may be a bit hard to see, but there *are* stable circular orbits for high angular momentum values, as well as non-circular bound orbits. In the following subsections, we will discuss circular orbits, non-circular bound orbits, and open orbits—in that order.[17]

**Exercise 6.34** Using a computer and assuming several values for $l$, plot the effective potential for the Schwarzschild spacelike geodesics and discuss their properties. What types of "orbits" do you observe?

**Exercise 6.35** Duplicate the calculations in this subsection for the Robertson–Walker space-time (5.102). Approach the problem generally at first, then apply your results to the three cases of $k$. If you need to, you may assume $a \sim t^n$ where $n$ is any arbitrary real constant. For appropriate choices of constants, plot and interpret the effective potentials. You may use the results of Exercise 6.30.

---

[17]A detailed comparison between the Schwarzschild and the Newtonian cases may be found in [13].

## 6.2.6   Circular Orbits

The circular orbits of the Schwarzschild spacetime correspond to the case $r = R = \text{con-}$ stant. Since the possible $r = R$ orbits correspond to extrema of the effective potential curves, they may be found by applying the first derivative test on $V_{eff}$:

$$\frac{d}{dr} V_{eff} \bigg|_{r=R} = \epsilon \frac{GM}{R^2} - \frac{l^2}{R^3} + 3\frac{GMl^2}{c^2 R^4} = 0, \tag{6.112}$$

which upon rearranging gives

$$\epsilon GMR^2 - l^2 R + 3\frac{GMl^2}{c^2} = 0. \tag{6.113}$$

For lightlike geodesics $\epsilon = 0$, there is only one circular orbit: the so-called "photon orbit" with radius

$$R_{\text{photon}} = 3\frac{GM}{c^2} = \frac{3}{2}r_s. \tag{6.114}$$

This of course applies to any massless particle, photons being the most famous example. For massive particles $\epsilon = 1$ eqn (6.113) is quadratic in $R$; hence there are two possible orbits corresponding to

$$R_\pm = \frac{l^2}{2GM} \pm \frac{1}{2GM}\sqrt{l^4 - \frac{12G^2M^2l^2}{c^2}}. \tag{6.115}$$

Requiring that the argument in the square root is greater than or equal to zero sets a lower bound on the angular momentum per unit mass:

$$l \geq 2\sqrt{3}\frac{GM}{c}. \tag{6.116}$$

To find out whether these orbits are stable requires taking the second derivative and substituting the solutions (6.114) and (6.115) as usual. This is clearly rather grueling; however, a look at Fig.[6.9a] will convince the reader that the photon orbits are clearly unstable, as they correspond to a maximum rather than a minimum. So a lightlike particle, given appropriate initial conditions, may be stuck in a circular orbit forever unless perturbed slightly, in which case it will fly off, either away to radial infinity or toward the gravitational center $r = 0$. In the timelike case, the situation depends on the value of the angular momentum $l$. For large values of $l$ there are two circular orbits which may be found by taking the limit $l \to \infty$ in (6.115):

$$\lim_{l \to \infty} R_\pm = \frac{l^2}{2GM} \pm \frac{l^2}{2GM}\left(1 - \frac{12G^2M^2}{c^2 l^2}\right)^{\frac{1}{2}} \tag{6.117}$$

$$\approx \frac{l^2}{2GM} \pm \frac{l^2}{2GM}\left(1 - \frac{6G^2M^2}{c^2 l^2}\right), \tag{6.118}$$

where we have used (4.138) in the second line. The two solutions then correspond to a stable outer orbit with radius

$$R_+ \approx \frac{l^2}{GM} - 3\frac{GM}{c^2} \approx \frac{l^2}{GM}, \tag{6.119}$$

and an unstable inner orbit that corresponds exactly to the photon orbit

$$R_- \approx 3\frac{GM}{c^2} = \frac{3}{2}r_s. \tag{6.120}$$

**Exercise 6.36** Explicitly derive every equation in this subsection. In other words, start at the beginning and fill in all calculational gaps.

**Exercise 6.37** Find the radii of any circular spacelike Schwarzschild orbits that may exist. You can use the results of Exercise 6.34.

**Exercise 6.38** Explicitly study the stability of the circular orbits found in this subsection for both massless and massive particles by applying the second derivative test to $V_{eff}$.

**Exercise 6.39** Discuss and interpret circular geodesics in the Robertson–Walker metric (5.102). Approach the problem generally at first, then apply your results to the three cases of $k$. If you need to, you may assume $a \sim t^n$ where $n$ is any arbitrary real constant. You may wish to use the results of Exercise 6.35.

## 6.2.7  Interlude

Generally speaking, the effective potential method may be used to construct a formal solution to the geodesic equations in the following way. Starting with (6.110) we can solve for $d\lambda$ and "integrate"

$$\lambda(r) = \int \frac{dr}{\sqrt{2(\mathcal{E} - V_{eff})}}, \tag{6.121}$$

from which we can "invert" $\lambda(r)$ to find $r(\lambda)$. The coordinate time $t(\lambda)$ may then be found by "integrating" (6.100):

$$t(\lambda) = e \int \frac{d\lambda}{1 - \frac{r_s}{r(\lambda)}}, \tag{6.122}$$

and "finally"

$$\varphi(\lambda) = l \int \frac{d\lambda}{r^2(\lambda)} \tag{6.123}$$

from (6.103). The reader may have noticed that I have put some keywords in quotes above. This is to emphasize that all of this is easier said than done. Even with the use of first integrals, performing these formal calculations analytically is hardly a walk in the park. Computations performed when these problems were first introduced[18] typically used approximate methods. Modern-day approaches usually employ numerical computations on the computer to arbitrary degrees of accuracy.

**Exercise 6.40** Construct and solve the integrals (6.121), (6.122), and (6.123) to find a complete description of the motion of:

1. A radial photon.
2. A radial massive particle.

Compare your results with the analysis in §6.2.3.

**Exercise 6.41** Construct and solve the integrals (6.121), (6.122), and (6.123) to find the radii of circular orbits for:

1. A photon.
2. A massive particle.

Compare your results with the analysis in §6.2.6.

To give the reader a taste of how any of these computations are done, let's reformulate (6.110) and put it in a slightly more convenient form. To do this we first remove $\lambda$ from the picture:

$$\frac{dr}{d\varphi} = \frac{dr}{d\lambda}\frac{d\lambda}{d\varphi} = \frac{\dot{r}}{\dot{\varphi}} = \frac{r^2\dot{r}}{l}, \tag{6.124}$$

where the chain rule and (6.103) have been used. Next we substitute this in (6.110) to get

$$\left(\frac{dr}{d\varphi}\right)^2 = 2\frac{r^4}{l^2}(\mathcal{E} - V_{eff})$$

$$= \frac{2\mathcal{E}}{l^2}r^4 + 2\epsilon\frac{GM}{l^2}r^3 - r^2 + \frac{2GM}{c^2}r. \tag{6.125}$$

It is typical in the analogous Newtonian problem to use the substitution

$$u(r) = \frac{1}{r}. \tag{6.126}$$

Hence

$$\frac{du}{d\varphi} = \frac{du}{dr}\frac{dr}{d\varphi} = -\frac{1}{r^2}\frac{dr}{d\varphi} \tag{6.127}$$

and

$$\frac{dr}{d\varphi} = -\frac{1}{u^2}\frac{du}{d\varphi}. \tag{6.128}$$

---

[18]These issues also arise in non-relativistic mechanics and were known in some form or another to Newton himself.

Substituting in (6.125), we end up with

$$\left(\frac{du}{d\varphi}\right)^2 = 2\frac{\mathcal{E}}{l^2} + 2\epsilon\frac{GM}{l^2}u - u^2 + \frac{2GM}{c^2}u^3. \tag{6.129}$$

This is still too complicated; however, it is in a much better form now for our purposes.

## 6.2.8 Bound Non-Circular Orbits

Equation (6.129) differs from the analogous one in Newtonian mechanics only in the cubic term. The exact Newtonian formula is just

$$\left(\frac{du_0}{d\varphi}\right)^2 = 2\frac{\mathcal{E}}{l^2} + 2\frac{GM}{l^2}u_0 - u_0^2, \tag{6.130}$$

with $\epsilon = 1$, and we are calling the Newtonian solutions $u_0$. This eqn (6.130) has well-known exact solutions: They are the so-called "conic sections," which may be described by

$$u_0\left(\varphi\right) = \frac{1 - n\cos\varphi}{np}. \tag{6.131}$$

The constant $n$ is known as the "eccentricity" and has three possible values:

- $n < 1$ gives ellipses.
- $n > 1$ gives hyperbolas.
- $n = 1$ gives parabolas.

The quantity $p$ has units of distance and in the case of the ellipse yields the distance between the ellipse's minor axis and either one of its foci. Since we are interested in bound orbits, we require that $n < 1$. It is straightforward to show that (6.131) satisfies (6.130) if

$$p = \frac{l^2}{\sqrt{2\mathcal{E}l^2 + G^2M^2}}$$
$$n = \frac{\sqrt{2\mathcal{E}l^2 + G^2M^2}}{GM}, \tag{6.132}$$

such that the full Newtonian solution is

$$u_0\left(\varphi\right) = \frac{GM}{l^2}\left(1 - \frac{\sqrt{2\mathcal{E}l^2 + G^2M^2}}{GM}\cos\varphi\right). \tag{6.133}$$

**Exercise 6.42** Substitute (6.131) into (6.130) to show that it, satisfies it provided that (6.132) are true.

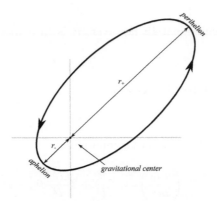

Fig. 6.10: Newtonian bound orbits are perfect ellipses.

Recalling the discussion around Fig. [6.8], we define the distance of closest approach (aphelion) to be $r = r_-$, and that of furthest (perihelion) to be $r = r_+$; see Fig. [6.10]. Solving (6.129) in full is way too complicated; it leads to a type of integrals known as "elliptic integrals," which are notoriously difficult to solve. In fact, they have no general analytical solutions. We can, however, solve it numerically: Under a reasonable choice of initial conditions we find that one of its solutions has the form shown in Fig. [6.11]. The orbits are not closed ellipses as in the Newtonian case, but rather continue to precess. We will make a rough estimate of the angle of precession $\delta\varphi$ in the remainder of this subsection.

**Exercise 6.43** Construct a computer program to numerically solve (6.129) with reasonable assumptions of the constants and the initial conditions. Plot the orbits that you find. You may do so using any computational technique and programming language or one of the commercially available symbolic manipulation software packages.

Inverting (6.129) leads to

$$\varphi(u) = \int \frac{du}{\sqrt{\frac{2\mathfrak{E}}{l^2} + 2\frac{GM}{l^2}u - u^2 + 2\frac{GM}{c^2}u^3}}, \tag{6.134}$$

where we have used $\epsilon = 1$ for timelike particles. Now the change in $\varphi$ as the distance changes from $u_+$ to $u_-$ over an entire revolution is the same as the change from $u_-$ to $u_+$; hence the total change is $2\,|\varphi(u_+) - \varphi(u_-)|$. Note that in the Newtonian solution this would be exactly $2\pi$; hence the angle of precession would be the difference

$$\delta\varphi = 2\,|\varphi(u_+) - \varphi(u_-)| - 2\pi, \tag{6.135}$$

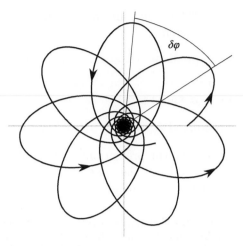

Fig. 6.11: Orbital precession in a Schwarzschild background (highly exaggerated).

which is

$$\delta\varphi = 2 \left| \int_{u_-}^{u_+} \frac{du}{\sqrt{\frac{2\mathfrak{E}}{l^2} + 2\frac{GM}{l^2}u - u^2 + 2\frac{GM}{c^2}u^3}} \right| - 2\pi. \tag{6.136}$$

In [6] Weinberg gives an approximate estimate of this integral, by far the most straightforward such calculation in the literature. We will not reproduce his work, but rather just report his result:

$$\delta\varphi \approx \frac{3\pi GM}{c^2}\left(u_+ + u_-\right) \text{ radians per revolution.} \tag{6.137}$$

The value is positive, which means that the precession proceeds in the same direction as the direction of motion of the object. Orbital precession is a very small effect; in fact, as mentioned earlier, the only planet in our solar system that exhibits a clearly observable precession is Mercury, by virtue of its closeness to the gravitational center, i.e. the Sun. The precession was first observed in 1859 by Urbain Jean Joseph Le Verrier[19] and for some time was assumed to be the result of the gravitational tug of a hitherto undiscovered planet in orbit between Mercury and the Sun. So sure was Le Verrier that such a planet existed that he even gave it a name: "Vulcan." Such a planet was, of course, never found, and Mercury's precession was only explained with the advent of general relativity. As we mentioned above, it is the extra cubic term in (6.108) that led to the series of arguments that finally explained orbital precession. If we plug the observed values of $u_-$ and $u_+$ as well as the Sun's mass $M$ into eqn

---

[19] French mathematician (1811–1877).

(6.137), we find $\delta\varphi = 0.1038''$ per revolution, and since Mercury makes 415 revolutions per century the total precession is:

$$\text{predicted } \delta\varphi = 43.03'' \text{ per century} \tag{6.138}$$

The observed value was found to be

$$\text{observed } \delta\varphi = 43.11 \pm 0.45'' \text{ per century}, \tag{6.139}$$

in astonishing agreement with the prediction.[20] Explaining the precession of Mercury was one of the major triumphs of the general theory of relativity and considerably helped staple it as a major advancement in our understanding of the universe.

**Exercise 6.44** Estimate $\delta\varphi$ for a planet in orbit around the supermassive black hole at the center of the M87 galaxy. You may take its aphelion at $r = 2r_s$ and its perihelion at $r = 2.5r_s$. The mass of the M87 black hole is given in Fig. [5.12]. If you have worked out Exercise 6.43 use the program you wrote to plot the orbit of this planet.

**Exercise 6.45** Consider a metric of the form

$$ds^2 = -c^2 dt^2 + \left(1 - ar^2\right) dr^2 + r^2 d\Omega^2, \tag{6.140}$$

where $a$ is a positive constant. Show that in this spacetime closed null orbits are ellipses.

### 6.2.9   Open Orbits

Our final example of computing the geodesics of the Schwarzschild metric is the case of a particle coming in from radial infinity. We will look at lightlike trajectories this time; in other words, $\epsilon = 0$ in (6.129):

$$\left(\frac{du}{d\varphi}\right)^2 = 2\frac{\mathfrak{E}}{l^2} - u^2 + \frac{2GM}{c^2}u^3. \tag{6.141}$$

As before, an exact analytical solution of (6.141) is not possible. Once again, using some reasonable initial parameters a numerical solution is found, and it gives the plot in Fig. [6.12].

**Exercise 6.46** Construct a computer program to numerically solve (6.141) with reasonable assumptions of the constants and the initial conditions. Plot the orbits that you find. You may do so using any computational technique and programming language or one of the commercially available symbolic manipulation software packages. If you have done Exercise 6.43, you should be able to adapt the code you already have to this exercise.

---

[20]Even though we have used the Schwarzschild metric to emulate the Sun's gravitational field, in itself an approximation.

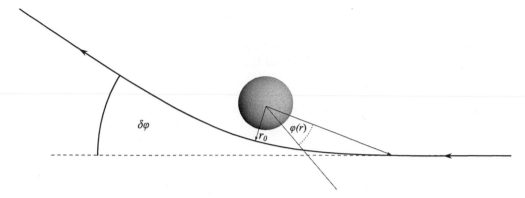

Fig. 6.12: The bending of light as it passes a massive object (highly exaggerated); also known as gravitational lensing.

As shown in the figure, we define the distance of closest approach by $r = r_0$. Now, the total change in $\varphi$ as $r$ decreases from infinity then increases again to infinity is twice its change from infinity to $r_0$. The Newtonian result would be exactly $\pi$ since lightlike particles do not deflect in Newton's theory. So the deflection we need to find would just be the difference:

$$\delta\varphi = 2\left|\varphi\left(r_0\right) - \varphi_\infty\right| - \pi. \tag{6.142}$$

Hence we have

$$\delta\varphi = 2\left|\int_{u_0}^{u_\infty} \frac{du}{\sqrt{\frac{2E}{l^2} - u^2 + 2\frac{GM}{c^2}u^3}}\right| - \pi. \tag{6.143}$$

We refer to Weinberg's approximation again, in which case he finds

$$\delta\varphi \approx \frac{4GMu_0}{c^2}. \tag{6.144}$$

The effect of the bending of light as it passes a massive gravitational source is very small. For example, if the beam is passing our Sun, the theory predicts (for a value of $r_0 = 6.95 \times 10^5$km) just $\delta\varphi = 1.75''$. It was, however, clear to scientists in the early twentieth century that such an effect, if observed, would constitute an even more important validation to the general theory than explaining the precession of Mercury, since in this case it predicts an entirely new phenomenon—one that doesn't even exist in the Newtonian picture. Following the end of World War I, Sir Arthur Stanley Eddington[21] led an expedition to photograph the total solar eclipse of May 29, 1919. The photographs showed the distant stars behind the eclipsed Sun shifted by an angle that conformed very closely with the prediction outlined above. The story of

---

[21]English astronomer and physicist (1882–1944).

this expedition is worth reading to interested readers, and they are referred to books of science history.[22] Since then this phenomenon, so-called **gravitational lensing**, has been observed and validated many times to high degrees of accuracy. For example, as recently as 1979 it was first observed that light from distant clusters of galaxies can bend around other clusters.

**Exercise 6.47** Estimate $\delta\varphi$ for a beam of light passing the supermassive black hole at the center of the M87 galaxy. You may take $r_0 = 1.2r_s$. The mass of the M87 black hole is given in Fig. [5.12]. If you have worked out Exercise 6.46 use the program you wrote to plot the path of this light beam.

## 6.3 Generally Covariant Electrodynamics

There really isn't much to say here. As discussed earlier, Maxwell's theory of electrodynamics is already fully specially covariant and requires no modifications to conform with special relativity. In addition, minimal substitution automatically makes it generally covariant. In other words, most of the equations in §4.8 carry over unscathed to curved spacetimes; the only difference is that indices are now raised and lowered by some curved metric tensor $g_{\mu\nu}$. For completeness, we relist these equations:

$$\text{4-potential:}\quad A^\alpha = \left(\frac{V}{c}, \mathbf{A}\right) \tag{6.145}$$

$$\text{4-current:}\quad J^\alpha = (c\rho, \mathbf{J}) \tag{6.146}$$

$$\text{Field tensor:}\quad F^{\alpha\beta} = \partial^\alpha A^\beta - \partial^\beta A^\alpha \tag{6.147}$$

$$\text{Maxwell's equations:}\quad \nabla_\alpha F^{\alpha\beta} = \mu_0 J^\beta \tag{6.148}$$

$$\partial_\alpha F_{\beta\gamma} + \partial_\gamma F_{\alpha\beta} + \partial_\beta F_{\gamma\alpha} = 0 \tag{6.149}$$

$$\text{Continuity equation:}\quad \nabla_\alpha J^\alpha = 0 \tag{6.150}$$

$$\text{d'Alembert's equation:}\quad \Box A^\alpha = \mu_0 J^\alpha \tag{6.151}$$

$$\text{Lorentz force:}\quad f^\mu = q F^\mu{}_\nu \frac{dx^\nu}{d\tau}, \tag{6.152}$$

where we recall that eqns (6.147) and (6.149) are already covariant and the d'Alembertian box operator is defined by

$$\Box = \nabla^\mu \nabla_\mu = g^{\mu\nu} \nabla_\mu \nabla_\nu = \frac{1}{\sqrt{-|g|}} \partial_\mu \left[\sqrt{-|g|}\partial^\mu\right., \tag{6.153}$$

where once again a minus sign is inserted inside the square root of the metric's determinant to make it positive, as pointed out back in §2.5.

**Exercise 6.48** Understanding the behavior of electromagnetic fields in a curved spacetime background requires substituting the metric tensor in the above equations. Do this for as

---

[22]For example, [14].

many of the equations of this section as you feel like, and for as many of the metric tensors we have studied as you can, particularly Schwarzschild (5.47) and Robertson–Walker (5.102). Simplify as much as possible.

**Exercise 6.49** This exercise may amount to a project that the reader may wish to undertake only if they have a few days to spare. Starting with any advanced textbook on electrodynamics, choose (or your instructor may choose for you) some of the solved examples that involve, say, finding electric or magnetic fields in a variety of situations, calculating the motion of charged particles in an electromagnetic field, or any other example that may be of interest. Construct the equivalent problem in the curved spacetime background of your choice (Schwarzschild being the most important we have presented) and attempt to redo the same problem, but this time in a curved background. Depending on which problem(s) you chose, you may find that an analytical solution is too complicated. In which case you may proceed to solve the equations numerically on the computer. Compare your results with the results of the original problem(s).

*Historical note*: Since the quantum-mechanical version of electromagnetic theory (Quantum Electrodynamics, or QED) has been known for some time, one can perform *quantum* electromagnetic calculations in a curved background similar to the non-quantum ones you will do in this exercise. Such calculations are dubbed "semiclassical," since only one side of the picture—the electromagnetic side—is quantum mechanical, while gravity is still unquantized. A calculation such as this was performed by Stephen Hawking in 1974 to deduce that due to quantum-mechanical effects near the event horizon of a black hole, it is possible for black holes to radiate electromagnetically. This discovery is now known as the **Hawking radiation**. Note that electromagnetic radiation traveling away from the event horizon carries away some of the energy, and therefore mass, of the black hole. This means that eventually black holes would run out of mass and "evaporate" completely. It is an extremely slow process, however. In fact, it was calculated that a black hole only as massive as our Sun would evaporate over a period of $10^{64}$ years! Supermassive black holes, like the ones at the centers of galaxies, would take a lot longer than this: anywhere up to $10^{106}$ years.

# 7

# Riemann and Einstein

*I had the decisive idea of the analogy between the mathematical problem of the [general] theory [of relativity] and the Gaussian theory of surfaces only in 1912 [...] after my return to Zurich, without being aware at that time of the work of Riemann, Ricci, and Levi-Civita. This was first brought to my attention by my friend Grossmann[1] when I posed to him the problem of looking for generally covariant tensors whose components depend only on derivatives of the coefficients $g_{\mu\nu}$ of the quadratic fundamental invariant $g_{\mu\nu}dx^\mu dx^\nu$.*

Albert Einstein

## Introduction

The general theory of relativity can be broken down into two major parts: Firstly, there is the principle of equivalence, which inevitably leads to the ideas of general covariance, free-fall geodesics, and particle trajectories. All of this is based on the concept of a metric, which we take to mean a coordinate chart on a preexisting spacetime manifold. This is what we have been occupied with thus far. The second major part is the ideas of Riemannian geometry, effectively applying exquisite rigor to the concepts discussed in the first part. In addition, and standing by itself as a major edifice of Einstein's genius, we come to the gravitational field equations formulated by Einstein in 1915 in a very close race with David Hilbert. The interesting story of how Einstein was led to his famous field equations is told in many sources,[2] but it might be important to note here that this discovery was not, by any stretch of the imagination, inevitable. In fact, one might argue that, like all similarly great discoveries and despite its basis in a solid foundation of geometry, the final form of the equations is simply based on intuition, physical insight, and pure unadulterated genius. Having familiarized ourselves with general covariance, we now turn our attention to a bit more rigor. We will review Riemann's arguments on the exact definitions of the terms "manifold," "metric," "curvature," and "connection." We will then follow this with a discussion of the Einstein field equations and use it to fill in the gaps of the previous chapters.

---

[1] Marcel Grossmann: Hungarian mathematician (1878–1936).
[2] For example, see the articles [15] and [16].

*Covariant Physics: From Classical Mechanics to General Relativity and Beyond.* Moataz H. Emam, Oxford University Press (2021). © Moataz H. Emam.
DOI: 10.1093/oso/9780198864899.003.0007

## 7.1 Manifolds

Mathematicians tend to make things as precise as possible, which is essential to the required rigor of mathematics itself. This inevitably leads to complicated notions and definitions. We will try to avoid this and present here a more intuitive, yet still rigorous, approach. One says that a **manifold** $\mathcal{M}$, more precisely an $n$-manifold, or $n$-fold $\mathcal{M}_n$, is a space of points in $n$ dimensions that is locally flat. Generally speaking, the manifold may or may not be described by a "chart" or a "map" (the grid which we call the "metric"). As discussed more than once before, the metric may cover every point on the manifold or have some defects, such as in Schwarzschild's original metric, but the manifold itself exists independently of metrics. The local "flatness" may be either Euclidean or Minkowskian/Lorentzian. In elementary mathematics we have the real number line, usually called $\mathbb{R}$. One says that a Euclidean $n$-space may be denoted by a number $n$ of orthogonal real lines, or $\mathbb{R}^n$. In the same vein, a Minkowskian/Lorentzian $n$-spacetime is $\mathbb{R}^{(n-1,1)}$, to denote that one of the $n$-dimensions has a different signature.[3] A manifold $\mathcal{M}_n$ may be **open** or **closed**. As the name implies, an open manifold is surrounded by a boundary, which in itself is also a manifold of dimension $n-1$. For example, a line curve is a one-dimensional shape: a 1-fold $\mathcal{M}_1$, whose boundary is two points, themselves zero-dimensional, i.e. 0-fold $\mathcal{M}_0$. Likewise, an ordinary surface such as a plane is a 2-fold $\mathcal{M}_2$, surrounded by a curve (the edge) that is one-dimensional, i.e. a 1-fold $\mathcal{M}_1$, and so on. In contrast, a closed manifold such as a circle or a sphere (the surface, *not* the interior) has no boundary. A common notation used to describe the "boundary" of a manifold is the partial derivative symbol, so we say that the boundary of a manifold $\mathcal{M}_n$ is $\partial \mathcal{M}_n = \mathcal{M}_{n-1}$. In this language, the boundary of a closed manifold is simply $\partial \mathcal{M}_n = 0$. Fig. [7.1] provides some visual examples.

It is perhaps interesting to note that the boundary of a boundary is always vanishing; for instance, the boundary of a plane is a closed curve completely surrounding the plane, so it is itself closed such that $\partial^2 \mathcal{M}_n = 0$. In more advanced discussions of these topics, the symbol $\partial$ is treated as an operator—a "sort of" derivative—and it has the property of being "nilpotent," which simply means that applying it twice always vanishes: $\partial^2 = 0$. A more intuitive approach as to why this $\partial$ symbol can in fact be thought of as a derivative is to note an example: The interior of a sphere is a three-dimensional manifold $\mathcal{M}_3$ whose volume is $\frac{4}{3}\pi r^3$. It is bounded by the two-dimensional surface of the sphere, i.e. $\mathcal{M}_2 = \partial \mathcal{M}_3$. Note that the surface of the boundary $\mathcal{M}_2$ has an area of $4\pi r^2$, which is *exactly* the derivative of the volume with respect to $r$! So one says that the "surface" of $\mathcal{M}_2$ is related to the "surface" of $\mathcal{M}_3$ by differentiation, implying the notation $\mathcal{M}_2 = \partial \mathcal{M}_3$. This example is not to be taken literally (the language $\mathcal{M}_{n-1} = \partial \mathcal{M}_n$ has a much broader meaning) but it explains the origins of using a derivative symbol to denote the relation between a manifold and its boundary.

Manifolds can be "embedded" in higher-dimensional manifolds. For instance, a circle drawn on a piece of paper is a 1-fold that is embedded in a 2-fold, which in turn

---

[3]Which in this book we take to be the negative one.

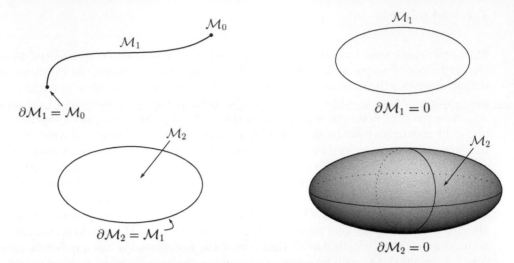

Fig. 7.1: Simple examples of open ($\partial\mathcal{M}_n = \mathcal{M}_{n-1}$) and closed ($\partial\mathcal{M}_n = 0$) $n$-folds. Higher-dimensional manifolds follow the same hierarchy, except that they are naturally difficult for us to visualize or plot. For example, although we live inside a 3-fold—our ordinary three-dimensional space—it is difficult, if at all possible, to visualize it as a "surface" with or without a boundary. It gets worse with four-dimensional spacetime, which is essentially a 4-fold with Lorentzian signature.

is embedded in a 3-fold (the three-dimensional space that surrounds it), and so on. In this case, the embedded manifold is known as a **submanifold**. We alluded to this property earlier when discussing the Robertson–Walker metric and pointed out that an important notion in Riemannian geometry is to think of each manifold *without* embedding. The properties of the manifold should be made intrinsic, without reference to any higher dimension, rather than extrinsic. This allows us physicists to talk about the expanding universe, for example, as an expanding manifold *without* asking what it is expanding into. Sure, it may be expanding into a higher-dimensional manifold—our universe may just be a submanifold—however, until this is observed and confirmed experimentally, we can make no assumptions about it. The concept of "metric" was developed specifically for this purpose. It charts a manifold without reference to any embedding. And from the metric everything else is derived. The mathematical theory of manifolds encompasses a lot of detail and generality. For example, there are non-smooth manifolds, non-differentiable manifolds, complex manifolds, quaternionic manifolds,[4] and other "exotic" classifications. However, our interest here is in manifolds that are real, smooth (i.e. containing no "kinks" or real singularities[5]), and differentiable (i.e. you can do calculus on them). It is true that every once in a while

---

[4] Quaternions are generalizations of ordinary complex numbers.

[5] We are of course led to spacetime singularities in many spacetimes; such as, the $r = 0$ point in the Schwarzschild spacetime, however we just take them as a sign of our ignorance of what exactly happens there.

"exotic" manifolds appear in physics,[6] but this is very rare, and in general relativity proper we have no need for them. Our manifolds here are real and smooth, and can be charted with a metric or more. These simpler versions are known as **Riemannian manifolds**. A subclass of this is **Lorentzian manifolds**, i.e. ones that are locally Minkowskian $\mathbb{R}^{(n-1,1)}$, particularly for $n = 4$ (our ordinary flat spacetime $\mathbb{R}^{(3,1)}$[7]).

I have said too much! We really do not need a lot of the language defined in the last two paragraphs. However, it does make a discussion easier to give things proper names, as well as hopefully whet the reader's appetite for more on this exciting topic. One last notion we need to define, however, is that of **tangent spaces**. As the name implies, and as far as our purposes go, a tangent space is a flat manifold that touches a curved manifold at a point. Yes, it is as simple as that. These tangent spaces can also be Lorentzian, endowed with a flat Minkowskian metric in the usual sense. In fact, mathematically speaking, these are exactly the spaces where our usual 4-vectors live! A curved manifold does not allow a simple definition of a vector, since by its very nature a vector is "straight" and cannot be "bent" onto a curved surface. Hence our ordinary 4-vectors live in a flat tangent space of their own, and saying it is flat is exactly the same thing as saying that close enough to the curved manifold it *looks* flat. In other words, the notion of a tangent space gives a precise mathematical definition to the concept of locally flat space, which in turn is the mathematical manifestation of the principle of equivalence. The notation usually used to describe a flat space (or spacetime) that is tangent to a curved manifold $\mathcal{M}_n$ at a specific point $x$ is $T_x\mathcal{M}_n$. We will use this notion again in Chapter 9 to develop more mathematics. In the meantime, Fig. [7.2] provides a visual of the concept of tangent spaces.

## 7.2   Calculus on Manifolds

The rather lengthy, and possibly dry, discussion in the previous section is somewhat essential as the foundation for a better understanding of what follows. An important notion to discuss, which we devote an entire section to, is the notion of doing calculus on the smooth differentiable Lorentzian manifolds we are interested in. In particular, how does one differentiate vectors, and more generally tensors, on a curved spacetime manifold? We have already developed a simple notion of doing so, the covariant derivative, which in itself is enough for most purposes. It pays to repeat the discussion here while making it a bit more rigorous at the same time.

Generally what we would like to do is define a derivative $\nabla$, covariant or not, as what mathematicians call a "map," or simply a "transformation," which takes a

---

[6]Such as in supersymmetric theories.

[7]Generally, a Riemannian manifold that is endowed with a metric that is *not* positive definite is called a **pseudo-Riemannian manifold**; Lorentzian manifolds are a specific case of those.

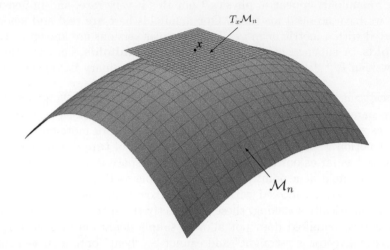

Fig. 7.2: A curved manifold $\mathcal{M}_n$ with a flat manifold $T_x\mathcal{M}_n$ tangent to it at a point.

tensor of rank $(p, q)$ to another tensor of rank $(p, q + 1)$ on any manifold. We say that it should satisfy the most basic rules of calculus; namely:[8]

1. Linearity: $\nabla \otimes (A + B) = \nabla \otimes A + \nabla \otimes B$, where $A$ and $B$ are tensors of any rank, including scalars. The symbol $\otimes$ here means just about any product you need it to mean: It could mean ordinary multiplication, dot product, cross product, etc, depending on what $A$ and $B$ are.

2. The Leibnitz[9] or product rule: $\nabla (A \otimes B) = A \otimes \nabla B + (\nabla A) \otimes B$.

3. Commutativity with contraction: $\nabla_\mu \left( A^{\alpha_1 \cdots \alpha_k}{}_{\beta_1 \cdots \beta_k} \right) = (\nabla A)_\mu{}^{\alpha_1 \cdots \alpha_k}{}_{\beta_1 \cdots \beta_k}$, which may seem trivial to us at this point.

4. Can be used as a **directional derivative**: Finds the change of some scalar field $f$ in the direction of a given vector $A^\mu$, or, in other words the, projection of the change of $f$ in the direction of the vector: $A^\mu \nabla_\mu f$. Note that $f$ may be a component of a higher-rank tensor; in other words, we also include things like $A^\mu \nabla_\mu B^\nu$, and so on.

5. Torsion free: $\nabla_\mu \nabla_\nu f = \nabla_\nu \nabla_\mu f$, where $f$ is some scalar field. This particular property is assumed true in general covariance (and all the sub-covariances discussed in this book), but in fact it may be dropped. Some modified theories of gravity do not require it (we will meet one in Chapter 10). As far as we can tell thus far, nature seems to be described by torsion-free physics.

---

[8]The reader may find some of these rules unknown to them as they usually arise in a much higher level of calculus than even multivariable calculus. The discussion here is spelled out even more in [17].

[9]Gottfried Wilhelm (von) Leibnitz or Leibniz: German mathematician (1646–1716). He is most famous for inventing differential and integral calculus, independently of Sir Isaac Newton. To this day there is some debate among historians on the question of priority. The modern language of derivatives, particularly the notations $\frac{d}{dx}$ and $\int dx$, etc, is due to him.

More advanced textbooks on the subject take a very cautious and rigorous approach to defining derivatives. They start by proving that derivatives that satisfy the above properties can in fact exist on general manifolds, then proceed to show that, once found, they are unique, in the sense that two differently defined derivatives have the *same* action on any tensor field, provided that both derivatives satisfy the five (or four if you drop the last one) properties above. They further show that two differently defined derivatives, denoted, say, by $\nabla$ and $\tilde{\nabla}$, *must* be related by

$$\nabla_\mu A^\nu = \tilde{\nabla}_\mu A^\nu + C^\nu_{\mu\alpha} A^\alpha. \tag{7.1}$$

The quantities $C^\nu_{\mu\alpha}$ are called the **connection**, or the **Levi-Civita connection**. At this point they are not *yet* the usual Christoffel symbols. They only become so when one takes $\tilde{\nabla}$ to be the ordinary partial derivative, i.e.

$$\nabla_\mu A^\nu = \partial_\mu A^\nu + \Gamma^\nu_{\mu\alpha} A^\alpha, \tag{7.2}$$

which the reader may recall we wrote down as a "guess" in §2.3. Let us next consider the covariant derivative as a directional derivative (property 4 above). Consider a curved line, i.e. just $\mathcal{M}_1$, embedded in a manifold $\mathcal{M}_{n>1}$ and described by a vector $x^\mu(\lambda)$ that is tangent to $\mathcal{M}_1$ everywhere.[10] Also consider any tensor field $B$ defined on this manifold (more precisely, in the tangent space $T_x\mathcal{M}_n$ of the manifold). For demonstrational purposes let's consider just a vector field $B^\nu$, but any tensor of any rank can be used, and $B^\nu$ is modified accordingly by simply adding the necessary indices to it. The gradient of the components of $B^\nu$, i.e. $\nabla_\mu B^\nu$, signifies the change of $B^\nu$ with respect to the $n$ coordinates. If one wishes to know how much $B^\nu$ is changing *along* the curve $\mathcal{M}_1$, then one projects $\nabla_\mu B^\nu$ on $x^\mu(\lambda)$. In other words, take the scalar product $x^\mu \nabla_\mu B^\nu$. This is called a **directional derivative**:[11]

$$DB^\nu = x^\mu \nabla_\mu B^\nu, \tag{7.3}$$

evaluated at any point on the curve $\mathcal{M}_1$. If one now requires that $B^\nu$ does *not* change anywhere along the curve, then we are effectively parallel transporting $B^\nu$ along $\mathcal{M}_1$. In other words, the directional derivative must vanish:

$$DB^\nu = x^\mu \nabla_\mu B^\nu = 0. \tag{7.4}$$

Equation (7.4) is thus known as the **parallel transport equation**. It makes the notion of parallel transport, which we just took intuitively back in §2.3, more precise. The reader may have noticed the similarity between (7.4) and the geodesic equation in the form (6.34). In fact, if one takes $B^\nu$ to be the 4-velocity $U^\nu$ and projects its gradient on itself, then (7.4) becomes *exactly* (6.34). What this means is that one can write the definition:

---

[10]Our interest is of course $n = 4$ with ($\mu = 0, 1, 2, 3$), but the discussion is general for any $n$.

[11]This is just the curved four-dimensional generalization of (1.86).

*"A geodesic is a curve whose tangent vectors parallel transport along themselves without change."*

The discussion thus far should (somewhat) convince the reader that as long as the above list of properties is satisfied, there are many different ways of defining a covariant derivative on a general manifold. If, however, one looks at our case of interest—Riemannian manifolds, i.e. manifolds endowed with a metric—then it turns out that a specific and unique definition becomes "natural." This arises from the requirement that scalar products, defined with a metric, do not change upon parallel transport. This is the same as saying that angles between vectors do not change if both vectors are parallel transported along a given curve, as of course they shouldn't. In other words, given two vectors, $B^\nu$ and $C^\mu$, we require that

$$D\left(B^\nu C_\nu\right) = D\left(g_{\mu\nu}B^\mu C^\nu\right) = x^\alpha \nabla_\alpha \left(g_{\mu\nu}B^\mu C^\nu\right) = 0. \tag{7.5}$$

Expanding this using the product rule (property 2 above) gives

$$g_{\mu\nu}C^\nu x^\alpha \nabla_\alpha B^\mu + g_{\mu\nu}B^\mu x^\alpha \nabla_\alpha C^\nu + B^\mu C^\nu x^\alpha \nabla_\alpha g_{\mu\nu} = 0. \tag{7.6}$$

The first two terms of (7.6) vanish because of (7.4), which applies to both of them, implying

$$\nabla_\alpha g_{\mu\nu} = 0, \tag{7.7}$$

which is the **metric compatibility condition** (2.92) we promised to prove back in §2.4. A uniqueness theorem may be proven (see [17]) demonstrating that for any specific metric tensor one and *only* one covariant derivative can be defined. This allows us to replicate the series of arguments that lead to (2.99), or in spacetime (4.33).

**Exercise 7.1** Prove that the connection $C^\nu_{\mu\alpha}$ must be symmetric in its lower indices if and only if property 5 is satisfied.

**Exercise 7.2** Verify that the Levi-Civita tensor is also metric compatible, i.e. satisfies $\nabla_\alpha \mathscr{E}_{\mu\nu\rho} = 0$. You may need the relation (8.111).

**Exercise 7.3** Explicitly verify the metric compatibility condition (7.7) for

1. Schwarzschild's exterior metric (5.47).
2. Schwarzschild's interior metric (5.61).
3. The Robertson–Walker metric (5.102).

You may also wish to practice with (5.87), (5.90), or any of the metrics listed in §8.6. Note that although (5.87) and (5.90) describe the same spacetime manifold as (5.47), the metric compatibility of the latter does not necessarily guarantee the compatibility of the former two.

**Exercise 7.4** Consider a sphere of radius $R$ centered around the origin. In spherical coordinates the curve $\theta = 90°$ on the sphere represents the equator (feel free to think of this sphere as an approximation of the globe). As a great circle, the equator is necessarily a geodesic on the sphere (as we showed in §3.1.1). However, any *other* circle of latitude $\theta \neq 90°$ is not. Let us show that parallel transporting a vector around any non-geodesic circle will return the vector pointing in a *different* direction. Define a 3-vector that lives on the sphere, or, more precisely, tangential to the sphere; even *more* precisely, it lives in $T_x \mathcal{M}_2$ where $\mathcal{M}_2$ is the sphere's surface. In spherical coordinates a vector tangential to a sphere would have no component in the radial direction, and is thus a 2-vector: $A^i = (A^\theta, A^\varphi)$. Parallel transport this vector around the circle of latitude $\theta$, and find the angle $\alpha$ it makes with its original self after it has returned to its starting point. Show that if the circle of parallel transport was the equator, then that angle is zero. The situation is described in Fig. [7.3]. Follow the steps:

1. Find the differential equations that arise from the parallel transport eqn (7.4). This is a calculation similar to the one we did to find the geodesic equations on the sphere back in §3.1.1, wherein you will find everything you need (e.g. the Christoffel symbols on the sphere).
2. The two equations you found are coupled first-order differential equations, meaning that they mix the components $A^\theta$ and $A^\varphi$. Uncouple them to find two uncoupled second-order differential equations. Note that $A^\theta$ and $A^\varphi$ are functions in $\varphi$ only, since $\theta$ has been fixed.
3. Solve the two equations you found (they should be familiar). You will need initial conditions to find the constants of integration. One convenient choice is to assume that the vector started out at $\varphi = 0$ with $A^\theta = 0$ and $A^\varphi = 1$. In other words, pointing directly due east and its length is unity.
4. What you have found are the components of the vector at *any* point $\varphi$ along the circle of parallel transport. When you substitute $\varphi = 0$ you should get the components of the vector *before* parallel transport, but when you substitute $\varphi = 360°$ you will find the components *after* a full-circle parallel transport. Find the angle between the original vector before parallel transport and the final vector after parallel transport. What would that angle be if $\theta = 90°$? Finish the problem symbolically first, then, if you wish plug in numbers; for instance, you may choose the circle of latitude $\theta = 66.56334°$ which corresponds to the current location of the Tropic of Cancer on the globe.

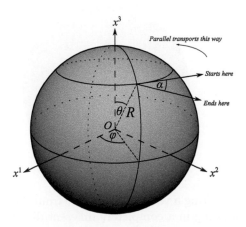

Fig. 7.3: Parallel transporting a vector around a circle of latitude $\theta$.

## 7.3 Curvature

So far in this book we have worked with a very loose, albeit intuitive, definition of curvature. We have essentially said, that given a specific manifold $\mathcal{M}_n$, if *all* of its geodesics *everywhere* are straight lines, then we call this manifold flat; notice the emphasis on "all" and "everywhere." While this makes sense, it is not rigorous enough, and would also require completely and thoroughly solving the geodesic equations for the given metric and not missing one possible solution, in order to confidently conclude whether or not a manifold is curved. We have already seen how difficult it is to solve the geodesic equations in the Schwarzschild case, which is really one of the simplest spacetime solutions known.[12] We will now describe what is today taken to be the canonical definition of curvature, based on the work by Riemann[13] in the nineteenth century. Other scientists who have contributed significantly to the modern theory of curvature are (in no particular order) Gauss, Ricci,[14] and Levi-Civita.

Riemann's definition of curvature is based on parallel transport and goes like this: Consider a vector $B^\sigma$ defined at a point $x$ on a manifold $\mathcal{M}_n$; in other words, it lives in $T_x\mathcal{M}_n$. Now parallel transport $B^\sigma$ around an arbitrary *closed* loop on the manifold, back to where it started. If it returns unchanged, i.e. pointing in the same direction, and this happens *everywhere* on the manifold, then the manifold is flat. If it *does* change then the manifold is curved, and the measure of the vector's change can be taken as the measure of curvature of the manifold. It is as simple as that. A clear example would be parallel transporting a vector on any closed loop on, say, a spherical surface, as can be seen by inspecting Fig. [7.4]. Also see Exercise 7.4.

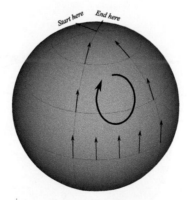

Fig. 7.4: Parallel transporting a vector on a sphere from the north pole back; note that the vector ends up pointing in a completely different direction.

---

[12]It was grueling even in the "simple" case of the sphere back in §3.1.1.

[13]Bernhard Riemann: German mathematician (1826–1866).

[14]Gregorio Ricci-Curbastro: Italian mathematician (1853–1925). Ricci is also known for the discovery of tensor calculus.

To derive the quantity we need,[15] recall that parallel transport follows the formula (7.4):

$$DB^\sigma = x^\mu \nabla_\mu B^\sigma = x^\mu \partial_\mu B^\sigma + x^\mu \Gamma^\sigma_{\mu\alpha} B^\alpha = 0, \tag{7.8}$$

which leads to

$$x^\mu \partial_\mu B^\sigma = -x^\mu \Gamma^\sigma_{\mu\alpha} B^\alpha, \tag{7.9}$$

or, if the vector $x^\mu$ is infinitesimal,[16]

$$dB^\sigma = dx^\mu \partial_\mu B^\sigma = -dx^\mu \Gamma^\sigma_{\mu\alpha} B^\alpha. \tag{7.10}$$

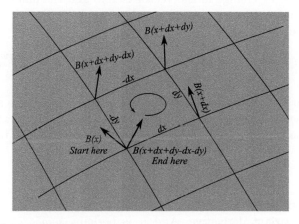

Fig. 7.5: Parallel transporting a vector around an infinitesimally small closed loop.

Now, for simplicity, let's choose an infinitesimal "square" loop, as shown in Fig. [7.5], and parallel transport $B^\sigma$ around it.[17] Transporting along the first leg of the loop gives a "new" vector with components:

$$B^\sigma \left(x^\alpha + dx^\alpha\right) = B^\sigma \left(x^\alpha\right) - dx^\mu \Gamma^\sigma_{\mu\alpha} B^\alpha. \tag{7.11}$$

Now transporting again along the second leg gives

$$B^\sigma \left(x^\alpha + dx^\alpha + dy^\alpha\right) = B^\sigma \left(x^\alpha + dx^\alpha\right) + dB^\sigma \left(x^\alpha + dx^\alpha\right), \tag{7.12}$$

which gives

$$B^\sigma \left(x^\alpha + dx^\alpha + dy^\alpha\right) = B^\sigma - dx^\mu \Gamma^\sigma_{\mu\alpha} B^\alpha$$
$$-dy^\mu \left(\Gamma^\sigma_{\mu\rho} + d\Gamma^\nu_{\mu\rho}\right) [B^\rho - dy^\kappa \Gamma^\rho_{\epsilon\kappa} B^\epsilon]. \tag{7.13}$$

---

[15] There are many ways to do the same calculation; the following is perhaps the simplest.

[16] Compare this with eqn (2.60).

[17] I put the word "square" in quotes because defining a square on a curved surface is not really straightforward, as we discovered back in §5.6.

The total change around the entire loop is

$$\Delta B^{\sigma} = B^{\sigma}\left(x^{\alpha} + dx^{\alpha} + dy^{\alpha}\right) - B^{\sigma}\left(x^{\alpha} + dy^{\alpha} + dx^{\alpha}\right), \qquad (7.14)$$

which after some algebra gives

$$\Delta B^{\sigma} = \left[\left(\partial_{\nu}\Gamma^{\sigma}_{\mu\rho}\right) - \left(\partial_{\rho}\Gamma^{\sigma}_{\mu\nu}\right) + \Gamma^{\alpha}_{\mu\rho}\Gamma^{\sigma}_{\alpha\nu} - \Gamma^{\alpha}_{\mu\nu}\Gamma^{\sigma}_{\alpha\rho}\right] B^{\mu} dx^{\nu} dy^{\rho}. \qquad (7.15)$$

The total change is then proportional to the vector itself and the area of the infinitesimal loop. The quantity in square brackets is the one we are looking for; if it vanishes, then the change in the vector is zero and the manifold is considered flat. This object can be shown to be a tensor[18] of rank $(1,3)$; it is called the **Riemann curvature tensor**:

$$R^{\sigma}_{\ \mu\nu\rho} = \left(\partial_{\nu}\Gamma^{\sigma}_{\mu\rho}\right) - \left(\partial_{\rho}\Gamma^{\sigma}_{\mu\nu}\right) + \Gamma^{\alpha}_{\mu\rho}\Gamma^{\sigma}_{\alpha\nu} - \Gamma^{\alpha}_{\mu\nu}\Gamma^{\sigma}_{\alpha\rho}. \qquad (7.16)$$

The derivation just outlined is *somewhat* rigorous, but there are others even more so. For example, it can be shown that the Riemann curvature tensor is a consequence of the non-commutativity of the covariant derivative when applied to lower-index vectors as follows:

$$\nabla_{\mu}\nabla_{\rho}B_{\nu} - \nabla_{\rho}\nabla_{\mu}B_{\nu} = R^{\sigma}_{\ \mu\rho\nu}B_{\sigma}. \qquad (7.17)$$

It is very straightforward to derive (7.16) from (7.17) by substituting the explicit form of the covariant derivative. And it is perhaps *somewhat* intuitive that parallel transport around a loop is equivalent to successive applications of the covariant derivative; however, an exact proof is rather tedious.

**Exercise 7.5** Prove that $R^{\sigma}_{\ \mu\nu\rho}$ transforms like a tensor.

**Exercise 7.6** Derive (7.16) from (7.17).

The reader is warned that there are several conventions to the ordering of indices on $R^{\sigma}_{\ \mu\nu\rho}$, so be careful consulting different texts. Also note that nowhere in the derivation, or any other derivation of $R^{\sigma}_{\ \mu\nu\rho}$, for that matter, was the metric explicit. In other words, the curvature tensor has the same form whether the manifold is metric compatible or not. If it isn't, then the Christoffel symbols $\Gamma^{\nu}_{\mu\alpha}$ are replaced by the more general connection $C^{\nu}_{\mu\alpha}$ defined in (7.1).

The Riemann curvature tensor has four indices; hence the total number of its components in $n$-dimensional space is $n^4$, which is 256 in our four-dimensional spacetime. However, there are symmetries of the tensor that greatly reduce the number of independent components. They are manifest once the upper index is lowered:

---

[18]Transforms via (2.36), as you can check, even though it is built of non-tensorial elements.

1. The antisymmetry of the first two indices: $R_{\sigma\mu\rho\nu} = -R_{\mu\sigma\rho\nu}$.
2. The antisymmetry of the last two indices: $R_{\sigma\mu\rho\nu} = -R_{\sigma\mu\nu\rho}$.
3. The symmetry of switching pairs: $R_{\sigma\mu\rho\nu} = R_{\rho\nu\sigma\mu}$.
4. The first Bianchi identity:[19] $R_{\sigma\mu\rho\nu} + R_{\sigma\nu\mu\rho} + R_{\sigma\rho\nu\mu} = 0$, where the last three indices are cyclically permutated.
5. The second Bianchi identity[20]: $\nabla_{\alpha}R_{\mu\nu\rho\sigma} + \nabla_{\rho}R_{\mu\nu\sigma\alpha} + \nabla_{\sigma}R_{\mu\nu\alpha\rho} = 0$.

**Exercise 7.7** Explicitly prove the symmetries of the Riemann curvature tensor.

These symmetries reduce the independent components of the Riemann tensor all the way down to 20 (in our $n = 4$), not all of which are necessarily non-vanishing. As an example, here are the non-vanishing components of the Riemann tensor for the Schwarzschild metric in the form (5.52):

$$R^t_{\ rrt} = \tfrac{1}{2}\frac{H''}{H}, \qquad\qquad R^t_{\ \theta\theta t} = \tfrac{1}{2}rH'$$

$$R^t_{\ \varphi\varphi t} = \tfrac{1}{2}r\sin^2\theta H', \qquad R^r_{\ trt} = \tfrac{1}{2}HH''$$

$$R^r_{\ \theta\theta r} = \tfrac{1}{2}rH', \qquad\qquad R^r_{\ \varphi\varphi r} = \tfrac{1}{2}r\sin^2\theta H'$$

$$R^\theta_{\ t\theta t} = \tfrac{1}{2r}HH', \qquad\qquad R^\theta_{\ r\theta r} = -\tfrac{1}{2r}\frac{H'}{H} \qquad\qquad (7.18)$$

$$R^\theta_{\ \varphi\varphi\theta} = (H-1)\sin^2\theta, \qquad R^\varphi_{\ t\varphi t} = \tfrac{1}{2r}HH'$$

$$R^\varphi_{\ r\varphi r} = -\tfrac{1}{2r}\frac{H'}{H}, \qquad\qquad R^\varphi_{\ \theta\varphi\theta} = 1 - H.$$

The remaining components can be found by using the symmetries; for example, $R^\theta_{\ tt\theta} = -R^\theta_{\ t\theta t}$ and so on. Notice that since $H = \left(1 - \frac{r_s}{r}\right)$, $H' = \frac{r_s}{r^2}$ and $H'' = -2\frac{r_s}{r^3}$ and none of the components of the Riemann tensor experience any trouble at the Schwarzschild radius $r = r_s$, even though they are derived from the metric (5.52) which is defective at that radius. On the other hand, all of them blow up at the true singularity $r = 0$; i.e. the curvature itself becomes infinite. The Riemann tensor is thus insensitive to coordinate singularities and is, in fact, the most rigorous way of telling the difference between them and a true manifold singularity. Sometimes, however, the question arises of whether this depends on the specific choice of the coordinate system used. For example, in the case of the Schwarzschild solution, would the components of the Riemann tensor still exhibit the same behavior at the various points if other metrics describing the same spacetime, such as (5.87) and (5.89), are used? Rather than recalculating the components of the Riemann tensor again for each metric, one can instead check the behavior of *invariants*. These are exactly what you would expect:

---

[19]Discovered, in fact, by Ricci.
[20]This one *was* discovered by the Italian mathematician Luigi Bianchi (1856–1928).

scalar quantities that, just by virtue of being scalar, do not change from one coordinate system to the other for the same manifold. One such quantity is

$$K = R_{\mu\nu\rho\sigma}R^{\mu\nu\rho\sigma}, \tag{7.19}$$

constructed from the full contraction of the Riemann tensor with itself. This $K$ is sometimes referred to as the **Kretschmann scalar**.[21] For the Schwarzschild case it is just

$$K = \frac{48G^2M^2}{c^4r^6}, \tag{7.20}$$

which clearly diverges at $r = 0$, vanishes at infinity $r \to \infty$, and has a finite value at the Schwarzschild radius, as can be easily checked. There are several other ways of constructing such invariants. We will define one more in eqn (7.23) below.

---

**Exercise 7.8** Derive (7.20).

**Exercise 7.9** Find the components of the Riemann tensor for the metric of Exercise 2.20, and hence show that it is not flat. Compute the Kretschmann scalar as well.

**Exercise 7.10** Find the components of the Riemann tensor for a sphere of radius $a$. The metric is $dl^2 = a^2\left(d\theta^2 + \sin^2\theta d\varphi^2\right)$. We found the Christoffel symbols earlier (3.37). Find the sphere's Kretschmann scalar.

**Exercise 7.11** Find all the non-vanishing components of the Riemann tensor for the following solutions and identify any true spacetime singularities they may have:

1. The interior Schwarzschild metric (5.61).
2. The Robertson–Walker metric (5.102).

**Exercise 7.12** Find the Kretschmann scalar for the metrics in Exercise 7.11 and verify your conclusions concerning singularities.

---

Because of the symmetries of the Riemann tensor it can be shown that all possible traces over its indices vanish except one, which is found by contracting the upper index of the Riemann tensor with the lower middle index. This is the **Ricci tensor**:

$$R_{\mu\nu} = R^{\rho}{}_{\mu\rho\nu}, \tag{7.21}$$

which can be calculated directly from the connection by applying (7.21) to (7.16)[22]

$$R_{\mu\nu} = \left(\partial_\nu\Gamma^{\rho}_{\mu\rho}\right) - \left(\partial_\rho\Gamma^{\rho}_{\mu\nu}\right) + \Gamma^{\alpha}_{\mu\rho}\Gamma^{\rho}_{\alpha\nu} - \Gamma^{\alpha}_{\mu\nu}\Gamma^{\rho}_{\alpha\rho}. \tag{7.22}$$

---

**Exercise 7.13** Prove that the Ricci tensor is symmetric in its two indices, i.e. $R_{\mu\nu} = R_{\nu\mu}$.

---

[21] Erich Justus Kretschmann: German physicist (1887–1973).

[22] The Ricci tensor is a second derivative of the metric tensor, just as the Newtonian tensor $R_{ij}$ was the second derivative of the gravitational potential $\Phi$ in §3.3. Since we already know that in general relativity the components $g_{\mu\nu}$ of the metric tensor act as "gravitational potentials," it is clear that $R_{ij}$ is the non-relativistic Newtonian approximation of $R_{\mu\nu}$. One can in fact explicitly show that eqn (3.89) arises as a low-curvature/weak gravity limit of (7.22).

**Exercise 7.14** Find the components of the Ricci tensor for the metric of Exercise 2.20. Calculate them once by applying eqn (7.21) to the results of Exercise 7.9 and a second time using (7.22) directly.

**Exercise 7.15** Find all the non-vanishing components of the Ricci tensor for a sphere of radius $a$. Calculate them once by applying eqn (7.21) to the results of Exercise 7.10 and a second time using (7.22) directly.

**Exercise 7.16** Find all the non-vanishing components of the Ricci tensor for the following solutions:

1. The exterior Schwarzschild metric (5.47).
2. The interior Schwarzschild metric (5.61).
3. The Robertson–Walker metric (5.102).

The trace of the Ricci tensor is known as the **Ricci scalar**:

$$R = R^{\mu}{}_{\mu} = g^{\mu\nu} R_{\mu\nu}, \tag{7.23}$$

a variant of which was found by Gauss in his earlier treatment of the subject of curvature.

**Exercise 7.17** Find the Ricci scalar for the metric of Exercise 2.20 using the result of Exercise 7.14.

**Exercise 7.18** Find the Ricci scalar for a sphere of radius $a$ using the result of Exercise 7.15.

**Exercise 7.19** Find the Ricci scalar for the following solutions using the results of Exercise 7.16:

1. The exterior Schwarzschild metric (5.47).
2. The interior Schwarzschild metric (5.61).
3. The Robertson–Walker metric (5.102).

Furthermore, by using the symmetries of the Riemann curvature, particularly the second Bianchi identity, it is possible to show that the *only* divergenceless second-rank tensor that can be defined from the Riemann tensor is the combination $R_{\mu\nu} - \frac{1}{2} g_{\mu\nu} R$. In other words,

$$\nabla^{\mu} \left( R_{\mu\nu} - \frac{1}{2} g_{\mu\nu} R \right) = 0. \tag{7.24}$$

This property is particularly important to us, as we will see; hence we define the so-called **Einstein tensor**:

$$G_{\mu\nu} = R_{\mu\nu} - \frac{1}{2} g_{\mu\nu} R, \tag{7.25}$$

such that

$$\nabla^{\mu} G_{\mu\nu} = 0. \tag{7.26}$$

**Exercise 7.20** Find the components of the Einstein tensor for the metric of Exercise 2.20 using the results of Exercises 7.14 and 7.17.

**Exercise 7.21** Find the components of the Einstein tensor for a sphere of radius $a$ using the results of Exercises 7.15 and 7.18.

**Exercise 7.22** Find all the non-vanishing components of the Einstein tensor for the following solutions using the results of Exercises 7.16 and 7.19:

1. The exterior Schwarzschild metric (5.47).
2. The interior Schwarzschild metric (5.61).
3. The Robertson–Walker metric (5.102).

Now, since the metric tensor *itself* is divergenceless, i.e. $\nabla^\mu g_{\mu\nu} = 0$, which follows from the metric compatibility condition (7.7), an even more general divergenceless quantity may be defined as follows:

$$\nabla^\mu \left( G_{\mu\nu} + \Lambda g_{\mu\nu} \right) = 0, \tag{7.27}$$

where $\Lambda$ is an arbitrary real constant. This particular constant is of great physical importance and at the same time constitutes a great mystery, as we will see.

**Exercise 7.23** Prove (7.26).

**Exercise 7.24** Prove that Killing vectors satisfy the following identities:

$$\nabla_\mu \nabla_\nu \xi^\rho = R^\rho{}_{\nu\mu\lambda} \xi^\lambda$$
$$\xi^\mu \nabla_\mu R = 0. \tag{7.28}$$

## 7.4   The Vacuum Einstein Equations

In the years leading up to his seminal papers of 1915, and well after he had developed the equivalence principle and its consequences, Einstein was searching for equations that would relate the matter/energy content of any given space with the curvature of said space (more precisely, spacetime). In so doing, he was effectively trying to find generalizations for the Poisson and Laplace equations of Newtonian mechanics: Recall that in Newton's theory the gravitational potential $\Phi$ (review the discussions in §3.3 and §5.3) leads to the Newtonian gravitational field $\vec{\mathfrak{g}}$ via

$$\vec{\mathfrak{g}} = -\vec{\nabla}\Phi. \tag{7.29}$$

Furthermore, the gravitational Gauss law (see footnote 18 in §5.3) requires

$$\vec{\nabla} \cdot \vec{\mathfrak{g}} = -4\pi G\rho, \tag{7.30}$$

where $\rho$ is the density of matter in kg/m$^3$. Equation (7.30) follows from the radial nature of the gravitational field, and in that regard is exactly analogous to the equation by the same name that one studies in Maxwell's theory of electrodynamics.[23] Substituting (7.29) into (7.30) leads to the Poisson equation:

$$\nabla^2\Phi = 4\pi G\rho. \tag{7.31}$$

Note that if the space under study is devoid of matter, then the Poisson equation reduces to the Laplace equation:

$$\nabla^2\Phi = 0. \tag{7.32}$$

The Poisson and Laplace equations are second-order partial differential field equations that relate the matter content of the region of space under study (the right-hand side) with the gravitational potential in the same region (the left-hand side). Since we have already established in §5.3 that the components of the metric tensor $g_{\mu\nu}$ play the role of the Newtonian gravitational potential, the general relativistic field equations should be second-order partial differential equations in $g_{\mu\nu}$, with the right-hand side involving something that describes matter as well as energy. It should also reduce to both (7.31) and (7.32) in the appropriate limit.

In this section we present, without motivation, the *vacuum* Einstein field equations, generalizing (7.32). We will explain later Einstein's reasoning for them.[24] They are simply the requirement that the Ricci tensor vanishes:

$$R_{\mu\nu} = 0. \tag{7.33}$$

How many differential equations are these? Let's count: The Ricci tensor has 16 components in four-spacetime dimensions; however, it can be shown that it is symmetric (see Exercise 7.13). Hence it has only 10 independent components, which means that (7.33) constitutes 10 independent differential equations in the likewise 10 independent metric components. If the metric is diagonal, then so is the Ricci tensor and

---

[23]If none of this sounds familiar, don't worry about it; it is not essential to what follows, it merely provides a background.

[24]We will discuss at least *some* reasonable motivation for them. What Einstein's actual reasoning was is somewhat of a debate among science historians.

we end up with only four equations; namely $R_{00} = R_{11} = R_{22} = R_{33} = 0$. Generally (7.33) are non-linear differential equations, which means that if two solutions are found, their sum is *not* necessarily a solution, in contrast with the analogous non-relativistic Laplace eqn (7.32). This makes the problem of finding exact solutions to (7.33) particularly difficult. There are techniques for "constructing" solutions, the simplest of which is using the metric's symmetries. A famous example of this is the Schwarzschild solution, which exploits spherical symmetry.

It is straightforward to show that (7.33) gives (7.32) in the weak field limit we have applied before in §5.3. Using (5.35) and (5.36) in (7.33) gives

$$R_{\mu\nu} = \frac{1}{2} \left[ \delta^{ij} \left( \partial_i \partial_\nu h_{j\mu} \right) + \delta^{ij} \left( \partial_j \partial_\mu h_{i\nu} \right) - \eta^{\alpha\beta} \left( \partial_\mu \partial_\nu h_{\alpha\beta} \right) - \eta^{\alpha\beta} \left( \partial_\alpha \partial_\beta h_{\mu\nu} \right) \right] = 0. \tag{7.34}$$

Letting $(\mu = \nu = 0)$ and looking at the static case only—in other words, all time derivatives vanish—we end up with

$$R_{00} = -\frac{1}{2} \delta^{ij} \partial_i \partial_j h_{00} = -\frac{1}{2} \nabla^2 h_{00} = 0. \tag{7.35}$$

But in §5.3 we have already established that $h_{00} = -\frac{2}{c^2} \Phi$; hence

$$R_{00} = \frac{1}{c^2} \nabla^2 \Phi = 0, \tag{7.36}$$

which leads to (7.32). Let's go a bit further and finish the derivation of the Newtonian metric (5.46). The spatial components of (7.34) are

$$R_{kl} = \frac{1}{2} \left[ \delta^{ij} \left( \partial_i \partial_l h_{jk} \right) + \delta^{ij} \left( \partial_j \partial_k h_{il} \right) - \eta^{\alpha\beta} \left( \partial_k \partial_l h_{\alpha\beta} \right) - \eta^{\alpha\beta} \left( \partial_\alpha \partial_\beta h_{kl} \right) \right] = 0. \tag{7.37}$$

Contract both sides of (7.37) with $\delta^{kl}$ and simplify

$$\delta^{kl} R_{kl} = \frac{1}{2} \left[ \nabla^2 h_{00} + 2 \partial^i \partial^j h_{ij} - 2\delta^{ij} \nabla^2 h_{ij} \right] = 0, \tag{7.38}$$

where $\nabla^2 = \delta^{ij} \partial_i \partial_j$ and we have also allowed all time derivatives to vanish. Using the spherically symmetric Newtonian metric (5.40), where $h_{ij} = f(r) \delta_{ij}$, we end up with

$$\delta^{kl} R_{kl} = \frac{1}{2} \left[ \nabla^2 h_{00} - 4 \nabla^2 f \right] = 0, \tag{7.39}$$

which implies

$$f = \frac{1}{4} h_{00} + b, \tag{7.40}$$

where $b$ is an arbitrary integration constant. We can choose $b = 0$, since the Einstein vacuum equations are differential equations in the metric components and an added constant does not affect the physics. We can also absorb the factor of 4 in the definition of $f$. Hence we end up with $f = h_{00}$, which we have promised to derive back in §5.3.

As another demonstration, let's, at long last, see how Schwarzschild found his famous metric. Recall that he was simply trying to construct a spherically symmetric solution to (7.33) in the region of space *outside* a sphere of a given radius, i.e. in vacuum. There are several ways of doing this; the most obvious is to write the *ansatz*:[25]

$$ds^2 = -c^2 A\left(r\right) dt^2 + B\left(r\right) dr^2 + C\left(r\right) r^2 d\Omega^2, \tag{7.41}$$

where we require that the unknown functions $A$, $B$, and $C$ become unity for $r \to \infty$, since the metric should reduce to the flat Minkowski metric at infinity. Next, it is possible to simplify (7.41) by making the coordinate redefinition

$$\tilde{r} = \sqrt{C\left(r\right)}r, \tag{7.42}$$

which, after some algebra, leads to

$$dr = \frac{d\tilde{r}}{\sqrt{C}}\left(1 - \frac{\tilde{r}}{2}\frac{d}{d\tilde{r}}\ln C\right). \tag{7.43}$$

The metric then becomes

$$ds^2 = -c^2 A\left(\tilde{r}\right) dt^2 + \frac{B}{C}\left(1 - \frac{\tilde{r}}{2}\frac{d}{d\tilde{r}}\ln C\right)^2 d\tilde{r}^2 + \tilde{r}^2 d\Omega^2. \tag{7.44}$$

Note that the angular components no longer have an unknown function next to them, which was the entire purpose of this redefinition of the radial coordinate. In fact, the quantity $\frac{B}{C}\left(1 - \frac{\tilde{r}}{2}\frac{d}{d\tilde{r}}\ln C\right)^2$ is an arbitrary function in $\tilde{r}$ which we can rename

$$\frac{B}{C}\left(1 - \frac{\tilde{r}}{2}\frac{d}{d\tilde{r}}\ln C\right)^2 \to B\left(\tilde{r}\right). \tag{7.45}$$

We can also rename $\tilde{r} \to r$ such that the metric becomes just

$$ds^2 = -c^2 A\left(r\right) dt^2 + B\left(r\right) dr^2 + r^2 d\Omega^2. \tag{7.46}$$

This reduction is based on the spherical symmetry of the metric, and no others can be made. The next step is to substitute the metric (7.46) in (7.33) to get

---

[25] *Ansatz* is a German word that means "the initial placement of a tool at a work piece," with the plural *ansätze*. In physics and mathematics the term is used to mean an educated guess that is verified later by its results. This definition is due to [18]. "Ansatz" is used heavily in today's physics literature, and seems to be one of several words that German physicists have introduced in the language of theoretical physics. We have already met "gedanken," and in §9.6 we will meet some more.

$$R_{tt} = \frac{1}{2}\frac{A''}{B} - \frac{1}{4}\frac{A'^2}{AB} - \frac{1}{4}\frac{A'B'}{B^2} + \frac{1}{r}\frac{A'}{B} = 0 \tag{7.47}$$

$$R_{rr} = -\frac{1}{2}\frac{A''}{A} + \frac{1}{4}\frac{A'^2}{A^2} + \frac{1}{4}\frac{A'B'}{AB} + \frac{1}{r}\frac{B'}{B} = 0 \tag{7.48}$$

$$R_{\theta\theta} = 1 + \frac{r}{2}\frac{B'}{B^2} - \frac{r}{2}\frac{A'}{AB} - \frac{1}{B} = 0 \tag{7.49}$$

$$R_{\varphi\varphi} = \left(1 + \frac{r}{2}\frac{B'}{B^2} - \frac{r}{2}\frac{A'}{AB} - \frac{1}{B}\right)\sin^2\theta = 0, \tag{7.50}$$

where a prime denotes a derivative with respect to $r$. Due to the spherical symme-
try, the $\varphi$ equation is just $R_{\varphi\varphi} = R_{\theta\theta}\sin^2\theta = 0$ and does not contribute any new
information; hence it can be dropped. By inspection let's multiply $R_{tt}$ by $B/A$ and
add

$$\frac{B}{A}R_{tt} + R_{rr} = 0, \tag{7.51}$$

which leads to

$$\frac{A'}{A} = -\frac{B'}{B}, \tag{7.52}$$

which is easily solvable to

$$A = KB^{-1}, \tag{7.53}$$

where $K$ is an arbitrary integration constant. If we once again make a coordinate
redefinition $\tilde{t} = K^{1/2}t$, then we can absorb the constant in the time definition and
rename back to $t$, just as we did before for $r$ and $\tilde{r}$. Hence

$$A = B^{-1} = H, \tag{7.54}$$

which already brings us to (5.52). Using (7.54), the $\theta$ eqn (7.49) becomes

$$1 - rH' - H = 0. \tag{7.55}$$

It is easy to check that this can be written as follows:

$$1 - \frac{d}{dr}(rH) = 0, \tag{7.56}$$

which is easily integrated via separation of variables thus

$$d(rH) = dr$$
$$rH = r + b$$
$$H(r) = 1 + \frac{b}{r}, \tag{7.57}$$

with $b$ being an arbitrary integration constant. Clearly this result satisfies the asymptotic flatness condition in that $H$ becomes unity at $r \to \infty$. In addition, we recall that
in order for this metric to reduce to the Newtonian metric (5.46), which in turn leads

to the Newtonian expression for a spherically symmetric potential, the constant must be

$$b = -\frac{2GM}{c^2}, \tag{7.58}$$

leading to the full Schwarzschild solution (5.47). The other major example of a metric that we have found, namely the Robertson–Walker metric, including the Friedmann eqns (5.112), cannot be verified using the vacuum eqns (7.33) because in that case the space it describes—i.e. the entire universe—is not devoid of matter and energy. For this we have to use the full Einstein equations, but we have to make one more definition before we can get to that.

**Exercise 7.25** Consider a general spherically symmetric metric where time evolution is allowed; in other words:

$$ds^2 = -c^2 A\,(t,r)\,dt^2 + B\,(t,r)\,dr^2 + r^2\left(d\theta^2 + \sin^2\theta d\varphi^2\right). \tag{7.59}$$

Show that the vacuum Einstein equations imply that this metric is not a reasonable solution. This is a result of **Birkhoff's theorem**,[26] which proves that any spherically symmetric solution of the vacuum field equations must be static and asymptotically flat. This further implies that the Schwarzschild solution is not just *a* spherically symmetric solution, but rather *the* only solution. This means that any other spherically symmetric static solution can be transformed to the Schwarzschild metric. We already know of two such metrics: (5.87) and (5.89). Birkhoff's theorem is *only* true, by the way, in four-dimensional spacetime. Higher dimensional versions of general relativity allow for "rotationally" symmetric spacetime solutions that are different, in the sense that they cannot be transformed into each other and as such describe different manifolds.

**Exercise 7.26** This exercise may amount to a project that the reader may wish to undertake only if they have a few days to spare. Following Schwarzschild's argument let's construct a general cylindrically symmetric static spacetime:

$$ds^2 = -c^2 A\,(\rho)\,dt^2 + B\,(\rho)\,d\rho^2 + \rho^2 d\varphi^2 + C\,(\rho)\,dz^2. \tag{7.60}$$

The fact that the $d\varphi^2$ term doesn't include an unknown function is based on a reparameterization similar to (7.42) in the spherically symmetric case.

1. Find the differential equations in the unknown functions $A$, $B$, and $C$ that arise from the vacuum field equations. Make an attempt at solving them. If you do succeed, you should find not just one solution, but a general class of solutions parameterized by constants.
2. Discuss and interpret the solutions: Do they have singularities? Real or coordinate? Do you think these solutions represent real objects that may be found in nature? Why or why not?

**Exercise 7.27** If you have done exercise/project (7.26) and still thirst for more, go ahead and construct the geodesic equations for the solution(s) you have found. Find the metric symmetries and conserved quantities and use them to simplify the geodesic equations. If you are still looking for something to do, analyze and catalog the metric's geodesics using the effective potential method. Not quite enough yet? Find and plot all the lightlike, timelike, and spacelike geodesics. Congratulations, you have done as much work as is possibly publishable in a research paper (or two).

---

[26] George David Birkhoff: American mathematician (1884–1944).

**Exercise 7.28** Another possible project-like exercise is finding a "gravitational wave" solution. Start with the metric

$$ds^2 = -2dudv + a^2\left(u\right)dx^2 + b^2\left(u\right)dy^2. \tag{7.61}$$

This is a light cone type metric similar to the one we studied in Exercise 4.5. The parameters $u$ and $v$ are defined by

$$u = \frac{1}{\sqrt{2}}\left(ct - z\right)$$

$$v = \frac{1}{\sqrt{2}}\left(ct + z\right). \tag{7.62}$$

1. Substitute in the vacuum field equations and find the differential equations that the unknown functions $a$ and $b$ must satisfy.
2. You should get only one partial differential equation in $a$ and $b$; solve it using the method of separation of variables.
3. Explain why the solution you found is a gravitational "wave."
4. Construct the geodesic equations of this spacetime. Simplify using the metric isometries.
5. Analyze the geodesics using the effective potential method.
6. Find and plot all the lightlike, timelike, and spacelike geodesics.

*Historical note*: It was realized by Einstein himself as early as 1916 that gravitational waves are allowed solutions of general relativity. In electrodynamics, it has already been known for many decades that any accelerating charged object must produce electromagnetic waves; in fact, this is the main principle of radio transmission and reception. Similarly, accelerating masses should create a wavelike disturbance in the spacetime background. Unlike electromagnetic waves, gravitational waves are not waves propagating *in* spacetime, but are the propagation *of* spacetime disturbances. For many decades their existence was strongly believed by most scientists; however, it was also known that they must be an extremely weak effect indeed; so weak that it would require cosmic events of catastrophic proportions to generate waves that are barely strong enough to be detectable by our technology. On September 2015, almost an entire century after Einstein predicted them, these waves were finally detected for the first time (see the discussion toward the end of §5.7) as the result of an international scientific effort under the heading of the Laser Interferometer Gravitational-Wave Observatory (LIGO) scientific collaboration. Only two years after that discovery, three of the leaders of the project—Rainer Weiss,[27] Kip Thorne,[28] and Barry Barish[29]—were awarded the 2017 Nobel Prize in Physics for the discovery of gravitational waves. Usually Nobel prizes are awarded several years after the discovery they honor. The fact that the Nobel came so quickly this time is a testament to how excited the scientific community was for this major achievement.

---

[27] American astrophysicist (b. 1932).

[28] American theoretical physicist (b. 1940). Thorne is also known for having served as executive producer for the 2014 movie *Interstellar*, for which he also acted as scientific consultant. Because of his efforts the movie has one of the most accurate depictions of black holes and their effects in cinema. Shortly after, he wrote a tie-in book explaining the science behind the movie to the general public [19].

[29] American experimental physicist (b. 1936).

## 7.5 The Stress Energy Momentum Tensor

As frequently mentioned, the complete Einstein field equations are generalizations of the Newtonian Poisson eqn (7.31). The left-hand side of the Poisson equation is comprised of second-order derivatives of the potential; hence the Einstein equations should also have second-order derivatives of the metric tensor. This was obvious in the vacuum equations discussed in the previous section. Now the right-hand side of the Poisson equation is made up of the *source* of the gravitational field; in the Newtonian case this is just the mass density $\rho$. Relativistically, however, the sources must also include energy density and momentum flow, since we already learned from special relativity that they are all really just different sides of the same thing; e.g. see eqn (4.149). Furthermore, one cannot use the 4-momentum $P^\mu$ to do this job, since it is defined for point particles only. What we really need is an equivalent quantity to specify the energy and momentum of continuous matter and/or energy distributions. One would also expect it to be a four-dimensional second-rank tensor to match the left-hand side. The Cauchy stress tensor discussed in §3.4 presents itself as a possible candidate. All one needs to do is to generalize it to four dimensions by adding a zeroth row and column, as well as make sure that all of its components are fully relativistic. So we define the **stress energy momentum tensor** $T_{\mu\nu}$ as a symmetric second-rank tensor that describes the density and flux of energy/matter and momentum *in* spacetime.[30] In reference to the Cauchy tensor, one may write

$$[T_{\mu\nu}] = \begin{bmatrix} T_{00} & T_{0i} \\ T_{i0} & \sigma_{ij} \to T_{ij} \end{bmatrix}. \tag{7.63}$$

Similarly to the Cauchy tensor, the exact values of the components of $T_{\mu\nu}$ depend on what type of energy and/or matter we assume to be present; as such, we must draw from various fields of physics. The components $T_{ij}$ are the relativistic generalization of the components $\sigma_{ij}$ of the Cauchy tensor. They represent the flux of the $i^{\text{th}}$ component of the field's momentum across surfaces of constant $x^j$; specifically, the diagonal spatial components $\mathcal{P}_i = T_{ii}$ (no sum over $i$) are the stresses normal to the $i^{\text{th}}$ surfaces, also sometimes known as the "pressure," and the components $\mathcal{S}_{ij} = T_{ij}$ $(i \neq j)$ are the stresses parallel to the surfaces, also known as the "shearing" or "shear" stresses (see Fig. [3.5] and review the discussion in §3.4 if necessary). The $T_{00}$ component is the energy of the field per unit volume $\mathcal{E}$. Since $E = mc^2$, $\mathcal{E} = \rho c^2$ where $\rho$ is ordinary matter density in $\text{kg/m}^3$. The energy content may be of pure fields as well; for example, for an electromagnetic field $\mathcal{E} = \left(\frac{1}{2}\varepsilon_0 E^2 + \frac{1}{2\mu_0}B^2\right)$.[31] The flux of said energy/mass across surfaces of constant $x^i$ are the components of the field's momentum density $\mathcal{M}_i = T_{i0} = T_{0i}$. All in all,

---

[30]Frequently referred to as the "energy momentum tensor," "stress energy tensor," or just "stress tensor;" we will use the latter because it is the shortest form. Not to be confused with its three-dimensional non-relativistic counterpart.

[31]Here $E$ and $B$ are the magnitudes of the electric and magnetic fields in vacuum respectively.

$$T_{\mu\nu} = \begin{bmatrix} \mathcal{E} & \mathcal{M}_1 & \mathcal{M}_2 & \mathcal{M}_3 \\ \mathcal{M}_1 & \mathcal{P}_1 & \mathcal{S}_1{}^2 & \mathcal{S}_1{}^3 \\ \mathcal{M}_2 & \mathcal{S}_2{}^1 & \mathcal{P}_2 & \mathcal{S}_2{}^3 \\ \mathcal{M}_3 & \mathcal{S}_3{}^1 & \mathcal{S}_3{}^2 & \mathcal{P}_3 \end{bmatrix}. \tag{7.64}$$

As noted earlier, the exact values of these quantities are found from various fields of physics. As such, here are the most common, and the simplest, examples:

- **Dust**: Consider a region of space filled with dust, defined as a distribution of matter/energy whose density $\rho$, which may be non-uniform and time-dependent, is so sparse it exerts no pressure. This has a very simple stress tensor of the form

$$T_{\mu\nu} = \rho U_\mu U_\nu, \tag{7.65}$$

where $U^\mu$ represents the 4-velocity of the dust (treated here as a velocity *field*). If we consider the local rest frame of the fluid, then $U^\mu = (c, \mathbf{0})$, giving the particularly simple form

$$T_{\mu\nu} = \begin{bmatrix} \rho c^2 & 0 & 0 & 0 \\ 0 & 0 & 0 & 0 \\ 0 & 0 & 0 & 0 \\ 0 & 0 & 0 & 0 \end{bmatrix}. \tag{7.66}$$

- **A perfect fluid**: A fluid in equilibrium filling out all of space is defined as a distribution of matter/energy whose pressure can no longer be ignored, but its viscosity, i.e. the off-diagonal shear stress components, is vanishing. The stress tensor is given by

$$T_{\mu\nu} = \left( \rho + \frac{p}{c^2} \right) U_\mu U_\nu + p g_{\mu\nu}, \tag{7.67}$$

where $U^\mu$ represents the 4-velocity of the fluid.[32] If we consider the local Minkowski-Cartesian rest frame of the fluid, then we find

$$T_{\mu\nu} = \begin{bmatrix} \rho c^2 & 0 & 0 & 0 \\ 0 & p & 0 & 0 \\ 0 & 0 & p & 0 \\ 0 & 0 & 0 & p \end{bmatrix}, \tag{7.68}$$

or in Minkowski non-Cartesian coordinates

$$T_{\mu\nu} = \begin{bmatrix} \rho c^2 & 0 & 0 & 0 \\ 0 & p g_{11} & 0 & 0 \\ 0 & 0 & p g_{22} & 0 \\ 0 & 0 & 0 & p g_{33} \end{bmatrix}, \tag{7.69}$$

where $g_{ij}$ may be, for example, spherical coordinates, i.e. $g_{11} = 1$, $g_{\theta\theta} = r^2$, and $g_{\varphi\varphi} = r^2 \sin^2 \theta$.

---

[32] Compare with (3.101).

- **The electromagnetic field**: In a region of space filled only with electromagnetic fields, Maxwell's theory implies a generalization of (3.107):

$$T_{\mu\nu} = \frac{1}{\mu_0} \left( F_\mu{}^\alpha F_{\nu\alpha} - \frac{1}{4} g_{\mu\nu} F^{\alpha\beta} F_{\alpha\beta} \right), \tag{7.70}$$

where $F_{\mu\nu}$ is the field strength tensor defined by (4.185).

- **The free particle**: Although rarely used, here is the stress tensor representing a freely moving point particle with energy $E$:

$$T_{\mu\nu} (x^\mu) = \frac{E}{c^2} \dot{x}_\mu \dot{x}_\nu \delta^4 \left( x^\mu - x_p^\mu \right), \tag{7.71}$$

which we include in the list just in case you were curious. $x^\mu$ is the position 4-vector of the spacetime point where $T_{\mu\nu}$ is evaluated, while $x_p^\mu (\lambda)$ is the 4-position of the particle itself and an over dot is the usual derivative with respect to $\lambda$. Finally, $\delta^4 \left( x^\mu - x_p^\mu \right)$ is the four-dimensional Dirac delta function. If you don't know what that is; just move on, it's not important as far as our purposes are concerned.

**Exercise 7.29** Explicitly write down the components of the electromagnetic $T_{\mu\nu}$ (7.70) in Minkowski-Cartesian coordinates. Verify that

$$T_{00} = \frac{1}{2} \left( \varepsilon_0 E^2 + \frac{1}{\mu_0} B^2 \right), \tag{7.72}$$

which is the energy density in Joules per volume of electric and magnetic fields, as may be checked in any textbook on electromagnetism. Demonstrate that the spatial components $T_{ij}$ give (3.107). Finally, show that the components $T_{i0}$, or alternatively $T_{0i}$, yield the components of the so-called **Poynting vector**:[33]

$$\mathbf{S} = \frac{1}{\mu_0} \left( \mathbf{E} \times \mathbf{B} \right), \tag{7.73}$$

which represents the magnitude and direction of the flow of electromagnetic energy in units of Joules per area per second.

It is crucial for our purposes to note that the stress tensor satisfies a conservation law of the form

$$\nabla^\mu T_{\mu\nu} = 0. \tag{7.74}$$

In local coordinates, i.e. $\nabla \rightarrow \partial$, eqn (7.74) implies conservation of energy and momentum in the usual sense. Globally, however, it doesn't quite represent energy-momentum conservation. This is yet another side of the problem alluded to earlier in our discussion of the energy-momentum of a particle moving in a curved background. Nevertheless, eqn (7.74) is true whether we give it an interpretation or not.

[33]Discovered by the English physicist John Henry Poynting: (1852–1914), and also independently by his countryman the physicist Oliver Heaviside (1850–1925).

This is as much as we need to learn about the stress tensor and its properties. As far as our purposes are concerned, we do not need to know the exact details of how the above examples are constructed. The reader interested in finding new solutions to the Einstein field equations given a specific matter/energy situation should just consult the literature for the appropriate stress tensor. However, we will give a prescription later on on how to derive the stress tensor if a quantity called the "action" is known. More on that in Chapter 8.

**Exercise 7.30** Apply (7.74) on the perfect fluid stress tensor (7.67) in Minkowski-Cartesian spacetime and write down the resulting equations. Interpret your results.

**Exercise 7.31** The following is known as the **weak energy condition**:

$$T_{\mu\nu}a^{\mu}a^{\nu} \geq 0, \tag{7.75}$$

where $a^{\mu}$ is any timelike vector. What does the weak energy condition imply for the case of the perfect fluid?

**Exercise 7.32** The following is known as the **strong energy condition**:

$$T_{\mu\nu}a^{\mu}a^{\nu} \geq \frac{1}{2}T_{\alpha}^{\alpha}a^{\beta}a_{\beta}, \tag{7.76}$$

where $a^{\mu}$ is any timelike vector. What does the strong energy condition imply for the case of the perfect fluid?

**Exercise 7.33** Stress tensors of situations other than the ones we reviewed can be generated from known ones. For example, the stress tensor for an ideal fluid (7.68) is in the fluid's rest frame. Construct the stress tensor of a group of particles moving with a uniform velocity $v$ in the $x$ direction by Lorentz boosting (7.68).

**Exercise 7.34** If $T_{\mu\nu}$ transforms as a 4-tensor, while its components $T_{i0}$ and $T_{0i}$ transform like 3-vectors and $T_{ij}$ transforms like a 3-tensor, prove that $T_{00}$ is invariant under spatial rotations.

**Exercise 7.35** Apply (7.74) to the stress tensor of a free particle (7.71). If you do this correctly, the result should surprise you. Or does it? Only attempt this exercise if Dirac delta functions don't scare you.

## 7.6   The Einstein Field Equations

As noted earlier, the story of how Einstein found his field equations, in a close race with Hilbert, is quite interesting and is found in many sources. Suffice it here to note that in his search he was (most likely) guided by Poisson's eqn (7.31). What Einstein needed was an equation that had the form

$$\text{Curvature stuff} \sim \text{Energy/matter stuff.} \tag{7.77}$$

To figure out the right-hand side of (7.77), consider (7.65). Contracting both sides with $U^\mu U^\nu$ gives

$$U^\mu U^\nu T_{\mu\nu} = \rho U^\mu U_\mu U^\nu U_\nu = \rho c^4, \tag{7.78}$$

which implies the correspondence

$$T_{\mu\nu}|_{\text{GR}} \leftrightarrow \rho|_{\text{Newton}} . \tag{7.79}$$

Hence the right-hand side should include the stress tensor, which certainly makes sense. What about the curvature side? We already know that in general relativity the metric tensor plays the same role the gravitational potential $\Phi$ played in Newtonian gravity. Hence the left-hand side of (7.77) had to have second derivatives of $g_{\mu\nu}$, just as the Poisson eqn (7.31) had second derivatives of $\Phi$. This reduces the pool of choices considerably. In fact, of the possibilities listed in §7.3, one choice that immediately comes to mind is the Ricci tensor $R_{\mu\nu}$. Another way of arriving at the same conclusion is by considering the following argument: Assume there are two particles moving on two separate, but initially parallel, geodesics. It can be shown that they will move away or toward each other with an "acceleration" that is proportional to $R^\sigma{}_{\mu\rho\nu} U^\mu U^\nu$, where $R^\sigma{}_{\mu\rho\nu}$ is the Riemann tensor and $U^\mu$ is the 4-velocity of the particles.[34] On the other hand, if one calculates the same problem in Newtonian gravity, where in this case the two particles are following their free-fall paths in a gravitational field, we find that the acceleration between the particles is proportional to $\nabla^i \nabla^j \Phi$ (as was discussed in §3.3 and the exercises within). This implies the correspondence

$$R^\sigma{}_{\mu\rho\nu}|_{\text{GR}} \leftrightarrow \nabla^i \nabla^j \Phi|_{\text{Newton}} . \tag{7.80}$$

With the knowledge that $R^{ij} = \nabla^i \nabla^j \Phi$ is the Newtonian version of the Ricci tensor, we then arrive at

$$R_{\mu\nu} U^\mu U^\nu = R^\sigma{}_{\mu\sigma\nu} U^\mu U^\nu \propto T_{\mu\nu} U^\mu U^\nu, \tag{7.81}$$

which implies the "field equation:"

$$R_{\mu\nu} = \kappa T_{\mu\nu} \quad \text{Wrong!} \tag{7.82}$$

where $\kappa$ is some arbitrary constant to be found later. This was Einstein's first "guess" at the field equations. However, he soon realized that (7.82) suffers from a *serious* problem: As mentioned above, the stress tensor must satisfy eqn (7.74), which, if

---

[34]See [17] for an exact analysis of this situation.

applied on (7.82), implies that $\nabla^\mu R_{\mu\nu} = 0$. Taken with (7.24), this leads to $\nabla^\mu R = 0$. Now consider taking the trace of (7.82) to find $g^{\mu\nu} R_{\mu\nu} = R = \kappa g^{\mu\nu} T_{\mu\nu}$. This means that $\nabla^\mu \left( g^{\mu\nu} T_{\mu\nu} \right) = 0$ as well. *However*, if we consider the simplest situation (7.65), we end up with $\nabla^\mu \rho = 0$. What this means is that in order for (7.82) to be consistent with (7.74) the density of matter has to be constant at every point in space. This is a completely unphysical restriction on any realistic matter distribution. Hence Einstein discarded (7.82). The final answer to this problem is included in the problem itself, since (7.24) means that the Einstein tensor (7.25) is truly divergenceless *without* the need to impose any unnecessary conditions on $T_{\mu\nu}$. Hence the next choice Einstein made was

$$G_{\mu\nu} = R_{\mu\nu} - \frac{1}{2} R g_{\mu\nu} = \kappa T_{\mu\nu}. \tag{7.83}$$

This is, in fact, the correct choice. The only remaining bit is to figure out the explicit value of the constant $\kappa$. This is easily done by attempting to reduce (7.83) to the Poisson equation. First, contract both sides of (7.83) with the inverse metric tensor $g^{\mu\nu}$, i.e. find the trace. Doing so and rearranging gives

$$R = -\kappa T, \tag{7.84}$$

where $T = g^{\mu\nu} T_{\mu\nu}$. Now using this back in (7.83) gives

$$R_{\mu\nu} = \kappa \left( T_{\mu\nu} - \frac{1}{2} T g_{\mu\nu} \right). \tag{7.85}$$

This is still exact and perfectly equivalent to (7.83). Now, since we are attempting to derive the Newtonian limit, we may use (5.35) as before. We can also assume that the matter density is that of dust, i.e. (7.65), since only $\rho$ appears in the Poisson equation. Hence we are led to only the 00 component of (7.85):

$$R_{00} = \kappa \rho c^2 \left[ 1 - \frac{1}{2} \left( -1 + h_{00} \right) \left( -1 - h^{00} \right) \right] = \frac{1}{2} \kappa \rho c^2, \tag{7.86}$$

where we have used $T = g^{00} T_{00} = \left( -1 - h^{00} \right) \rho c^2$ and $\left( -1 + h_{00} \right) \left( -1 - h^{00} \right) \approx 1$. Finally, we have already established that in the low-curvature limit, $R_{00}$ is given by (7.36); hence

$$\nabla^2 \Phi = \frac{1}{2} \kappa \rho c^4. \tag{7.87}$$

Comparing (7.87) with (7.31) leads to

$$\kappa = \frac{8\pi G}{c^4}. \tag{7.88}$$

This constant is sometimes referred to as the **Einstein constant**. Hence the full and correct (first) version of the **Einstein field equations** is

$$G_{\mu\nu} = R_{\mu\nu} - \frac{1}{2}Rg_{\mu\nu} = \frac{8\pi G}{c^4}T_{\mu\nu}. \tag{7.89}$$

The forms (7.83) and (7.85) of the field equations can still be used along with (7.88). Of course, the full equations reduce to the vacuum version (7.83), as can be seen by setting $T_{\mu\nu} = 0$ in (7.85). There is a second version of the full equations that we will also discuss. Based on (7.27), the reader may have wondered the following: If satisfying the divergenceless condition on both sides of the equation is of such importance, then is using the expression $G_{\mu\nu} + \Lambda g_{\mu\nu}$ in the left-hand side of (7.89) warranted? Einstein did in fact consider this, and there are cases where this new constant "$\Lambda$" comes into play, as we will describe in the next section.

In the meantime, let us demonstrate the use of the full Einstein field equations by applying it to the cosmological Robertson–Walker metric (5.102). Recall that this metric was constructed based on the conditions of homogeneity and isotropy. The next step in such a construction is to plug it into (7.89) to figure out what constraints need be applied on the scale factor $a\,(t)$. Using (5.102) for flat space ($k = 0$), the components of the Einstein tensor are

$$G_{tt} = 3\frac{\dot{a}^2}{a^2}$$

$$G_{rr} = -2\frac{\ddot{a}}{a} - \frac{\dot{a}^2}{a^2}$$

$$G_{\theta\theta} = a^2 r^2 \left( -2\frac{\ddot{a}}{a} - \frac{\dot{a}^2}{a^2} \right) = a^2 r^2 G_{rr}$$

$$G_{\varphi\varphi} = a^2 r^2 \sin^2\theta \left( -2\frac{\ddot{a}}{a} - \frac{\dot{a}^2}{a^2} \right) = a^2 r^2 \sin^2\theta G_{rr}, \tag{7.90}$$

where the over dot here represents a derivative with respect to the coordinate time $t$. Modeling the contents of the universe as a perfect fluid means using the stress tensor (7.69) in Robertson–Walker coordinates:

$$T_{tt} = \rho c^2$$
$$T_{rr} = a^2 p, \quad T_{\theta\theta} = a^2 r^2 p, \quad T_{\varphi\varphi} = a^2 r^2 \sin^2\theta p. \tag{7.91}$$

Clearly only the $tt$ and $rr$ components of the Einstein equation give independent information. This is due to the assumed isotropy of the universe. Equations (7.89) then give the Friedmann eqns (5.112) as promised. It is possible of course to find a version for the Friedmann equations for the remaining values of $k = (+1, -1)$ via the same process.

**Exercise 7.36** Derive the Friedmann equations for an arbitrary $k$. You will need the results of either Exercise 7.16 or Exercise 7.22.

**Exercise 7.37** To find the spacetime metric inside a sphere of a uniform mass distribution $M$ and radius $a$, Schwarzschild used the stress tensor (7.65), where (5.60) relates $\rho$ with $M$. Show that with this choice the interior solution (5.61) satisfies the Einstein field eqns (7.89), or alternatively (7.85). You will need the results of either Exercise 7.16 or Exercise 7.22.

**Exercise 7.38** Another way of deriving the spatial part of the Robertson–Walker metric is by exploiting the fact that 3-spaces with constant curvature have Riemann tensor components of the form

$$R^i{}_{jkl} = k \left( \delta^i_k g_{jl} - \delta^i_l g_{jk} \right), \tag{7.92}$$

where $k$ is an arbitrary constant representing the constant curvature. Starting with a general spherically symmetric metric:

$$dl^2 = A\left(r\right) dr^2 + r^2 \left( d\theta^2 + \sin^2\theta d\varphi^2 \right), \tag{7.93}$$

exploit property (7.92) to show that it is satisfied if and only if the function $A\left(r\right)$ is the one appearing in (5.102).

**Exercise 7.39** Working in geometrized units; i.e. $G = c = 1$, consider the Schwarzschild metric (5.47).

1. Define the transformation $t = u + r + 2M \ln\left(\frac{r}{2M} - 1\right)$, where $u$ is a lightlike coordinate. Show that this leads to the metric

$$ds^2 = -\left(1 - \frac{2M}{r}\right) du^2 - 2dudr + r^2 \left( d\theta^2 + \sin^2\theta d\varphi^2 \right). \tag{7.94}$$

This is called the "retarded" or "outgoing" Schwarzschild metric. Alternatively, we could have defined $t = v - r - 2M \ln\left(\frac{r}{2M} - 1\right)$, where $v$ is a lightlike coordinate. This leads to the "advanced" or "ingoing" Schwarzschild metric

$$ds^2 = -\left(1 - \frac{2M}{r}\right) dv^2 + 2dudr + r^2 \left( d\theta^2 + \sin^2\theta d\varphi^2 \right). \tag{7.95}$$

2. Now, letting the mass $M$ depend on *either* $u$ in the retarded case or $v$ in the advanced case, we find

$$ds^2 = -\left[1 - \frac{2M\left(u\right)}{r}\right] du^2 - 2dudr + r^2 \left( d\theta^2 + \sin^2\theta d\varphi^2 \right) \text{ and}$$

$$ds^2 = -\left[1 - \frac{2M\left(v\right)}{r}\right] dv^2 + 2dvdr + r^2 \left( d\theta^2 + \sin^2\theta d\varphi^2 \right). \tag{7.96}$$

This, in either version, is called the **Vaidya spacetime**.[35] It describes a spherically symmetric object that either emits (retarded/outgoing) or absorbs (advanced/ingoing) dust, thereby increasing or decreasing its mass $M$. It represents the simplest dynamic generalization to the Schwarzschild solution. Find the Ricci tensor, Ricci scalar, and Einstein tensor for the retarded/outgoing case.

3. Substitute in the Einstein field eqns (7.89) and show that the stress tensor must have the form

$$T_{\mu\nu} = -\frac{\left(\partial_u M\right)}{4\pi r^2} \left(\partial_\mu u\right)\left(\partial_\nu u\right). \tag{7.97}$$

[35]Prahalad Chunnilal Vaidya: an Indian physicist and mathematician (1918–2010).

4. Show that the retarded Validya metric can be recast in the form

$$ds^2 = \frac{2M(u)}{r}du^2 - dt^2 + dr^2 + r^2\left(d\theta^2 + \sin^2\theta d\varphi^2\right),\tag{7.98}$$

which may *appear* five-dimensional but is really not: the coordinate $u$ is not independent of $t$ and $r$.

5. Freely study the properties of the Vaidya solution in any of its incarnations. Its effective potential? Orbits? Geodesics? Its properties as a possible black hole? Your pick.

**Exercise 7.40** Using the approximate metric (5.35), one can derive the so-called **linearized Einstein field equations**, which are applicable to situations with low curvature such as the Newtonian metric (5.46). Show that they are:

$$(\partial_\mu\partial_\nu h) + (\partial^\alpha\partial_\alpha h_{\mu\nu}) - (\partial_\mu\partial_\rho h_\nu^\rho) - (\partial_\nu\partial_\rho h_\mu^\rho) - \eta_{\mu\nu}\left[(\partial^\alpha\partial_\alpha h) - \left(\partial_\alpha\partial_\beta h^{\alpha\beta}\right)\right] = -2\kappa T_{\mu\nu},\tag{7.99}$$

where $h = h_\alpha^\alpha$.

## 7.7 Einstein's Greatest Blunder

As discussed briefly in §5.8, all possible solutions to the Friedmann equations, for all possible values of $k$, are dynamic, in the sense that the scale factor of the universe cannot be a constant. This is obvious by simply looking at the first of eqns (5.112); $\dot{a}$ only vanishes for a universe completely devoid of matter or energy: $\rho = 0$. In Einstein's time, before Hubble discovered that the universe is indeed expanding, the common belief was that the universe must be static. In order to guarantee this, something must exist on the right-hand side of the first Friedmann equation to cancel the effect of the density $\rho$ and give a vanishing $\dot{a}$. Working backward, Einstein quickly figured out that the presence of a $\Lambda$ term in his field equations would accomplish this objective. In other words, he now considered field equations of the form

$$G_{\mu\nu} + \Lambda g_{\mu\nu} = R_{\mu\nu} - \frac{1}{2}Rg_{\mu\nu} + \Lambda g_{\mu\nu} = \frac{8\pi G}{c^4}T_{\mu\nu}.\tag{7.100}$$

This is the second and final version of Einstein's field equations. The so-called **cosmological constant** $\Lambda$ is assumed to be a positive real number. It can be considered as part of the matter/energy content by moving it to the right-hand side:

$$G_{\mu\nu} = R_{\mu\nu} - \frac{1}{2}Rg_{\mu\nu} = \frac{8\pi G}{c^4}T_{\mu\nu} - \Lambda g_{\mu\nu}.\tag{7.101}$$

It can be shown that this term generates a negative pressure and thus acts as a "repulsive" force, counteracting the attractive gravitational effect of the universe's matter/energy. If it cancels it *exactly*, one gets the static universe that Einstein sought in this revision of his equations. However, it can further be shown that such a static universe is also unstable, in the sense that any small change, or perturbation, to this delicate balance would cause the universe to start expanding or contracting uncontrollably. In the years after Einstein introduced (7.100/7.101), Edwin Hubble discovered that the universe *is* in fact expanding ($\dot{a} > 0$) and Einstein realized that he did not need the cosmological constant. He went on to strike the cosmological constant from his equations, calling it his "greatest blunder."[36] Had he trusted his initial instinct, he would have theoretically predicted the expansion of the universe.

It is of great interest to note that what Einstein called his greatest blunder turned out not to have been a blunder after all. In 1998, evidence from observing type IA supernovae [21] indicated that the universe is not only expanding, but is also *accelerating* in its expansion, i.e. $\ddot{a} > 0$! This was surprising, since one would assume that the acceleration of the universe $\ddot{a}$ is negative, as the matter density would gravitationally "pull in" the universe rather than "push" it outward. This observation was confirmed several times over, and it is clear now that there exists a non-vanishing positive value for the cosmological constant that should be accounted for in the gravitational field equations; in other words, eqn (7.100) should be used when considering the universe as a whole.[37] The source of this constant is somewhat unclear. The most common assumption is that it arises from what is called **dark energy**, an unknown type of energy that pervades the entire universe and has a repulsive effect on it as a whole. It is sometimes also referred to as the **vacuum energy**, since it is not related to the matter content and would still be there if the vacuum were devoid of matter and other types of known energy. Because the field equations can be written in the form (7.101), the cosmological constant term can be thought of as part of the matter/energy right-hand side of the Einstein equations, rather than the geometric left-hand side.

It gets better. There is, in fact, such an energy of the vacuum that arises from studying particle physics at a very small scale. It has been known since the 1960s that quantum field theory[38] allows for a residual energy of the vacuum to exist, even when no other types of matter or energy are present. This is not just an object of theoretical interest; it has been experimentally verified and measured (see the so-called **Casimir effect**[39]). If one calculates this quantum vacuum energy in units of energy density, one finds find

$$\rho|_{\text{Quantum Vacuum}} \sim 10^{112} \text{erg/cm}^3, \tag{7.102}$$

which is a *very* large number. On the other hand, the observed value of the dark energy density required for the current rate of acceleration of the universe is roughly

$$\rho_\Lambda \sim 10^{-8} \text{erg/cm}^3. \tag{7.103}$$

---

[36] As told in [20].

[37] One wonders what Einstein would have said had he lived to know about this.

[38] The special relativistic version of quantum mechanics.

[39] Hendrik Brugt Gerhard Casimir: Dutch physicist (1909–2000).

The quantum $\rho|_{\text{Quantum Vacuum}}$ exceeds the observed $\rho_\Lambda$ by a catastrophic 120 orders of magnitude!! This is known as the **cosmological constant problem** and is sometimes referred to as the "worst prediction in the history of physics." If the source of dark energy and the cosmological constant is indeed this quantum vacuum energy, then something *else* must be canceling it: something huge! But to this day no one knows what that is. If the two are not related, then the source of dark energy, and consequently the cosmological constant, is completely unknown.

There are other ways of trying to solve this problem; for example, some modified theories of gravity (generalizations to Einstein's relativity) have terms in their field equations that may provide an explanation to the large-scale acceleration of the universe based on geometry rather than an unknown energy content. Most of these, however, also make other predictions that are not observed, and despite this (huge!) problem, Einstein's general theory of relativity remains our best understanding of the geometry of the universe at large as well as small (but not too small) scales.

**Exercise 7.41** Find the metric of a spherically symmetric mass distribution $M$ with radius $a$ in the presence of a cosmological constant $\Lambda$ but nothing else. In other words, solve (7.100) for vanishing $T_{\mu\nu}$. Clearly the metric is still (5.52), except that $H$ is no longer defined by (5.51). It should, however, reduce to the Schwarzschild form for vanishing $\Lambda$. This "black hole with cosmological constant" solution is well known, of course, but let's pretend it's not.

**Exercise 7.42** Study the solution you found in Exercise 7.41. Does it have an event horizon? Use the method of the effective potential to see what you can figure out concerning its orbital geodesics.

# 8

# Least Action and Classical Fields

*I have tried to read philosophers of all ages and have found many illuminating ideas but no steady progress toward deeper knowledge and understanding. Science, however, gives me the feeling of steady progress: I am convinced that theoretical physics is actual philosophy. It has revolutionized fundamental concepts and has taught us new methods of thinking which are applicable far beyond physics.*

Max Born

## Introduction

At this point you, dear reader, can claim that you know the fundamentals of the general theory of relativity, and have considerable insight into the idea of covariance in general. You certainly have acquired enough knowledge to perform some major calculations in the topic. More advanced texts on the subject (see the bibliography) could now be read with "relative" ease. In this chapter and the next one, we will discuss the basics of some advanced material that might be particularly useful for students planning to read modern research papers in theoretical physics. To begin with, we will discuss the all-important principle of least action and its application on the motion of point particles, both classically and relativistically. We will also take this opportunity to introduce the concept of *fields*, which may already be somewhat familiar to you. The concepts discussed here appear in various branches of physics, and you may have already been exposed to them; especially if you have studied advanced classical mechanics and electrodynamics. We will, however, make no assumptions and will present everything from scratch.

## 8.1 The Principle of Least Action

The so-called **principle of least action** is a time-honored concept that has been shown over and over again to be of great importance to many branches of physics. It has a long history of development with roots in the ancient writings of Euclid and

*Covariant Physics: From Classical Mechanics to General Relativity and Beyond.* Moataz H. Emam, Oxford University Press (2021). © Moataz H. Emam.
DOI: 10.1093/oso/9780198864899.003.0008

Hero of Alexandria.[1] In more modern times it was developed in various forms by (in no particular order): Leibnitz, Fermat,[2] Euler, Maupertuis,[3] Lagrange,[4] Jacobi, and Gauss, all very notable mathematicians and physicists. The formulation that the principle is most useful for, however, is due to Hamilton,[5] hence it is best known today as **Hamilton's principle of least action**, or just **Hamilton's principle**.

The principle can be stated as follows: For each case of the motion of a specific particle or the behavior of a given physical field there exists an integral known as the **action**, usually denoted by the letter $S$, such that its vanishing *variation* $\delta S = 0$ (explained below) necessarily leads to the equations that govern the behavior of said particle and/or field. The integral is sometimes performed over time, in which case the integrand is known as the **Lagrangian function** $L$, while sometimes it is performed over both time and space, i.e. spacetime, in which case it is performed over the so-called **Lagrangian density function** $\mathcal{L}$. The dimensions of the action are those of energy × time. There are many things here that require an explanation, and this is perhaps best done by delving straight into specific applications.

### 8.1.1 The Actions of Non-Relativistic Particles

In classical non-relativistic mechanics, the Lagrangian function $L$ is simply the difference between the kinetic energy of the system under study and its potential energy; thus

$$L\left(q_a, \dot{q}_a; t\right) = T - V, \tag{8.1}$$

such that the action is

$$S = \int_{t_i}^{t_f} dt\, L\left(q_a, \dot{q}_a; t\right) = \int_{t_i}^{t_f} dt\, \left(T - V\right), \tag{8.2}$$

where $T$ is the total kinetic energy of a system of particles, $V$ is the sum of all potential energies present, and the integral is performed between the starting time of the particles' motion $t_{i[nitial]}$ and the ending time $t_{f[inal]}$. The functions $q_a\left(t\right)$ are called the **generalized coordinates**, and $\dot{q}_a\left(t\right)$ are their rates of change with respect to time, known as the **generalized velocities**. The action is a "function" in these functions, technically termed a **functional**. Now if there are $N$ point particles in the problem under study, each particle is described by at most three generalized coordinates in three-dimensional space; hence the indices count over $a = 1, 2, 3, \cdots, 3N$. The generalized coordinates $q_a$ may be degrees of freedom with dimensions of length, such as

---

[1] Hero of Alexandria also known as Heron of Alexandria: an Egyptian-Greek mathematician and engineer (*c.* 10CE–*c.* 70CE).

[2] Pierre de Fermat: French lawyer and mathematician (1607–1665).

[3] Pierre Louis Moreau de Maupertuis: French mathematician (1698–1759).

[4] Joseph-Louis Lagrange: Italian mathematician and astronomer (1736–1813).

[5] Sir William Rowan Hamilton: Irish mathematician (1805–1865).

the usual $x$, $y$, $z$, $\rho$, and $r$; in which case $\dot{q}_a$ are the usual components of velocity $\dot{x}$, $\dot{r}$, etc. They may also be with dimensions of radians,[6] such as the angles $\theta$ or $\varphi$; in which case $\dot{q}_a$ are angular speeds $\dot{\theta}$, and so on. Clearly then, if the $q$'s are the usual spatial coordinates, one generally needs three for each particle. Some of these coordinates may not be independent of each other, however, which means that the generalized coordinates may be reduced to less than the maximum of $3N$. For example, there are two time-dependent coordinates $x(t)$ and $y(t)$ needed to describe the location of the bob of a simple pendulum in the plane, as in Fig. [8.3]. However, they are constrained by the length of the pendulum $l$ such that $l^2 = x^2 + y^2$. The existence of this constraint reduces the number of degrees of freedom by one, which is usually taken to be the angle $\theta$ that the pendulum makes, with the vertical defined by $\tan\theta = x/y$. Hence in this case there is only one generalized coordinate, $q = \theta$, even though the problem is essentially two-dimensional.

Note that in (8.1) and (8.2) the time parameter $t$ is placed after a semicolon to emphasize that the Lagrangian, and consequently the action, while explicitly dependent on $q_a$ and $\dot{q}_a$, is usually only implicitly dependent on $t$; i.e. $t$ does not explicitly appear in $L$. There are cases where $t$ may appear explicitly in the Lagrangian; however, these cases are rare, and they represent physical situations where energy is not conserved. We will not be concerned with these cases and will always take

$$\frac{\partial L}{\partial t} = 0. \tag{8.3}$$

Let us also point out that it is possible to define the so-called **generalized momenta** $\pi_a$, associated with each of the generalized coordinates, as follows:

$$\pi_a = \frac{\partial L}{\partial \dot{q}_a}. \tag{8.4}$$

This gives an expression for the linear momentum $\pi_a = m\dot{q}_a$ if $q_a$ is a coordinate with dimensions of length, and it yields *angular* momentum if its associated $q_a$ is an angle. There are cases where the generalized momentum is not as straightforward as this, as we will see. Sometimes the generalized momentum is also referred to as the **canonical momentum**.

As a simple example, consider the Lagrangian for a point particle of mass $m$ constrained to move along the $x$ axis only and attached to a spring with a spring constant $k$, as shown in Fig. [8.1]. In this case there is only one generalized coordinate, $q = x$, and one generalized velocity, $\dot{q} = \dot{x}$, such that

$$L(x, \dot{x}; t) = T - V = \frac{1}{2}m\dot{x}^2 - \frac{1}{2}kx^2, \tag{8.5}$$

---

[6]Technically radians is not a dimension, but rather a dimensionless measure of an angle.

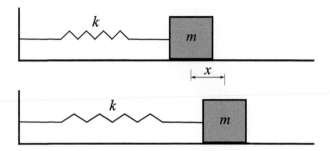

Fig. 8.1: Hooke's law problem. The top diagram shows the box $m$ at equilibrium. The bottom diagram shows it in motion. The distance $x$ is the extension (or contraction) of the box from equilibrium.

with

$$S = \int_{t_i}^{t_f} dt \left( \frac{1}{2}m\dot{x}^2 - \frac{1}{2}kx^2 \right). \tag{8.6}$$

The particle will follow the path described by Newton's second law, which in this case leads to the equation of motion $m\ddot{x} + kx = 0$, whose solution is a path $x(t)$ described by the usual sines and cosines of simple harmonic motion; but we will pretend that we do not know that. In fact, the whole idea is to replace Newton's formulation of mechanics with a new, (much) more general, one. Here is how it goes: If one plots the motion of the particle $x(t)$ between its starting and ending times, there are *a priori* an infinite number of possibilities (remember we are pretending we don't know the answer), as shown in Fig.[8.2]. If we can calculate the action integral $S$ for each and every one of these infinite possibilities, then the path that the particle will actually take will have the lowest value for $S$. Hence the term "least action principle." But how does one calculate $S$ for an infinite number of possible functions $x(t)$? It turns out we do not need to. The problem is similar to the problem of finding minima and maxima of functions in ordinary calculus. Recall that, given a function $f(x)$, finding its minimum and maximum points is simply the matter of finding the points $x_1$, $x_2$, etc. that satisfy the condition

$$\frac{df}{dx} = 0. \tag{8.7}$$

The principle of least action is a generalization of this concept. Instead of (8.7) we have the condition

$$\delta S = 0, \tag{8.8}$$

which technically finds both minima and maxima of $S$, just like the first derivative test. In physics, however, it is always a minimum that exists for $S$, so we don't have to worry about that. Equation (8.8) is a generalization of the first derivative test, *except* that instead of a function $f$ we have a *functional* $S$, and instead of finding points $x$ we are finding *functions* $q(t)$, which in the case of (8.6) is just $x(t)$! The condition

(8.7) gives an algebraic equation with numbers as its solutions, while (8.8) gives a *differential* equation with *functions* as its solutions! In the context of the mechanics of point particles, the differential equations that (8.8) leads to are the second-order differential equations of motion of the particle or particles. In the box on a spring case it is just $m\ddot{x} + kx = 0$, as we will see.

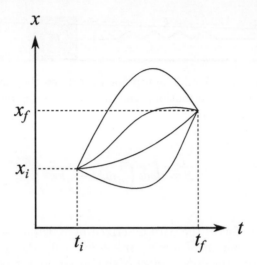

Fig. 8.2: The principle of least action: The actual path that a particle takes among the infinite possibilities is the one that guarantees the least possible value for the action $S$.

Now the operator $\delta$ is short hand for *variation of*, meaning that it represents the change of the action $S$, treated as a whole, just as $d/dx$ represented the change of a function $f$. In terms of performing the actual calculation (8.8), the variation $\delta$ is a differential operator very much like $d$ or $d/dx$ in that it satisfies all the rules of differential calculus, such as the product and chain rules.[7] If the concept of least action is rather hard to explain and to conceptually understand, the calculation itself is quite simple and replicates itself for most known cases of actions in physics. Let's see how it works for the box on a spring situation:

$$\delta S = \delta \int_{t_i}^{t_f} dt \left[ \frac{1}{2} m\dot{x}^2 - \frac{1}{2} kx^2 \right]$$

$$= \int_{t_i}^{t_f} dt \left[ \frac{1}{2} m\delta\left(\dot{x}^2\right) - \frac{1}{2} k\delta\left(x^2\right) \right] = \int_{t_i}^{t_f} dt \left[ m\dot{x}\delta\dot{x} - kx\delta x \right]. \tag{8.9}$$

---

[7]There is an entire branch of mathematics known as the **calculus of variations** that was developed via the efforts of many mathematicians, and most notably Euler, that deals with the meaning and properties of variations. You *will* come across it in the course of your studies if you haven't already. In this context, Hamilton's principle can be regarded as a special case, or more precisely an application, of the calculus of variations.

Note that since $\delta$ is an independent operator it passed through the integral and acted directly on the functions $x$ and $\dot{x}$. The first integral alone is

$$m \int_{t_i}^{t_f} dt \frac{dx}{dt} \delta \left( \frac{dx}{dt} \right) = m \int_{t_i}^{t_f} dt \frac{dx}{dt} \frac{d(\delta x)}{dt} = m \frac{dx}{dt} \delta x \Big|_{t_i}^{t_f} - m \int_{t_i}^{t_f} dt \delta x \frac{d^2 x}{dt^2}, \qquad (8.10)$$

where we have used integration by parts in the last step. The term $m \frac{dx}{dt} \delta x \big|_{t_i}^{t_f}$ is known as the **boundary term**; it evaluates the variation of $x$ at the *fixed* initial and starting points. We take it as vanishing, since all the possible paths of the particle must share precisely these two points. In fact, if it doesn't vanish we can always go back and "fix" the principle such that it does.[8] The resulting total condition is thus

$$\delta S = -\frac{1}{2} \int_{t_i}^{t_f} dt \, (m\ddot{x} + kx) \, \delta x = 0. \qquad (8.11)$$

If we now require that this condition is satisfied for any arbitrary variation $\delta x$, we are then led to the conclusion that the quantity inside the parentheses *must* vanish for the actual path of the particle; in other words,

$$m\ddot{x} + kx = 0, \qquad (8.12)$$

which is *exactly* what we would have found had we applied Newton's version of this calculation via $F = ma$! As a final check that the method works, let's also find the momentum associated with $x$ using (8.4):

$$\pi_x = \frac{d}{d\dot{x}} \left( \frac{1}{2} m\dot{x}^2 - \frac{1}{2} kx^2 \right) = m\dot{x}, \qquad (8.13)$$

as one would expect. Notice that in taking the derivative, $\dot{x}$ is treated as a parameter independent of $x$. This is a crucial concept in this approach. The reader may think that using Newton's version to figure out the motion of a particle on a spring is much easier, and in fact they would be absolutely correct. What we have done here is applied a very powerful principle on a very simple case: cracking a walnut with a jackhammer, if you will. It certainly does the job, but could be applied to much more complicated situations with a lot more ease than the increasingly complicated methods of Newtonian mechanics. An even easier case is that of a free particle, $V = 0$, whose action is just

$$S = \int_{t_i}^{t_f} dt \frac{1}{2} m\dot{x}^2, \qquad (8.14)$$

---

[8] As was pointed out in [22].

leading to the equation of motion of a free particle,

$$\ddot{x} = 0, \tag{8.15}$$

or, by inserting the mass back to recover the ordinary momentum $\pi = p = m\dot{x}$,

$$\frac{dp}{dt} = 0. \tag{8.16}$$

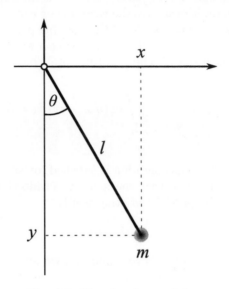

Fig. 8.3: The simple pendulum.

Another "walnut" example is that of figuring out the motion of a simple pendulum of length $l$ and mass $m$, as shown in Fig. [8.3]. An elementary calculation of the kinetic and potential energy terms leads to the Lagrangian and action

$$L = \frac{1}{2}ml^2\dot{\theta}^2 - mgl\,(1 - \cos\theta)$$

$$S = \int_{t_i}^{t_f} dt\,\left[\frac{1}{2}ml^2\dot{\theta}^2 - mgl\,(1 - \cos\theta)\right], \tag{8.17}$$

which leads to

$$\delta S = \int_{t_i}^{t_f} dt\,\left[\frac{1}{2}ml^2\delta\left(\dot{\theta}^2\right) + mgl\delta\,(\cos\theta)\right]$$

$$= \int_{t_i}^{t_f} dt\,\left[ml^2\dot{\theta}\delta\left(\dot{\theta}\right) - mgl\sin\theta\delta\theta\right]. \tag{8.18}$$

Once again, we integrate the first integral by parts

$$ml^2 \int\limits_{t_i}^{t_f} dt \frac{d\theta}{dt} \delta\left(\frac{d\theta}{dt}\right) = ml^2 \int\limits_{t_i}^{t_f} dt \frac{d\theta}{dt} \frac{d\delta\theta}{dt}$$

$$= ml^2 \frac{d\theta}{dt} \delta\theta \Big|_{t_i}^{t_f} - ml^2 \int\limits_{t_i}^{t_f} dt \frac{d^2\theta}{dt^2} \delta\theta, \tag{8.19}$$

and throw away the boundary term like before. Putting everything back together, we find

$$\delta S = \int\limits_{t_i}^{t_f} dt \left(-\ddot{\theta} - \frac{g}{l} \sin\theta\right) \delta\theta = 0, \tag{8.20}$$

from which we conclude that for arbitrary $\delta\theta$ we must have

$$\ddot{\theta} + \frac{g}{l} \sin\theta = 0, \tag{8.21}$$

exactly as we would have found had we used Newton's second law. As mentioned above, the generalized coordinate in this case is just $q = \theta$: the angle the pendulum makes with the vertical. Note that (8.4) gives

$$\pi_\theta = \frac{d}{d\dot{\theta}} \left[\frac{1}{2} ml^2 \dot{\theta}^2 - mg\left(1 - \cos\theta\right)\right] = ml^2 \dot{\theta}, \tag{8.22}$$

where we recognize the quantity $ml^2$ as the moment of inertia $I$ of the mass $m$ about the fixed point, and $\dot{\theta}$ is just the angular speed, frequently called $\omega$. The readers will then recognize the familiar expression for the angular momentum of the pendulum $\pi_\theta = I\omega$; which is a special case of (3.68).

Finally, we make note of a matter of notation. In the above examples the variation always ended up looking like this:

$$\delta S = \int\limits_{t_i}^{t_f} dt \,(\text{some second-order differential expression})\, \delta q = 0, \tag{8.23}$$

from which we conclude that since this works for any arbitrary $\delta q$, the expression in the parentheses must be equal to zero. It is a common notation to divide all sides of (8.23) by $\delta q$ and write

$$\frac{\delta S}{\delta q} = \int dt \,(\text{some second-order differential expression}) = 0, \tag{8.24}$$

from which one can argue that an equivalent form of the least action principle (8.8) is

$$\frac{\delta S}{\delta q} = 0, \tag{8.25}$$

renamed "variation with respect to $q$." Clearly (8.8) and (8.25) are trivially equivalent, but we will be using the later form when there arises an upfront need to define what the variation will be with respect to.

The principle of least action in non-relativistic mechanics replaces Newton's formulation of mechanics with an equivalent, but perhaps more elegant, formulation. It certainly also carries a certain "mystique" to it. In Richard Feynman's excellent,[9] as well as legendary, lectures on physics [22], he describes the principle of least action as follows: "Suppose you have a particle (in a gravitational field, for instance) which starts somewhere and moves to some other point by free motion. You throw it, and it goes up and comes down. It goes from the original place to the final place in a certain amount of time. Now, you try a different motion. Suppose that to get from here to there, it went [on a different path] but got there in just the same amount of time. [ . . ] If you calculate the kinetic energy at every moment on the path, take away the potential energy, and integrate it over the time during the whole path, you'll find that the number you'll get is bigger than that for the actual motion. In other words, the laws of Newton could be stated not in the form $F = ma$ but in the form: The average kinetic energy less the average potential energy is as little as possible for the path of an object going from one point to another."[10]

---

**Exercise 8.1** For the following cases in ordinary mechanics, formulate the Lagrangian, identify the generalized coordinates, then apply $\delta S = 0$ to find the equations of motion:

1. A particle of mass $m$ falling vertically in the uniform gravitational field of close to Earth's surface.
2. A particle of mass $m$ falling radially in the Newtonian gravitational potential (5.30).
3. A particle of mass $m$ fired from a canon in the uniform gravitational field of close to Earth's surface. This is the projectile motion problem of introductory physics.
4. The double pendulum shown in Fig. [8.4a].
5. A particle mass $m$ constrained to slide without friction on the walls of the cone shown in Fig. [8.4b].

---

### 8.1.2 The Euler–Lagrange Equations

The reader may have noticed that the calculation (8.8), which we have performed twice so far,[11] always follows the same exact route: Vary the action, which leads to variations in $\delta q$ and $\delta \dot{q}$, integrate the term in $\delta \dot{q}$ by parts and throw away the boundary

---

[9]Richard Phillips Feynman: American theoretical physicist (1918–1988).

[10]In this section of the lectures, Feynman is actually quoting his high school physics teacher Mr. Bader, who first got him interested in the topic of least action. Feynman went on to win a Nobel Prize for work that was heavily based on the principle of least action.

[11]More if you did any parts of Exercise 8.1.

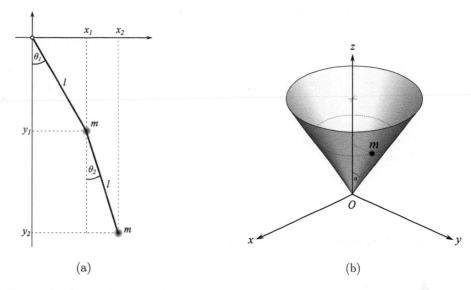

(a)                                                        (b)

Fig. 8.4: Examples for applying the least action principle on the motion of point masses.

term, and whatever is left must be vanishing for arbitrary $\delta q$. It is possible, then, to perform this process for arbitrary actions upfront to find an equation that depends on derivatives of the Lagrangian. Let's do this. Start with

$$S = \int_{t_i}^{t_f} dt L\left(q_a, \dot{q}_a; t\right) \tag{8.26}$$

and vary it:

$$\delta S = \int_{t_i}^{t_f} dt \delta L\left(q_a, \dot{q}_a; t\right) = 0. \tag{8.27}$$

Then apply the chain rule on $\delta L$:

$$\delta S = 0 = \int_{t_i}^{t_f} dt \left(\frac{\partial L}{\partial q_a}\delta q_a + \frac{\partial L}{\partial \dot{q}_a}\delta \dot{q}_a + \frac{\partial L}{\partial t}\delta t\right). \tag{8.28}$$

Next, we note that

$$\frac{\partial L}{\partial t} = \frac{\partial T}{\partial t} - \frac{\partial V}{\partial t}. \tag{8.29}$$

As pointed out earlier, we take this to vanish. The reason is two-fold: Firstly, the kinetic energy never depends on time explicitly; hence the first term of (8.29) automatically vanishes. Secondly, the second term is non-vanishing only for non-conservative potentials; i.e. potentials that do not satisfy energy conservation. We do not need to concern ourselves with those cases, so we take the second term to vanish as well.[12] From the remaining two terms, let's look at the second one and integrate it by parts:

$$\int\limits_{t_i}^{t_f} dt \frac{\partial L}{\partial \dot{q}_a} \delta \dot{q}_a = \int\limits_{t_i}^{t_f} dt \frac{\partial L}{\partial \dot{q}_a} \frac{d \delta q_a}{dt} = \left. \frac{\partial L}{\partial \dot{q}_a} \delta q_a \right|_{t_i}^{t_f} - \int\limits_{t_i}^{t_f} dt \frac{d}{dt} \left( \frac{\partial L}{\partial \dot{q}_a} \right) \delta q_a. \tag{8.30}$$

Throwing away the boundary term as usual and plugging back into (8.28), we finally end up with

$$\delta S = \int\limits_{t_i}^{t_f} dt \left[ \frac{\partial L}{\partial q_a} - \frac{d}{dt} \left( \frac{\partial L}{\partial \dot{q}_a} \right) \right] \delta q_a = 0, \tag{8.31}$$

which for arbitrary $\delta q_a$ implies

$$\frac{\partial L}{\partial q_a} - \frac{d}{dt} \left( \frac{\partial L}{\partial \dot{q}_a} \right) = 0. \tag{8.32}$$

This is a set of $3N$ second-order differential equations known as the **Euler–Lagrange equations**. We can use them directly on the Lagrangian to find the equations of motion, since the step (8.8) has already been performed. To confirm this, here is how they are applied to the two examples we have seen in the previous subsection:

**Box on a spring:** The Lagrangian is given by (8.5). Applying (8.32) using the single generalized coordinate $q = x$, we find

$$\frac{\partial L}{\partial x} = -kx$$

$$\frac{d}{dt} \left( \frac{\partial L}{\partial \dot{x}} \right) = \frac{d}{dt} (m\dot{x}) = m\ddot{x}, \tag{8.33}$$

from which (8.12) immediately arises.

**Simple pendulum:** From (8.17) the Lagrangian is

$$L = \frac{1}{2} m l^2 \dot{\theta}^2 - mgl \left( 1 - \cos \theta \right), \tag{8.34}$$

from which

$$\frac{\partial L}{\partial \theta} = -mgl \sin \theta, \tag{8.35}$$

$$\frac{d}{dt} \left( \frac{\partial L}{\partial \dot{\theta}} \right) = \frac{d}{dt} \left( \frac{1}{2} m l^2 \dot{\theta}^2 \right) = m l^2 \ddot{\theta}. \tag{8.36}$$

leading to (8.21) as expected.

---

[12]Textbooks in classical mechanics may be consulted for how to deal with the non-conservative cases; for example, see [12].

It is easy to see that the Euler–Lagrange equation is equivalent to Newton's second law by noting that the term inside the derivative in (8.32) is exactly the generalized momentum (8.4), and that a (conservative) force is always defined by $\mathbf{F} = -\vec{\nabla}V$; whose components in the generalized coordinates may be written $F_a = -\frac{\partial V}{\partial q_a}$. Going back to (8.32) we write

$$\frac{\partial(T-V)}{\partial q_a} - \frac{\partial \pi_a}{\partial t} = 0, \tag{8.37}$$

and recalling that the kinetic energy does not depend on the coordinates we are led to

$$F_a = \frac{\partial \pi_a}{\partial t}, \tag{8.38}$$

which is Newton's second law in component form, i.e. exactly $\mathbf{F} = d\mathbf{p}/dt$. In many cases, using the Euler–Lagrange equations directly on the Lagrangian is a lot more economical than applying (8.8) every time. We will pick and choose which to use depending on the situation.

**Exercise 8.2** Apply the Euler–Lagrange equations on the Lagrangians you derived in Exercise 8.1 and show that you get the same equations of motion.

## 8.1.3   The Actions of Relativistic Particles

Let's now see how to adapt the principle of least action if the motion of the particle in question is relativistic. The special relativistic equivalent for a free particle equation of motion is

$$\frac{dP^\mu}{d\tau} = 0, \tag{8.39}$$

from (4.128). And if we use the massive particle definition $P^\mu = mU^\mu = m\,dx^\mu/d\tau$, then

$$\frac{d^2 x^\mu}{d\tau^2} = \frac{d^2 x^\mu}{dt^2} = 0. \tag{8.40}$$

To figure out the action responsible for these equations, we may start by generalizing the concept of the Lagrangian being the difference between the kinetic and potential energies of the relativistic particle and see where it leads. The real test is whether or not what we postulate gives (8.40) as well as the correct canonical momentum; which in this case should just be the relativistic 3-momentum $p = \gamma m v$ (4.134). For a free particle, the potential energy is zero and the kinetic energy is given by (4.145)

$$T = (\gamma - 1)\, mc^2. \tag{8.41}$$

So we postulate

$$S \stackrel{?}{=} mc^2 \int_{\tau_i}^{\tau_f} d\tau \, (\gamma - 1). \tag{8.42}$$

Recall that $v = dx/dt$ is the speed of the particle's rest frame with respect to some inertial observer, but the integral is with respect to the proper time $\tau$, so we use (4.89) and write

$$S = mc^2 \int_{t_i}^{t_f} \frac{dt}{\gamma} (\gamma - 1) = mc^2 \int_{t_i}^{t_f} dt \left( 1 - \frac{1}{\gamma} \right). \tag{8.43}$$

However, since the objective is to apply (8.8), then the "1" inside the parentheses shouldn't matter; we thus remove it:

$$S = -mc^2 \int_{t_i}^{t_f} \frac{dt}{\gamma}. \tag{8.44}$$

This is, in fact, the correct action for a free relativistic particle. However, the way we constructed it is rather misleading, and I can almost hear my colleagues gasping in horror. The reason is that (8.42) seems to imply that the Lagrangian for a free particle is

$$L = T = (\gamma - 1) \, mc^2. \quad \text{Wrong!} \tag{8.45}$$

This is in fact incorrect. Although it *does* lead to the correct action following the argument above, we cannot consider (8.45) to be true. The reason is the added requirement that the Lagrangian should lead to the generalized momentum via (8.4). We already know that the special relativistic 3-momentum is defined by eqn (4.134), so let's check:

$$\frac{d}{dv} \left[ mc^2 \, (\gamma - 1) \right] = mc^2 \frac{d\gamma}{dv} = -m \left( 1 - \frac{v^2}{c^2} \right)^{-\frac{3}{2}} v. \tag{8.46}$$

This is certainly *not* (4.134). If, however, we extract the Lagrangian from (8.44):

$$L = -\frac{mc^2}{\gamma}, \tag{8.47}$$

and apply (8.4),

$$\frac{d}{dv} \left[ -\frac{mc^2}{\gamma} \right] = -mc^2 \frac{d}{dv} \left( \frac{1}{\gamma} \right) = \gamma mv, \tag{8.48}$$

which is indeed (4.134). This is interesting: The wrong approach leads to the correct action, from which the wrong Lagrangian can be corrected. As such, we take (8.44) and (8.47) to be the correct action and Lagrangian respectively for a freely moving relativistic massive particle.

The action has forms other than (8.44) that will lead us to some interesting interpretations. Using (4.89) again, we rewrite (8.44):

$$S = -mc^2 \int_{\tau_i}^{\tau_f} d\tau, \qquad (8.49)$$

which is still a correct and equivalent form for the action. But wait! Using $ds^2 = -c^2 d\tau^2$ we can write $|ds| = cd\tau$, which leads to

$$S = -mc \int_{s_i}^{s_f} ds, \qquad (8.50)$$

where we have used the absolute value symbol to avoid the square root of a negative number, then removed it. This does *not* alter the resulting physics.[13] As mentioned, (8.44), (8.49), and (8.50) are all equivalent; however, the third of these is by far the most important. Consider what it means: The quantity $ds$ is the element of length on the *path* of the particle in spacetime; the so-called world line of the particle. This makes the action $S$ proportional to the total length of the world line, and minimizing it by using (8.8) is like saying that the free particle will follow the shortest distance between two points in spacetime (a geodesic), which most certainly makes sense.

Now, we still need to derive the equations of motion (8.39) from the action. This is most easily done if one writes the line element $ds$ explicitly in (8.50)[14]

$$S = -mc \int \sqrt{\eta_{\mu\nu} dx^\mu dx^\nu}, \qquad (8.51)$$

then multiply by $(d\tau/d\tau)^2$ inside the square root

$$S = -mc \int d\tau \sqrt{\eta_{\mu\nu} \frac{dx^\mu}{d\tau} \frac{dx^\nu}{d\tau}} = -mc \int d\tau \sqrt{\eta_{\mu\nu} U^\mu U^\nu}. \qquad (8.52)$$

Now we can apply (8.8):

$$\delta S = -mc \int d\tau \, \delta \left( \eta_{\mu\nu} \frac{dx^\mu}{d\tau} \frac{dx^\nu}{d\tau} \right)^{\frac{1}{2}} = -\frac{1}{2} mc \int \frac{d\tau}{\left( \eta_{\mu\nu} \frac{dx^\mu}{d\tau} \frac{dx^\nu}{d\tau} \right)^{\frac{1}{2}}} \eta_{\mu\nu} \delta \left( U^\mu U^\nu \right)$$

$$= -\frac{1}{2} mc \int \frac{d\tau}{ds} \left( \eta_{\mu\nu} U^\mu \delta U^\nu + \eta_{\mu\nu} U^\nu \delta U^\mu \right) = 0. \qquad (8.53)$$

---

[13]The reader, feeling cheated by this, will recall that it is perfectly fine to use the signature $(+,-,-,-)$ for the metric (4.16), which would set $ds^2 = +c^2 d\tau^2$ and avoid the issue altogether.

[14]We will henceforth drop the limits of the action integral for brevity.

The expression inside the parentheses can be dealt with as follows: Since the Minkowski metric is symmetric $\eta_{\mu\nu} = \eta_{\nu\mu}$, then we can switch the $\eta$ indices in the second term. Following that, we rename the dummy indices $\mu \to \nu$ and $\nu \to \mu$:

$$\eta_{\mu\nu} U^\mu \, \delta U^\nu + \eta_{\mu\nu} U^\nu \, \delta U^\mu = \eta_{\mu\nu} U^\mu \, \delta U^\nu + \eta_{\nu\mu} U^\nu \, \delta U^\mu$$
$$= \eta_{\mu\nu} U^\mu \, \delta U^\nu + \eta_{\mu\nu} U^\mu \, \delta U^\nu = 2\eta_{\mu\nu} U^\mu \, \delta U^\nu, \qquad (8.54)$$

to end up with

$$\delta S = -mc\eta_{\mu\nu} \int \frac{d\tau}{ds} U^\mu \, \delta U^\nu = 0. \qquad (8.55)$$

If at this point we use $mU^\nu = P^\nu$, then using the chain rule $\delta P^\nu = \frac{dP^\nu}{d\tau}\delta\tau$ we get

$$-c\eta_{\mu\nu} \int \frac{d\tau}{ds} U^\mu \, \delta P^\nu = -c\eta_{\mu\nu} \int \frac{d\tau}{ds} U^\mu \frac{dP^\nu}{d\tau}\delta\tau = 0, \qquad (8.56)$$

leading directly to (8.39). However, if we continue without inserting $P^\nu$ we get

$$-mc\eta_{\mu\nu} \int \frac{d\tau}{ds} U^\mu \, \delta U^\nu = -mc\eta_{\mu\nu} \int \frac{d\tau}{ds} \frac{dx^\mu}{d\tau} \delta\left(\frac{dx^\nu}{d\tau}\right). \qquad (8.57)$$

Integrating by parts,

$$\delta S = -mc\eta_{\mu\nu} \left[ \frac{dx^\mu}{d\tau} \delta x^\nu \Big|_{\tau_i}^{\tau_f} - \int \frac{d\tau}{ds} \delta x^\nu \frac{d^2 x^\mu}{d\tau^2} \right] = 0. \qquad (8.58)$$

Once again the boundary term is thrown out for the exact same reason as before, and we clearly end up with the first equality of (8.40), as desired. We can get the second equality if we multiply by $(dt/dt)^2$ inside the square root of (8.51) instead. The fact that the action represents the "length" of a four-dimensional curve in spacetime and that minimizing it is akin to finding the shortest distance between two points is very telling. This means that if $ds^2$ is not flat, but representing a line element on a curved manifold, then requiring that $\delta S = 0$ should lead to the geodesic equation! We will see that this guess is in fact true in §8.5.

Finally, we ask: Since the action has different forms, can the Lagrangian too be written in different forms corresponding to (8.49), (8.50), (8.51), and (8.52)? In order to understand this, recall that the Lagrangian that we read off from (8.44) gave the correct answer for the momentum. This Lagrangian was taken from an action that integrates over $dt$, *not* $d\tau$. We then define the Lagrangian to be whatever is inside the integrand, as long as the integration is carried over $dt$. Hence we cannot read off a Lagrangian from any of the action forms (8.49), (8.50), (8.51), and (8.52). However, recall that in writing (8.52) we multiplied by $(d\tau/d\tau)^2$ inside the square root. There is in fact nothing to stop us from multiplying by $(dt/dt)^2$ instead, such that now

$$S = -mc \int dt \sqrt{\eta_{\mu\nu} \frac{dx^\mu}{dt} \frac{dx^\nu}{dt}}, \tag{8.59}$$

and the Lagrangian of the free relativistic particle is thus

$$L = -mc \sqrt{\eta_{\mu\nu} \frac{dx^\mu}{dt} \frac{dx^\nu}{dt}}. \tag{8.60}$$

In the presence of a potential energy, $V$, one can still just subtract it in the usual way:

$$L = -\frac{mc^2}{\gamma} - V = -mc \sqrt{\eta_{\mu\nu} \frac{dx^\mu}{dt} \frac{dx^\nu}{dt}} - V. \tag{8.61}$$

This is taken as axiomatic, since the interpretation of the Lagrangian as $T - V$ is no longer correct in relativistic mechanics. The first term of (8.61) is *not* the kinetic energy and has no physical meaning other than the fact that it gives the correct answer (it still has units of energy, though). Another thing to note is that the action is Lorentz invariant, i.e. all inertial observers will agree on it. However, the Lagrangian is *not*, since its dependence is on the time $t$, not the proper time $\tau$.

**Exercise 8.3** The calculations in this subsection were grueling. In order to make sure you understood them correctly (for the sake of what's coming next) prepare a detailed presentation of the material in this subsection so far, record it on a video, and hand it in for grading.[15]

For the sake of demonstration, let's take a look at the case of the relativistic motion of a particle under the influence of a constant force.[16] Consider that the particle is allowed to travel only along the $x$ direction. It is subjected to a constant force, $-F$, such that the potential energy is just $V = Fx$. If we further use $a = F/m = $ constant, then we can write the Lagrangian

$$L = -\frac{mc^2}{\gamma} - max = -mc^2 \sqrt{1 - \frac{\dot{x}^2}{c^2}} - max. \tag{8.62}$$

---

[15]This exercise is based on the saying "If you want to master something, teach it," famously attributed to Richard Feynman, who was a notoriously great teacher. A similar saying: "If you want to learn something, read about it. If you want to understand something, write about it. If you want to master something, teach it" is attributed to the American-Indian Yogi Bhajan (1929–2004). Whoever first said it doesn't really matter. It is very true, and I am a firm believer in it. In fact, it is most definitely a good idea for you to apply not just onto this subsection but onto any material at all.

[16]We explored this case back in Exercise 4.27 using a different method. What follows was adapted freely from [12].

The generalized coordinate is $q = x$ and the generalized velocity is $\dot{q} = \dot{x}$. The Euler–Lagrange eqn (8.32) is then

$$\frac{\partial L}{\partial x} - \frac{d}{dt}\left(\frac{\partial L}{\partial \dot{x}}\right) = -ma + mc^2\frac{d}{dt}\left(\frac{\dot{x}}{c\sqrt{1 - \frac{\dot{x}^2}{c^2}}}\right) = 0, \tag{8.63}$$

or

$$\frac{d}{dt}\left(\frac{\dot{x}}{\sqrt{1 - \frac{\dot{x}^2}{c^2}}}\right) = a. \tag{8.64}$$

Integrating once,

$$\frac{\dot{x}}{\sqrt{1 - \frac{\dot{x}^2}{c^2}}} = at + b. \tag{8.65}$$

Now rearranging,

$$dx = c\frac{at + b}{\sqrt{c^2 + (at + b)^2}}dt,$$

$$x = c\int \frac{at + b}{\sqrt{c^2 + (at + b)^2}}dt + x_0. \tag{8.66}$$

Integrating again, we end up with the solution

$$x(t) = \frac{c}{a}\left[\sqrt{c^2 + (at + b)^2} - \sqrt{c^2 + b^2}\right] + x_0. \tag{8.67}$$

At $t = 0$ we find $x = x_0$, so $x_0$ is the initial position of the particle. Let's assume that to be the origin, $x_0 = 0$. What about the particle's initial velocity? Equation (8.65) may be rearranged:

$$\dot{x} = \frac{c(at + b)}{\sqrt{c^2 + (at + b)^2}}. \tag{8.68}$$

Hence the initial velocity is

$$v_0 = \frac{cb}{\sqrt{c^2 + b^2}}. \tag{8.69}$$

If we assume that the particle also starts from rest, then $v_0 = 0$, which leads to $b = 0$. Under these initial conditions the solution (8.67) can be rewritten as

$$\left(x + \frac{c^2}{a}\right)^2 - c^2 t^2 = \frac{c^4}{a^2},\tag{8.70}$$

which is the equation of a *hyperbola* in the $x$–$t$ plane! The reader is encouraged to show that the non-relativistic solution may be found from (8.67) by assuming $(at + b) << c$, leading to the expected parabolic formula

$$x(t) = \frac{1}{2}at^2 + v_0 t + x_0,\tag{8.71}$$

where in this limit you will recognize that $v_0 \approx b$.

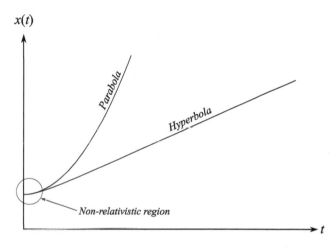

Fig. 8.5: Comparing the non-relativistic solution (top) with the relativistic solution (bottom) for a particle in motion under the influence of a constant force. Notice that initially both curves coincide. They start separating as time progresses and the speed, i.e. the slope of the curve, becomes large enough for relativistic effects to start showing. The hyperbola has an asymptote whose slope is exactly the speed of light; in other words it is the light cone.

As a final example, we give the Lagrangian of a particle under the influence of an electromagnetic field. This is

$$L = -\frac{mc^2}{\gamma} - qV + q\dot{\mathbf{r}} \cdot \mathbf{A},\tag{8.72}$$

where $V$ and $\mathbf{A}$ are the electric potential and magnetic vector potential respectively, and $q$ is the charge of the relativistic particle. The action is then

$$S = \int dt \left(-\frac{mc^2}{\gamma} - qV + q\dot{\mathbf{r}} \cdot \mathbf{A}\right).\tag{8.73}$$

Applying (8.8) gives the equation

$$\frac{d\mathbf{p}}{dt} = q\mathbf{E} + q\dot{\mathbf{r}} \times \mathbf{B},\tag{8.74}$$

where $\mathbf{p}$ is the relativistic 3-momentum (4.134) and the right-hand side is the Lorentz force that the particle will experience. The Lagrangian in explicitly covariant form (i.e. using the index–tensor language) is

$$L(\tau) = -mc\sqrt{\eta_{\mu\nu}\frac{dx^\mu}{dt}\frac{dx^\nu}{dt}} + q\frac{dx^\mu}{dt}A_\mu.\tag{8.75}$$

This Lagrangian is then expected to lead to eqn (4.192), with the definition

$$f^\mu = \frac{dP^\mu}{dt}.\tag{8.76}$$

**Exercise 8.4**  Derive (8.74) from (8.73).

**Exercise 8.5**  Explicitly show that (8.75) is equivalent to (8.73).

**Exercise 8.6**  Vary the action that arises from (8.75) to show that it does lead to (4.192) as claimed.

**Exercise 8.7**  Using (8.61), find the relativistic equations of motion of a particle mass $m$ in the following potential energies:

1. The simple harmonic oscillator $V = -\frac{1}{2}kx^2$. Compare your result with its non-relativistic counterpart.
2. The Newtonian gravitational potential energy (3.90). Compare your result with its non-relativistic counterpart.
3. An exponential potential $V = \alpha e^{\beta x}$, where $\alpha$ and $\beta$ are arbitrary constants.
4. The so-called **Yukawa potential** $V = \alpha\frac{e^{-\beta x}}{x}$, where $\alpha$ and $\beta > 0$ are arbitrary constants. This potential is used as an approximation of nuclear forces.[17]

**Exercise 8.8**  Here is another exercise that may act as a small project. Use your work from part 2 of Exercise 8.7 to show that the relativistic motion of a particle in a Newtonian gravitational field is a precessing ellipse! Calculate the precession of the perihelion of Mercury (whether approximately analytically or numerically) and show that it is much smaller than the value predicted by general relativity.

**Exercise 8.9**  Attempt solving the equations of motion you found in Exercise 8.7. If it proves too hard to do so analytically, then do it numerically on the computer with some different sets of initial conditions. Do not do the second one if you have already done Exercise 8.8.

---

[17]Discovered in 1935 by the Japanese theoretical physicist Hideki Yukawa (1907–1981).

## 8.2   Classical Field Theory

All of the examples of actions we have seen so far have been for particles. However, we claimed earlier that actions and Lagrangians can be written for fields, leading to *field* equations; rather than the equations of motion of an object. It is time to verify this claim, and in so doing we introduce the reader to the classical (i.e. non-quantum) theory of fields. Maxwell's theory of electrodynamics is of course the most well known of such field theories. It is, however, only one of many. Field theories in physics are sometimes referred to as **gauge theories**. The reason is that they all, including electrodynamics, have certain symmetries, known (for historical reasons) as **gauge symmetries**. These in turn are classified into "groups," a concept that has a very precise mathematical definition. We will not get into field theory from that perspective, instead leaving it to more specialized texts. For our purposes, we can classify field theories based on the rank of the tensor that describes its potential, or potentials. Electrodynamics, for instance, is defined by the the rank 1 tensor $A^\mu$: the 4-potential, as reviewed in §4.8 and §6.3. It is, then, an example of a **vector field theory**. We emphasize that the reference to vectors here is in the sense of the *potentials*, meaning that it is *not* the electric and magnetic fields that define the theory; rather it is $A^\mu$, from which the electromagnetic field can be found in the form of the second-rank field strength tensor $F_{\mu\nu}$ defined by (4.185).[18] Taking the derivative of the 4-potential raises its rank by 1; in other words, the rank of the field strength is always one higher than the potential that defines the theory. Similarly, one can define a **scalar field theory** by a scalar potential $\phi(x^\alpha)$, i.e. a rank zero tensor, and the corresponding field is found by taking the derivative; making the field strength itself a vector, e.g.

$$F_\mu = \partial_\mu \phi. \tag{8.77}$$

The definition (8.77) is automatically generally covariant since for a scalar field $\nabla_\mu \phi = \partial_\mu \phi$. Now we can go in the other direction and define higher-rank tensor field theories, where the potential is an arbitrary rank $n > 1$ tensor and the field strength is of a rank $n + 1$. For example, general relativity *itself* may be thought of as a rank 2 tensor field theory. The "potential" in this case is the metric tensor, while the field is its derivative; namely the connection or Christoffel symbols. But let's leave that case for later and explore the topics in order.

---

[18]There are other vector field theories defined by extensions of (4.185); e.g. theories where the components of the potential 4-vector do not commute with each other. The field strength tensor (not electromagnetic anymore) in cases like these would have a form such as $F^{\alpha\beta} = \partial^\alpha A^\beta - \partial^\beta A^\alpha - g\left(A^\alpha A^\beta - A^\beta A^\alpha\right)$, where $g$ is some, possibly complex, constant. Such theories are called **non-Abelian gauge theories** (compared to Maxwell's electrodynamics, which is an **Abelian gauge theory**); they are used to describe, among other things, the fields of the so-called **strong nuclear force** inside the nuclei of atoms. This is the force responsible for keeping the nucleus together against the electrostatic repulsion between the protons. As the name implies, it is much stronger than electric forces. The term "Abelian" is in honor of the Norwegian mathematician Niels Henrik Abel (1802–1829). Once again I find myself in a footnote that can easily turn into a book of its own. So I put an end to it and refer the interested reader to, for example, [10] and [23].

### 8.2.1 The Principle of Least Action for Scalar Fields

It turns out that scalar fields can be thought of as a natural extension of oscillatory fields in classical mechanics. Consider as a simple example an ordinary non-relativistic string, like that of a violin, of length $l$ that is stretched between two points in space; for example, two nails on a wall. Since this is a continuous material, it is best described by its mass per unit length $\mu = m/l$ rather than its total mass $m$. We will assume that the string is making very small oscillations about its equilibrium such that its length doesn't change by much from its rest length. Its kinetic energy per unit length, i.e. its kinetic energy *density*, is

$$\mathcal{T} = \frac{1}{2}\mu\dot{y}^2, \tag{8.78}$$

where $y(t,x)$ is the vertical displacement of the string, as shown in Fig. [8.6], and $\dot{y} = \partial y/\partial t$. The potential energy density responsible for the motion of the string, in this case in the $y$ direction, is given in terms of the tension of the string $f_T$, assumed constant throughout. This turns out to be

$$\mathcal{V} = \frac{1}{2}f_T y'^2, \tag{8.79}$$

where $y' = \partial y/\partial x$.[19]

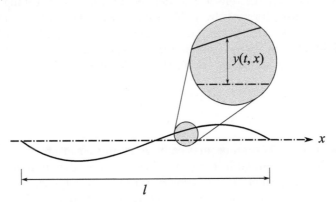

Fig. 8.6: A classical oscillating string.

Hence we can construct the **Lagrangian density**:

$$\mathcal{L} = \mathcal{T} - \mathcal{V} = \frac{1}{2}\left(\mu\dot{y}^2 - f_T y'^2\right) \tag{8.80}$$

in straightforward analogy with (8.1). The Lagrangian itself would be the integral of this density over the length $l$:

$$L = \int_{x_i}^{x_f} dx\mathcal{L}\left(\dot{y}, y'; t\right) = \int_0^l dx\frac{1}{2}\left(\mu\dot{y}^2 - f_T y'^2\right). \tag{8.81}$$

---

[19]See any textbook on mechanics for a detailed description of how $\mathcal{V}$ is derived.

Hence the full action would be

$$S = \int\limits_{t_i}^{t_f} dt L = \int\limits_{t_i}^{t_f} dt \int\limits_{x_i}^{x_f} dx \mathcal{L}\left(\dot{y}, y'; t\right) = \mu \int\limits_{t_i}^{t_f} \int\limits_{0}^{l} dt dx \frac{1}{2}\left(\dot{y}^2 - \frac{f_T}{\mu}y'^2\right). \qquad (8.82)$$

Applying (8.8) leads to the equation of motion of the string

$$y'' - \frac{\mu}{f_T}\ddot{y} = 0, \qquad (8.83)$$

which is the wave equation in one dimension with solutions representing oscillatory waves traveling on the string. Equation (8.83) may be written more generally for *any* one-dimensional wave motion as follows:

$$y'' - \frac{1}{v^2}\ddot{y} = 0, \qquad (8.84)$$

where it can be shown that $v$ (equivalent to $\sqrt{f_T/\mu}$ for the case of the string) is the speed of propagation of the wave. The action is then generally

$$S = \int\limits_{t_i}^{t_f} dt \int\limits_{x_i}^{x_f} dx \mathcal{L}\left(\dot{y}, y'\right) = \alpha \int\limits_{t_i}^{t_f} \int\limits_{0}^{a} dt dx \frac{1}{2}\left(\dot{y}^2 - v^2 y'^2\right), \qquad (8.85)$$

where $\alpha$ is some constant with units of kg/m. Note that the action is an integral over *both* time and space; effectively treating them on equal footing (hint hint)! One can take this further by looking at two-dimensional oscillations, such as on a stretched membrane like the surface of a percussion instrument described by the coordinates $x$ and $y$ and oscillating in the $z$ direction, in which case the action would be of the form

$$S = \alpha \int \int dt dx dy \frac{1}{2}\left[\dot{z}^2 - v^2\left(\frac{\partial z}{\partial x}\right)^2 - v^2\left(\frac{\partial z}{\partial y}\right)^2\right], \qquad (8.86)$$

where here the "rise" of the surface of the membrane is described by the function $z(t, x, y)$. This leads to the two-dimensional wave equation

$$\frac{\partial^2 z}{\partial x^2} + \frac{\partial^2 z}{\partial y^2} - \frac{1}{v^2}\frac{\partial^2 z}{\partial t^2} = 0. \qquad (8.87)$$

Finally, we generalize to a continuous three-dimensional function $\phi(t, \mathbf{r})$ such that

$$S = \alpha \int dt dV \frac{1}{2}\left[\dot{\phi}^2 - v^2\left(\frac{\partial \phi}{\partial x}\right)^2 - v^2\left(\frac{\partial \phi}{\partial y}\right)^2 - v^2\left(\frac{\partial \phi}{\partial z}\right)^2\right], \qquad (8.88)$$

where for brevity we did not write down the four integration symbols $\int$ and $dV$ represents the volume element $dxdydz$. This can be more compactly written:

$$S = \alpha \int dt dV \frac{1}{2} \left[ \dot{\phi}^2 - v^2 \left( \vec{\nabla}\phi \right) \cdot \left( \vec{\nabla}\phi \right) \right].$$ (8.89)

If one wishes for (8.89) to be coordinate invariant in three dimensions in the sense of §2.6, then we should insert the Jacobian $\sqrt{|g|}$ in the action. The equation of motion that (8.89) leads to is simply

$$\nabla^2 \phi - \frac{1}{v^2} \ddot{\phi} = 0.$$ (8.90)

This should begin to look familiar. The function $\phi$ describes a scalar field in three-dimensional space. If we now attempt to make (8.89) and (8.90) relativistic by making the speed of propagation of the $\phi$ waves equal to the speed of light ($v = c$), then (8.90) becomes

$$\nabla^2 \phi - \frac{1}{c^2} \ddot{\phi} = 0,$$ (8.91)

which can now be written as a relativistic field equation for $\phi(x^\alpha)$:

$$\eta^{\mu\nu} \partial_\mu \partial_\nu \phi = 0,$$ (8.92)

or equivalently

$$\Box \phi = \partial^\mu \partial_\mu \phi = 0,$$ (8.93)

where $\Box$ is the d'Alembertian operator. As we pointed out briefly following eqn (4.31), this is a wave equation whose solutions represent the propagation of waves of the field $\phi$. In the current context it is the special relativistic equation for a *free* scalar field. In quantum field theory, the scalar field $\phi$ is represented by the free motion of particles, called **scalar particles**, that are traveling at the speed of light and are hence massless. In contrast, the theory of quantum electrodynamics represents the vector field $A^\mu$ by a massless **vector particle**: the so-called photon. The fully (special) relativistic action from which (8.92/8.93) arises is found by adapting (8.89):

$$S = -\alpha \int d^4 x \frac{1}{2} \eta^{\mu\nu} \left( \partial_\mu \phi \right) \left( \partial_\nu \phi \right),$$ (8.94)

where $d^4 x = dx^0 dx^1 dx^2 dx^3 = cdt dx^1 dx^2 dx^3$ and the minus sign is added in as a matter of convention. If one wishes to make these relationships covariant under any coordinate system (in flat or curved spacetimes), then the minimal substitutions $\partial \to \nabla$ and $\eta \to g$ are used, as well as inserting a Jacobian (with the appropriate minus sign) in the action

$$S = -\alpha \int d^4 x \sqrt{-|g|} \frac{1}{2} g^{\mu\nu} \left( \nabla_\mu \phi \right) \left( \nabla_\nu \phi \right)$$
$$g^{\mu\nu} \nabla_\mu \nabla_\nu \phi = \nabla_\mu \nabla^\mu \phi = 0,$$ (8.95)

or in terms of the scalar field strength (8.77)

$$S = -\alpha \int d^4 x \sqrt{-|g|} \frac{1}{2} F_\mu F^\mu$$
$$\nabla_\mu F^\mu = 0.$$ (8.96)

The term $(\nabla_\mu \phi)(\nabla_\nu \phi)$ in the action is usually called the **kinetic term**.[20] It has the form of the square of a derivative just as ordinary non-relativistic theory has a kinetic term in $v^2$, which is also the square of a derivative, except that the time derivative becomes a derivative with respect to all of the spacetime dimensions; i.e. $d/dt \to \partial_\mu, \nabla_\mu$. As noted, the theory represents massless fields (or massless particles upon quantization). We can add a "mass" term, one that, upon quantization in quantum field theory, leads to the equations of motion of a relativistic quantum massive particle. Such a term is usually of the form $\frac{1}{2}m^2\phi^2$, such that

$$S = \alpha \int d^4x \sqrt{-|g|}\,\frac{1}{2}\left(-F_\mu F^\mu - m^2\phi^2\right), \tag{8.97}$$

leading to the so-called **Klein–Gordon equation**:[21]

$$\nabla_\mu \nabla^\mu \phi + m^2 \phi = 0. \tag{8.98}$$

Note the similarity between (8.97) and the action of a non-relativistic mass on a spring (8.5); in fact, it seems that an additional minimal substitution principle is in action here: $k \to m$ and $x \to \varphi$. The same can be seen when one compares (8.98) with (8.12). It is as if the massive scalar field is modeled upon a large number $N \to \infty$ of massive particles connected together with springs in space. This is very telling, and in quantum field theory the mathematics that arises is very similar, in fact identical, to having a (quantum) simple harmonic oscillator at every point in spacetime.[22]

We conclude this subsection by giving the Euler–Lagrange equation for scalar fields:

$$\frac{\partial \mathcal{L}}{\partial \phi} - \partial_\mu \frac{\partial \mathcal{L}}{\partial (\partial_\mu \phi)} = 0. \tag{8.99}$$

One can easily see that (8.99) is related to (8.32) by the minimal substitutions $L \to \mathcal{L}$, $q \to \phi$, and $d/dt \to \partial_\mu$.

---

**Exercise 8.10** Vary the actions (8.85), (8.86), (8.89), (8.94), (8.95), (8.97) with respect to their degrees of freedom $y$, $z$, and $\phi$, and show that the variation leads to their respective field equations.

---

[20] Once again, because $\phi$ is scalar field, $(\nabla_\mu \varphi)(\nabla_\nu \varphi)$ is the same thing as $(\partial_\mu \phi)(\partial_\nu \phi)$; however, $\nabla_\mu \nabla^\mu \phi$ is *not* the same thing as $\partial_\mu \partial^\mu \phi$.

[21] Due to the Swedish and German physicists Oskar Benjamin Klein (1894–1977) and Walter Gordon (1893–1939).

[22] The reader who has taken a course in ordinary non-relativistic quantum mechanics may recall that the ground state, or the lowest energy state, of a quantum harmonic oscillator is not zero, but has a minimum value. This means that, contrary to their non-quantum counterparts, quantum oscillators cannot sit still, even at their lowest energy state. As noted here, quantized relativistic fields (scalar or otherwise) behave like quantum harmonic oscillators at *every* point in space. What this means is that their lowest energy level is not zero! It has a small, but cumulative, residual value that exists even when there is nothing there. In fact, this is *exactly* the postulated source of the vacuum energy that we discussed in the previous chapter in the context of the cosmological constant problem.

**Exercise 8.11** Derive the Euler–Lagrange equation for scalar fields (8.99) by varying, i.e. applying $\delta S = 0$, to the general form scalar field action:

$$S = \alpha \int d^4x \mathcal{L}\left[\phi, (\partial_\mu \phi) ; x^\mu\right],$$
(8.100)

where $\alpha$ is some arbitrary constant.

**Exercise 8.12** Apply the Euler–Lagrange eqn (8.99) to the actions (8.89), (8.94), (8.95), (8.97), and show that it leads to their respective field equations.

## 8.2.2   The Principle of Least Action for Maxwell Fields

In the previous section we gave the Lagrangian/action describing the motion of a charged particle in an electromagnetic field. But what is the Lagrangian/action that describes the fields themselves? In other words, what is the action whose vanishing variation (8.8) leads to Maxwell's equations? We will present this in the explicitly covariant form reviewed in §4.8 and §6.3. The electrodynamic Lagrangian density is

$$\mathcal{L} = -\frac{1}{4\mu_0} F^{\mu\nu} F_{\mu\nu} - A_\mu J^\mu.$$
(8.101)

The action is hence

$$S = \int d^4x \sqrt{-|g|} \left(-\frac{1}{4\mu_0} F^{\mu\nu} F_{\mu\nu} - A_\mu J^\mu\right),$$
(8.102)

where $g_{\mu\nu}$ is used to raise and lower indices. The Euler–Lagrange equation that leads to the Maxwell equations is found by a similar minimal substitution to the one used for scalar fields. Explicitly, $L \to \mathcal{L}$, $q \to A^\mu$, and $d/dt \to \partial_\mu$. Hence

$$\frac{\partial \mathcal{L}}{\partial A^\mu} - \partial_\nu \frac{\partial \mathcal{L}}{\partial (\partial_\nu A^\mu)} = 0.$$
(8.103)

This actually leads to *only* (4.188). The second of the Maxwell eqns (4.189) is just a direct consequence of the *definition* (4.185), as can be checked by direct substitution.

**Exercise 8.13** Starting with the general form action:

$$S = \alpha \int d^4x \mathcal{L}\left[A^\mu, (\partial_\nu A^\mu) ; x^\mu\right],$$
(8.104)

derive the Euler–Lagrange eqn (8.103). You may want to attempt Exercise 8.11 first.

**Exercise 8.14** Vary the action (8.102) to find (4.188).

**Exercise 8.15**  Apply (8.103) to (8.101) to find (4.188).

**Exercise 8.16**  Consider the Minkowski-Cartesian action:

$$S = \alpha \int d^4x \left[ \frac{1}{2} \varepsilon^{\mu\nu\rho} A_\mu \left( \partial_\nu A_\rho \right) - A_\mu J^\mu \right], \tag{8.105}$$

where $\alpha$ is an arbitrary constant. This action belongs to a category of theories that go by the name **Chern–Simons theories**,[23] which are used mostly in string-theoretic type models. Derive the field equations that arise in this theory in terms of the field strength tensor $F_{\mu\nu}$, which you may take as still defined by (4.185).

## 8.3   The Stress Tensor from the Action

In §7.5 we gave the explicit form of the stress energy tensor (7.70) in the case where the region of space under study is filled with electromagnetic fields. This can be calculated from the definitions of each component. It turns out that there is a general recipe for calculating $T_{\mu\nu}$ for *any* type of field content, scalar or otherwise, directly from the action; this method is due to Hilbert and defines the stress tensor as the variation of the field action with respect to the metric tensor. It is

$$T_{\mu\nu} = -\frac{2}{\sqrt{-|g|}} \frac{\partial}{\partial g^{\mu\nu}} \left( \sqrt{-|g|} \mathcal{L}_M \right), \tag{8.106}$$

where we call the Lagrangian density that contains all those fields simply $\mathcal{L}_M$ ($M$ for matter). It is perhaps important to note that this method does not apply to the case of the perfect fluid stress tensor (7.67), as there doesn't seem to be a way of writing an action to describe perfect fluids. It only applies to the Lagrangian densities of fields. The expression (8.106) can be further simplified by applying the product rule

$$T_{\mu\nu} = -\frac{2}{\sqrt{-|g|}} \left[ \sqrt{-|g|} \frac{\partial \mathcal{L}_M}{\partial g^{\mu\nu}} + \mathcal{L}_M \frac{\partial}{\partial g^{\mu\nu}} \sqrt{-|g|} \right]. \tag{8.107}$$

To evaluate the second term we must draw upon a theorem from linear algebra known as **Jacobi's formula**, which states that the differential of the determinant of a square matrix $M$ is given by

$$d\,|M| = |M|\, \mathrm{Tr} \left[ M^{-1} dM \right]. \tag{8.108}$$

[23] Named after the Chinese-American mathematician Shiing-Shen Chern (1911–2004), pronounced "Chen," and the American mathematician James Harris Simons (b. 1938). However, the theories were introduced into physics by the Russian-American theoretical physicist Albert Solomonovich Schwarz (b. 1934) and further developed by the American theoretical physicist Edward Witten (b. 1951).

Applying this to $M \to g_{\mu\nu}$ and replacing $d$ with $\delta$ gives

$$\delta |g| = |g| g^{\mu\nu} \delta g_{\mu\nu} = -|g| g_{\mu\nu} \delta g^{\mu\nu}, \tag{8.109}$$

where we have used the easily verifiable relation

$$\delta g_{\mu\nu} = -g_{\mu\alpha} g_{\nu\beta} \delta g^{\alpha\beta} \tag{8.110}$$

in the last step. Then

$$\delta\sqrt{-|g|} = -\frac{1}{2\sqrt{-|g|}} \delta |g| = \frac{|g|}{2\sqrt{-|g|}} g_{\mu\nu} \delta g^{\mu\nu} = -\frac{1}{2}\sqrt{-|g|} g_{\mu\nu} \delta g^{\mu\nu}. \tag{8.111}$$

Hence we end up with

$$T_{\mu\nu} = -2\frac{\partial \mathcal{L}_M}{\partial g^{\mu\nu}} + g_{\mu\nu}\mathcal{L}_M. \tag{8.112}$$

Let us apply this to a scalar field Lagrangian density that includes both the kinetic term (8.95) and an arbitrary potential $V(\phi)$. The action is

$$S = \alpha \int d^4x \sqrt{-|g|} \left[ -\frac{1}{2}g^{\mu\nu} (\partial_\mu \phi)(\partial_\nu \phi) + V(\phi) \right]. \tag{8.113}$$

Hence

$$\mathcal{L}_M = \alpha \left[ -\frac{1}{2}g^{\mu\nu} (\partial_\mu \phi)(\partial_\nu \phi) + V(\phi) \right]. \tag{8.114}$$

Applying (8.112):

$$\begin{aligned}
T_{\mu\nu} &= -2\alpha \frac{\partial}{\partial g^{\mu\nu}} \left[ -\frac{1}{2}g^{\alpha\beta} (\partial_\alpha \phi)(\partial_\beta \phi) + V(\phi) \right] + g_{\mu\nu}\mathcal{L}_M \\
&= \alpha \frac{\partial g^{\alpha\beta}}{\partial g^{\mu\nu}} (\partial_\alpha \phi)(\partial_\beta \phi) + g_{\mu\nu}\mathcal{L}_M \\
&= \alpha \delta_\mu^\alpha \delta_\nu^\beta (\partial_\alpha \phi)(\partial_\beta \phi) + \alpha g_{\mu\nu} \left[ -\frac{1}{2}g^{\alpha\beta} (\partial_\alpha \phi)(\partial_\beta \phi) + V(\phi) \right] \\
&= -\frac{\alpha}{2}g_{\mu\nu} (\partial^\alpha \phi)(\partial_\alpha \phi) + \alpha (\partial_\mu \phi)(\partial_\nu \phi) + \alpha g_{\mu\nu} V(\phi),
\end{aligned} \tag{8.115}$$

where we have used

$$\frac{\partial g^{\alpha\beta}}{\partial g^{\mu\nu}} = \delta_\mu^\alpha \delta_\nu^\beta \text{ or } \overset{=}{} \delta_\nu^\alpha \delta_\mu^\beta. \tag{8.116}$$

---

**Exercise 8.17** What terms in addition to (8.115) would appear if we used the massive scalar field action (8.97) instead of (8.113)?

**Exercise 8.18** Derive the stress tensor (7.70) from the Maxwell action (8.102).

**Exercise 8.19** Find the stress tensor of the Chern–Simons field theory of Exercise 8.16.

## 8.4 General Relativistic Actions

We are finally in a position to write an action that describes physics in a curved spacetime background. In a sense, we have already done half the job: The relativistic Lagrangians and actions of the previous sections are exactly the same in curved background if one makes sure that minimal substitution is used appropriately everywhere. There is only one piece of the picture that is left to discuss: What is the action that gives the Einstein field equations? It should be an action with two terms: The first is the purely geometrical term that gives the left-hand side of (7.89), while the second should give the right-hand side. The latter is easy to guess: It is simply the pieces that describe whatever fields are present; for example, (8.95) for scalar fields, (8.102) for electromagnetic fields or any others we haven't discussed or all together; if more than one type of field is present. The action term involving the curvature side of the Einstein field equations was discovered by Hilbert in 1915 and is known today simply as the **Einstein–Hilbert action**:

$$S_{EH} = \frac{1}{2\kappa} \int d^4x \sqrt{-|g|} R, \tag{8.117}$$

where $\kappa$ is the Einstein constant (7.88) and $R$ is the Ricci scalar (7.23). The Lagrangian density is thus simply

$$\mathcal{L}_{EH} = \frac{1}{2\kappa} R. \tag{8.118}$$

As such, the Lagrangian density and action of a full model with matter/field content can be written as

$$\mathcal{L} = \frac{1}{2\kappa} R + \mathcal{L}_M, \tag{8.119}$$

$$S = \int d^4x \sqrt{-|g|} \left( \frac{1}{2\kappa} R + \mathcal{L}_M \right). \tag{8.120}$$

Varying (8.120) with respect to the metric tensor $g_{\mu\nu}$ or its inverse $g^{\mu\nu}$ (slightly easier) should yield the full Einstein equations in the form (7.89). In other words, (8.25) becomes

$$\frac{\delta S}{\delta g^{\mu\nu}} = 0. \tag{8.121}$$

Let's go ahead and apply (8.121) to (8.120):

$$\frac{\delta S}{\delta g^{\mu\nu}} = 0 = \int d^4x \left[ \frac{1}{2\kappa} \frac{\delta}{\delta g^{\mu\nu}} \left( \sqrt{-|g|} R \right) + \frac{\delta}{\delta g^{\mu\nu}} \left( \sqrt{-|g|} \mathcal{L}_M \right) \right],$$

$$\delta S = 0 = \int d^4x \sqrt{-|g|} \left[ \frac{1}{2\kappa} \left( \frac{\delta R}{\delta g^{\mu\nu}} + \frac{R}{\sqrt{-|g|}} \frac{\delta \sqrt{-|g|}}{\delta g^{\mu\nu}} \right) \right.$$

$$\left. + \frac{1}{\sqrt{-|g|}} \frac{\delta}{\delta g^{\mu\nu}} \left( \sqrt{-|g|} \mathcal{L}_M \right) \right] \delta g^{\mu\nu}. \tag{8.122}$$

As usual, we make the assertion that this should be true for any arbitrary variation $\delta g^{\mu\nu}$; the field equations *should* then be

$$\frac{\delta R}{\delta g^{\mu\nu}} + \frac{R}{\sqrt{-|g|}}\frac{\delta\sqrt{-|g|}}{\delta g^{\mu\nu}} = -2\kappa\frac{1}{\sqrt{-|g|}}\frac{\delta}{\delta g^{\mu\nu}}\left(\sqrt{-|g|}\mathcal{L}_M\right). \tag{8.123}$$

The right-hand side of (8.123) is exactly the definition of the stress tensor (times $\kappa$) we proposed in (8.106). As such, the left-hand side must lead to the Einstein tensor (7.25). Let's look at it more closely:

$$\frac{\delta R}{\delta g^{\mu\nu}} + \frac{R}{\sqrt{-|g|}}\frac{\delta\sqrt{-|g|}}{\delta g^{\mu\nu}} = \frac{\delta\left(g^{\alpha\beta}R_{\alpha\beta}\right)}{\delta g^{\mu\nu}} + \frac{R}{\sqrt{-|g|}}\frac{\delta\sqrt{-|g|}}{\delta g^{\mu\nu}}$$

$$= g^{\alpha\beta}\frac{\delta R_{\alpha\beta}}{\delta g^{\mu\nu}} + R_{\alpha\beta}\frac{\delta g^{\alpha\beta}}{\delta g^{\mu\nu}} + \frac{R}{\sqrt{-|g|}}\frac{\delta\sqrt{-|g|}}{\delta g^{\mu\nu}}. \tag{8.124}$$

The middle term of (8.124) is the simplest: it is just

$$R_{\alpha\beta}\frac{\delta g^{\alpha\beta}}{\delta g^{\mu\nu}} = R_{\alpha\beta}\delta^{\alpha}_{\mu}\delta^{\beta}_{\nu} = R_{\mu\nu}. \tag{8.125}$$

We have already evaluated the third term of (8.124) in (8.111). We are then left with only the first term of (8.124) to figure out. This term is the toughest, and the most subtle, to deal with. Using brute force on (7.22), it can be shown that the variation of the Ricci tensor is equivalent to the divergence of a vector $v_\kappa$:

$$g^{\alpha\beta}\delta R_{\alpha\beta} = \left(\nabla^\kappa v_\kappa\right), \tag{8.126}$$

where

$$v_\kappa = \nabla^\alpha\left(\delta g_{\alpha\kappa}\right) - g^{\alpha\beta}\nabla_\kappa\left(\delta g_{\alpha\beta}\right). \tag{8.127}$$

As such, the full first term of the action is

$$\frac{1}{2\kappa}\int d^4x\sqrt{-|g|}g^{\alpha\beta}\delta R_{\alpha\beta} = \frac{1}{2\kappa}\int d^4x\sqrt{-|g|}\nabla^\kappa v_\kappa. \tag{8.128}$$

The integral is that of a total derivative, which means that using the generalized form of **Stokes' theorem** it should yield a boundary term at infinity, i.e. essentially the value of the vector $v_\kappa$ evaluated at the *boundary* of the spacetime manifold. If the reader is not familiar with Stokes' theorem, we will discuss it in §9.5. Suffice it to note here that it is a mathematical theorem that relates the derivative of a vector on a manifold with the value of that vector on the boundary of said manifold. Hence eqn (8.123) becomes

$$K + R_{\mu\nu} - \frac{1}{2}g_{\mu\nu}R = \kappa T_{\mu\nu}, \tag{8.129}$$

where $K$ is the unwanted boundary term. If the spacetime manifold is closed (see §7.1), then this term automatically vanishes, as there is no boundary. If it is open, then the boundary is at infinity and we can just *require* that the variation of the metric at infinity vanishes, leading to $K = 0$. However, it turns out that $K$ depends not only on the variation of the metric $\delta g_{\mu\nu}$ but also on its first derivative, as you can see in (8.127), which may or may not vanish on the boundary. As such, the quantity $K$ may not easily go away, and we do not quite have the Einstein equations in (8.129). In his book [17], Wald showed that in this (rare) case the Einstein–Hilbert action itself may just be adjusted by adding a term designed specifically to cancel $K$. But, as pointed out, this is a rare situation indeed and for most intents and purposes $K$ may safely be taken to vanish and (8.117) is assumed the correct action.

One last thing to mention here: What about the cosmological constant? What action can be used to give eqns (7.100)? This can be guessed to be

$$S = \frac{1}{2\kappa} \int d^4x \sqrt{-|g|}\,(R - 2\Lambda),\tag{8.130}$$

as the reader can easily check. In conclusion, the most complete action of any classical field theory in a curved spacetime background is

$$S = \int d^4x \sqrt{-|g|}\left[\frac{1}{2\kappa}\,(R - 2\Lambda) + \mathcal{L}_M\right].\tag{8.131}$$

**Exercise 8.20** Show explicitly that the term "$-2\Lambda$" added to the Einstein–Hilbert action (8.130) does indeed give the correct form (7.100) of the Einstein field equations.

## 8.5 The Geodesic Equation One More Time

I would be doing you a great disservice dear reader, if I did not show you how to *derive* the geodesic equation from an action. We do this in its own section for its obvious importance. The idea is simple: The action describing the motion of a relativistic particle in *any* spacetime background is just (8.50). We have argued that since the action in this form represents the length of a particle's world line, it must be true that applying (8.8) necessarily leads to the geodesic equation itself. We have seen this in special relativity, but therein varying the action just led to the trivial case of a straight line. In curved backgrounds we have to use (8.50), with $ds$ being any arbitrary generally covariant metric; ergo

$$S = -mc \int \sqrt{g_{\alpha\beta}dx^\alpha dx^\beta} = -mc \int d\lambda \sqrt{g_{\alpha\beta}\frac{dx^\alpha}{d\lambda}\frac{dx^\beta}{d\lambda}},\tag{8.132}$$

which is just (8.51) after the minimal substitution $\eta \to g$, and we have used the affine parameter for the utmost generalization. The Euler–Lagrange equation is

$$\frac{\partial \mathcal{L}}{\partial x^\mu} - \frac{d}{d\lambda}\left[\frac{\partial \mathcal{L}}{\partial (dx^\mu/d\lambda)}\right] = 0,\tag{8.133}$$

where

$$\mathcal{L} = -mc\sqrt{g_{\alpha\beta}\frac{dx^\alpha}{d\lambda}\frac{dx^\beta}{d\lambda}}. \tag{8.134}$$

Now the first term of (8.133) is

$$\frac{\partial\mathcal{L}}{\partial x^\mu} = -\frac{mc}{\sqrt{g_{\alpha\beta}\frac{dx^\alpha}{d\lambda}\frac{dx^\beta}{d\lambda}}}\frac{1}{2}\partial_\mu\left(g_{\alpha\beta}\frac{dx^\alpha}{d\lambda}\frac{dx^\beta}{d\lambda}\right) = \frac{m^2c^2}{2\mathcal{L}}\partial_\mu\left(g_{\alpha\beta}\frac{dx^\alpha}{d\lambda}\frac{dx^\beta}{d\lambda}\right)$$

$$= \frac{m^2c^2}{2\mathcal{L}}\left[g_{\alpha\beta}\frac{dx^\alpha}{d\lambda}\partial_\mu\left(\frac{dx^\beta}{d\lambda}\right) + g_{\alpha\beta}\partial_\mu\left(\frac{dx^\alpha}{d\lambda}\right)\frac{dx^\beta}{d\lambda} + (\partial_\mu g_{\alpha\beta})\frac{dx^\alpha}{d\lambda}\frac{dx^\beta}{d\lambda}\right]$$

$$= \frac{m^2c^2}{2\mathcal{L}}\left[g_{\alpha\beta}\frac{dx^\alpha}{d\lambda}\frac{d\left(\partial_\mu x^\beta\right)}{d\lambda} + g_{\alpha\beta}\frac{d\left(\partial_\mu x^\alpha\right)}{d\lambda}\frac{dx^\beta}{d\lambda} + (\partial_\mu g_{\alpha\beta})\frac{dx^\alpha}{d\lambda}\frac{dx^\beta}{d\lambda}\right]$$

$$= \frac{m^2c^2}{2\mathcal{L}}\left[g_{\alpha\beta}\frac{dx^\alpha}{d\lambda}\frac{d\delta^\beta_\mu}{d\lambda} + g_{\alpha\beta}\frac{d\delta^\alpha_\mu}{d\lambda}\frac{dx^\beta}{d\lambda} + (\partial_\mu g_{\alpha\beta})\frac{dx^\alpha}{d\lambda}\frac{dx^\beta}{d\lambda}\right]$$

$$= \frac{m^2c^2}{2\mathcal{L}}(\partial_\mu g_{\alpha\beta})\frac{dx^\alpha}{d\lambda}\frac{dx^\beta}{d\lambda}, \tag{8.135}$$

where we have used $\partial_\mu x^\alpha = \delta^\alpha_\mu$. The next derivative is

$$\frac{\partial\mathcal{L}}{\partial\left(dx^\mu/d\lambda\right)} = \frac{m^2c^2}{2\mathcal{L}}g_{\alpha\beta}\left[\frac{dx^\alpha}{d\lambda}\left(\frac{\partial}{\partial\left(dx^\mu/d\lambda\right)}\frac{dx^\beta}{d\lambda}\right) + \left(\frac{\partial}{\partial\left(dx^\mu/d\lambda\right)}\frac{dx^\alpha}{d\lambda}\right)\frac{dx^\beta}{d\lambda}\right]$$

$$= \frac{m^2c^2}{2\mathcal{L}}g_{\alpha\beta}\left(\frac{dx^\alpha}{d\lambda}\delta^\beta_\mu + \delta^\alpha_\mu\frac{dx^\beta}{d\lambda}\right)$$

$$= \frac{m^2c^2}{2\mathcal{L}}\left(g_{\alpha\mu}\frac{dx^\alpha}{d\lambda} + g_{\mu\beta}\frac{dx^\beta}{d\lambda}\right) = \frac{m^2c^2}{\mathcal{L}}g_{\alpha\mu}\frac{dx^\alpha}{d\lambda}. \tag{8.136}$$

The final derivative is

$$\frac{d}{d\lambda}\left[\frac{\partial\mathcal{L}}{\partial\left(dx^\mu/d\lambda\right)}\right] = \frac{m^2c^2}{\mathcal{L}}\left[g_{\alpha\mu}\frac{d^2x^\alpha}{d\lambda^2} + \frac{dg_{\alpha\mu}}{d\lambda}\frac{dx^\alpha}{d\lambda}\right]. \tag{8.137}$$

Now applying the chain rule on the derivative of the metric tensor,

$$\frac{dg_{\alpha\mu}}{d\lambda} = \frac{\partial g_{\alpha\mu}}{\partial x^\beta}\frac{dx^\beta}{d\lambda} = (\partial_\beta g_{\alpha\mu})\frac{dx^\beta}{d\lambda}, \tag{8.138}$$

leads to

$$\frac{d}{d\lambda}\left[\frac{\partial L}{\partial\left(dx^\mu/d\lambda\right)}\right] = \frac{m^2c^2}{\mathcal{L}}g_{\alpha\mu}\frac{d^2x^\alpha}{d\lambda^2} + \frac{m^2c^2}{\mathcal{L}}(\partial_\beta g_{\alpha\mu})\frac{dx^\beta}{d\lambda}\frac{dx^\alpha}{d\lambda}$$

$$= \frac{m^2c^2}{\mathcal{L}}g_{\alpha\mu}\frac{d^2x^\alpha}{d\lambda^2} + \frac{m^2c^2}{2\mathcal{L}}(\partial_\beta g_{\alpha\mu})\frac{dx^\beta}{d\lambda}\frac{dx^\alpha}{d\lambda} + \frac{m^2c^2}{2\mathcal{L}}(\partial_\alpha g_{\beta\mu})\frac{dx^\beta}{d\lambda}\frac{dx^\alpha}{d\lambda}. \tag{8.139}$$

Putting everything together, the Euler–Lagrange equation is

$$\frac{1}{2}\left(\partial_\mu g_{\alpha\beta}\right)\frac{dx^\alpha}{d\lambda}\frac{dx^\beta}{d\lambda} - g_{\alpha\mu}\frac{d^2x^\alpha}{d\lambda^2} - \frac{1}{2}\left(\partial_\beta g_{\alpha\mu}\right)\frac{dx^\alpha}{d\lambda}\frac{dx^\beta}{d\lambda} - \frac{1}{2}\left(\partial_\alpha g_{\beta\mu}\right)\frac{dx^\alpha}{d\lambda}\frac{dx^\beta}{d\lambda} = 0$$

$$\rightarrow \quad g_{\alpha\mu}\frac{d^2x^\alpha}{d\lambda^2} + \frac{1}{2}\left[\left(\partial_\beta g_{\alpha\mu}\right) + \left(\partial_\alpha g_{\beta\mu}\right) - \left(\partial_\mu g_{\alpha\beta}\right)\right]\frac{dx^\alpha}{d\lambda}\frac{dx^\beta}{d\lambda} = 0.$$

$$(8.140)$$

Contracting with $g^{\mu\nu}$ and using the definition (4.33) of the Christoffel symbols

$$\frac{d^2x^\nu}{d\lambda^2} + \frac{1}{2}g^{\mu\nu}\left[\left(\partial_\alpha g_{\beta\mu}\right) + \left(\partial_\beta g_{\alpha\mu}\right) - \left(\partial_\mu g_{\alpha\beta}\right)\right]\frac{dx^\alpha}{d\lambda}\frac{dx^\beta}{d\lambda} = 0,$$

$$\text{we find:} \quad \frac{d^2x^\nu}{d\lambda^2} + \Gamma^\nu_{\alpha\beta}\frac{dx^\alpha}{d\lambda}\frac{dx^\beta}{d\lambda} = 0, \qquad (8.141)$$

as required. It is clear that this calculation reduces to the Minkowski-Cartesian result (8.40) if we set $g \rightarrow \eta$ and $\partial g = 0$ in every step.

**Exercise 8.21** Derive the geodesic equation by varying the action (8.132); in other words, use $\delta S = 0$.

## 8.6   Some More Spacetimes

As one last detailed example of a spacetime, we choose a solution that uses some of the new stuff we have learned in this chapter. At the end of the section we will list, without derivation, two more examples for completeness. Let's write down the generally covariant Lagrangian density (8.119) and action (8.120) describing an electromagnetic field in a curved spacetime background. Using (8.102), this is:

$$\mathcal{L}_M = \frac{1}{2\kappa}R - \frac{1}{4\mu_0}F^{\mu\nu}F_{\mu\nu} - A_\mu J^\mu, \qquad (8.142)$$

$$S = \int d^4x\sqrt{-|g|}\left(\frac{1}{2\kappa}R - \frac{1}{4\mu_0}F^{\mu\nu}F_{\mu\nu} - A_\mu J^\mu\right). \qquad (8.143)$$

Varying this with respect to $A^\mu$ leads to the Maxwell eqn (4.188), while, as noted earlier, (4.189) is a direct consequence of (4.185). Varying the action with respect to $g^{\mu\nu}$ leads to the Einstein field equations with the Maxwell stress tensor (7.70):

$$R_{\mu\nu} - \frac{1}{2}Rg_{\mu\nu} = \kappa T^{\text{Maxwell}}_{\mu\nu}$$

$$= \frac{\kappa}{\mu_0}\left(F_\mu{}^\alpha F_{\nu\alpha} - \frac{1}{4}g_{\mu\nu}F^{\alpha\beta}F_{\alpha\beta}\right). \qquad (8.144)$$

Let's look for a solution that represents a static uniform solid sphere with total mass $M$ and electric charge $Q$. Because of the spherical symmetry, our starting point is the ansatz (7.46), just as in the Schwarzschild case; although this time we expect the unknown functions $A$ and $B$ to end up being something other than what we found for Schwarzschild. We *do* expect them, however to reduce to the Schwarzschild case in the limit $Q \to 0$, as well as going to unity at $r \to \infty$. In addition, the Schwarzschild metric was a solution of the vacuum equations; here we have to use the full Einstein eqns (7.89). We elect to use version (7.85) of the field equations, since it is generally easier to calculate the trace of $T_{\mu\nu}$ than it is to compute the Einstein tensor; in fact, in this case the trace of the Maxwell tensor is vanishing:

$$T = g^{\mu\nu} T_{\mu\nu} = \frac{1}{\mu_0} \left( F^{\nu\alpha} F_{\nu\alpha} - F^{\alpha\beta} F_{\alpha\beta} \right) = 0. \tag{8.145}$$

Hence the field eqns (7.85) are just

$$R_{\mu\nu} = \kappa T_{\mu\nu}^{\text{Maxwell}} = \frac{\kappa}{\mu_0} \left( F_\mu{}^\alpha F_{\nu\alpha} - F^{\alpha\beta} F_{\alpha\beta} \right) \tag{8.146}$$

and the components of the Ricci tensor are read off from (7.47), (7.48), (7.49), and (7.50). This time it is not only the Einstein equations that need to be solved; one must also make sure that the Maxwell equations are satisfied for this configuration. In other words, the solution we are looking for must satisfy every *single* equation that results from $\delta S = 0$. Now because of spherical symmetry, the Faraday tensor is just

$$[F^{\mu\nu}] = \begin{bmatrix} 0 & -E_r/c & 0 & 0 \\ E_r/c & 0 & 0 & 0 \\ 0 & 0 & 0 & 0 \\ 0 & 0 & 0 & 0 \end{bmatrix}, \tag{8.147}$$

in the coordinates defined by (7.46). Hence

$$F^{tr} = -F^{rt} = -\frac{E_r}{c}. \tag{8.148}$$

Consequently finding the bits and pieces of the right-hand side of (8.146) is straightforward:

$$F^{\alpha\beta} F_{\alpha\beta} = F^{tr} F_{tr} + F^{rt} F_{rt} = 2F^{tr} F_{tr} = 2g_{tt} g_{rr} F^{tr} F^{tr} = \frac{2}{c^2} ABE_r^2,$$

$$F_t{}^\alpha F_{t\alpha} = F_t{}^r F_{tr} = g_{tt}^2 g_{rr} F^{tr} F^{tr} = \frac{1}{c^2} A^2 BE_r^2,$$

$$F_r{}^\alpha F_{r\alpha} = F_r{}^t F_{rt} = g_{tt} g_{rr}^2 F^{rt} F^{rt} = -\frac{1}{c^2} AB^2 E_r^2. \tag{8.149}$$

Putting the right- and left-hand sides of (8.146) together:

$$R_{tt} = \frac{\kappa}{2\mu_0 c^2} A^2 B E_r^2,$$

$$R_{rr} = -\frac{\kappa}{2\mu_0 c^2} A B^2 E_r^2,$$

$$R_{\theta\theta} = \frac{\kappa r^2}{2\mu_0 c^2} A B E_r^2,$$

$$R_{\varphi\varphi} = \frac{\kappa r^2}{2\mu_0 c^2} \sin^2 \theta A B E_r^2. \tag{8.150}$$

Before we proceed, let's note the following: Firstly, because of spherical symmetry the $R_{\varphi\varphi}$ components do not contribute anything new, just like in the Schwarzschild case. Secondly, one can see from the first two equations of (8.150) that (7.51) is true here as well, leading to

$$A = B^{-1}, \tag{8.151}$$

also the same as the Schwarzschild case. Hence we can set $A = B^{-1} = H$ and write

$$ds^2 = -c^2 H(r) dt^2 + H(r)^{-1} dr^2 + r^2 d\Omega^2, \tag{8.152}$$

similar to (5.52). Now, before proceeding let's find the explicit form of $E_r$ by solving the Maxwell equations. One can check that (4.189) is automatically satisfied, since it represents the no-monopoles law as well as the vanishing curl of the electric field; both of which are satisfied here.[24] Equation (4.188) needs to be constructed carefully. Since we are looking for the field *outside* the spherical charge distribution, $J^\mu = 0$. And in order to use the covariant derivative we also need the Christoffel symbols for (8.152), but these are the same as (6.87), keeping in mind that the function $H(r)$ is *not* the Schwarzschild one. We then write

$$\nabla_\alpha F^{\alpha\beta} = \partial_\alpha F^{\alpha\beta} + \Gamma^\alpha_{\alpha\mu} F^{\mu\beta} + \Gamma^\beta_{\alpha\mu} F^{\alpha\mu} = 0, \tag{8.153}$$

where we use the spacetime version of (2.80). The Christoffel symbols are symmetric in their lower indices, while the field tensor is antisymmetric; hence the third term vanishes exactly by virtue of the theorem in the paragraph following eqns (1.107). Equation (8.153) is two equations. The first is for the free index, $\beta = r$:

$$\nabla_\alpha F^{\alpha r} = \nabla_t F^{tr} = \partial_t F^{tr} + \Gamma^t_{tt} F^{\mu r} = 0, \tag{8.154}$$

which vanishes identically: The first term vanishes because $E_r$ is static and the second term vanishes because $\Gamma^t_{tt} = 0$. The second equation is for $\beta = t$:

$$\nabla_\alpha F^{\alpha t} = \nabla_r F^{rt} = \partial_r F^{rt} + \Gamma^\alpha_{\alpha r} F^{rt}$$

$$= \partial_r F^{rt} + \left( \frac{1}{2} \frac{H'}{H} - \frac{1}{2} \frac{H'}{H} + \frac{1}{r} + \frac{1}{r} \right) F^{rt}$$

$$= \frac{dE_r}{dr} + \frac{2}{r} E_r = 0, \tag{8.155}$$

---

[24]Recall that a radial field has a vanishing curl.

which is an easy first-order differential equation in $E_r$ that can be integrated by separation of variables thus:

$$\frac{dE_r}{E_r} = -2\frac{dr}{r},$$

$$\ln E_r = -2\ln r + b = \ln r^{-2} + b,$$

$$E_r = \frac{e^b}{r^2} = \frac{k}{r^2}, \qquad (8.156)$$

where $b$ and $k$ are arbitrary constants. Since we have obviously stumbled upon the ordinary Coulomb law for a spherically symmetric charge, we take the constant to be of the standard MKS form; hence

$$E_r = \frac{Q}{4\pi\varepsilon_0 r^2}. \qquad (8.157)$$

We can use the integral form of Gauss' law in these coordinates to show that the constant $k$ is in fact as chosen. The only thing to be careful with here is that the parameter $r$ in (8.157) is just a radial *coordinate*. It does *not* necessarily measure the true radial *distance* from the origin to the field point. In fact, since we do not know what the metric *inside* the sphere looks like, there really is no way of finding the actual distance in meters. Recall the discussion in §5.6.

**Exercise 8.22** To solve $\nabla_\alpha F^{\alpha\beta} = 0$ we used the spacetime version of (2.80). Solve it again using the spacetime version of (2.109) applied to $F^{\alpha\beta}$, i.e.

$$\nabla_\alpha F^{\alpha\beta} = \frac{1}{\sqrt{-|g|}}\partial_\alpha\left(\sqrt{-|g|}F^{\alpha\beta}\right) = 0. \qquad (8.158)$$

We are now ready to finish the problem. As in (7.55), we have

$$R_{\theta\theta} = 1 - rH' - H = 1 - \frac{d}{dr}(rH). \qquad (8.159)$$

Equating this with the third equation of (8.150) and rearranging gives

$$\frac{d}{dr}(rH) = 1 - \frac{GQ^2}{4\pi\varepsilon_0 c^4}\frac{1}{r^2}, \qquad (8.160)$$

where we have used the explicit form of Einstein's constant (7.88) as well as (4.172). Integrating

$$rH = \int dr\left(1 - \frac{GQ^2}{4\pi\varepsilon_0 c^4}\frac{1}{r^2}\right)$$

$$= r + \frac{GQ^2}{4\pi\varepsilon_0 c^4}\frac{1}{r} + C, \qquad (8.161)$$

we end up with

$$H\left(r\right) = 1 + \frac{C}{r} + \frac{GQ^2}{4\pi\varepsilon_0 c^4}\frac{1}{r^2}. \tag{8.162}$$

Now, since we require that this solution reduces to Schwarzschild in the limit $Q \to 0$, it must be true that the constant of integration is equal to the negative of the Schwarzschild radius: $C = -2GM/c^2 = -r_s$. If we further define

$$r_Q^2 = \frac{GQ^2}{4\pi\varepsilon_0 c^4}, \tag{8.163}$$

then

$$H\left(r\right) = 1 - \frac{r_s}{r} + \frac{r_Q^2}{r^2}, \tag{8.164}$$

and the full solution is hence

$$ds^2 = -c^2 \left(1 - \frac{r_s}{r} + \frac{r_Q^2}{r^2}\right) dt^2 + \frac{dr^2}{\left(1 - \frac{r_s}{r} + \frac{r_Q^2}{r^2}\right)} + r^2 d\Omega^2. \tag{8.165}$$

This is known as the **Reissner–Nordström metric**.[25] Just like the Schwarzschild metric, it has an anomalous behavior if the quantity $1 - \frac{r_s}{r} + \frac{r_Q^2}{r^2}$ vanishes. Since this expression is quadratic in $r$, we conclude that the Reissner–Nordström solution has *two* event horizons, which we can call $r_+$ and $r_-$. They are located at

$$r_\pm = \frac{1}{2}r_s \pm \frac{1}{2}\sqrt{r_s^2 - 4r_Q^2}. \tag{8.166}$$

**Exercise 8.23** Derive (8.166).

As in the Schwarzschild case, if our massive sphere has a radius $a$ that is greater than the larger event horizon, then the Reissner–Nordström metric describes spacetime outside the sphere. If, however, the radius is smaller, then the sphere collapses into a charged black hole. Note that the two event horizons reduce to a single one if it so happens that

$$r_s^2 = 4r_Q^2, \tag{8.167}$$

or equivalently

$$M^2 = \frac{Q^2}{4\pi\varepsilon_0 G}, \tag{8.168}$$

in which case the single event horizon is located at half the value of the Schwarzschild radius. Now if condition (8.168) is satisfied we get what has been termed an **extremal**

---

[25]Originally found by Hans Jacob Reissner, a German aeronautical engineer (1874–1967), then independently rediscovered by the German mathematician Hermann Klaus Hugo Weyl (1885–1955), the Finnish theoretical physicist Gunnar Nordström (1881–1923), as well as the British mathematical physicist George Barker Jeffery (1891–1957).

**charged black hole**. The reason it is called that is that if $r_s^2 < 4r_Q^2$, or equivalently $M^2 < Q^2/4\pi\varepsilon_0 G$, then the square root in (8.166) becomes imaginary, signaling the vanishing of the horizon entirely and effectively exposing the curvature singularity at $r = 0$. Technically this is no longer a black hole but is now what can be called a **naked singularity**; i.e. a singularity not "clothed" by an event horizon. Naked singularities have been studied extensively in the theoretical physics literature. There is some evidence that such spacetimes cannot arise in nature; in other words, the condition (8.168) signals the maximum charge a black hole with mass $M$ can attain; hence the use of the term "extremal." There is no rigorous proof of this, mind you; however, all evidence seems to point in this direction. For example, what if we try to create a naked singularity by *overcharging* a black hole to achieve $r_s^2 < 4r_Q^2$? We can consider doing so by bombarding it with particles that carry more charge than mass/energy. Calculations seem to indicate that if we continue to throw in these charges, then eventually, and right *before* the extremal condition (8.168) is breached, the black hole's charge will become of such magnitude that its electrostatic repulsion will push the particles back and none can cross the horizon any more [24]. Gedanken experiments such as this, and others concerned with the collapse of determinism and/or causality around a naked singularity, have caused physicists to postulate the existence of the so-called **cosmic censorship**, which in its simplest forms simply says that nature abhors naked singularities and would not allow one to form.[26]

Charged black holes are theoretically possible; all that a Schwarzschild black hole needs to do is absorb an excess of electrically charged particles. However, the observational techniques used by astronomers to detect black holes (as discussed briefly in §5.7) do not include the remote detection of an electric field; hence the existence of a charged black hole, let alone an extremal one, is still up in the air. They are, however, extremely interesting from a theoretical perspective, and have been recognized as the simplest case of higher-dimensional (*theoretical*) objects that appear in superstring theory known as "BPS branes."

**Exercise 8.24** Finding an interior Reissner–Nordström metric—in other words, the metric describing spacetime inside a spherical mass with uniform distribution of both mass and electric charge—has been proven much harder than most physicists would have believed. There are many candidates, each suffering from some physical defect or another. Construct the field equations (both Einstein and Maxwell) that need to be solved to find such a solution.

**Exercise 8.25** Plot the light cone diagram for the Reissner–Nordström spacetime following the methods in §6.2.2.

To conclude this subsection, we list two more metrics without much discussion, just to whet the reader's appetite for further study. The first is the **Kerr metric**,[27] alluded to earlier, representing a rotating *neutral* spherical mass or black hole:

---

[26] In its original form, cosmic censorship was proposed by Sir Roger Penrose in 1969.

[27] Discovered in 1963 by the New Zealand mathematician Roy Kerr (b. 1934).

$$ds^2 = -c^2 \left(1 - \frac{r_s}{\Sigma}r\right) dt^2 + \frac{\Sigma}{\Delta}dr^2 + \Sigma d\theta^2 + \left(r^2 + a^2 + \frac{r_s a^2}{\Sigma}r \sin^2 \theta\right) \sin^2 \theta d\varphi^2$$
$$-c\frac{2r_s a}{\Sigma}r \sin^2 \theta dt d\varphi, \tag{8.169}$$

where $r_s$ is the usual Schwarzschild radius,

$$a = \frac{J}{Mc}, \tag{8.170}$$
$$\Sigma = r^2 + a^2 \cos^2 \theta, \tag{8.171}$$
$$\Delta = r^2 - r_s r + a^2, \tag{8.172}$$

and $J$ is the angular momentum of the sphere/black hole.[28] The appearance of the cross term $dt d\varphi$ is interesting, as it indicates that the coordinate system described by (8.169) is non-orthogonal. Note that for $J = 0$ this metric reduces to Schwarzschild *exactly*, as of course it should. If $M \to 0$, i.e. the massive object vanishes, the metric reduces to flat spacetime; however, *not* in spherical coordinates like its static counterpart, but rather in the so-called oblate spheroidal coordinates. In Fig. [8.7] we give a (semi-serious) visual comparison between the Schwarzschild and Kerr black holes. The Schwarzschild solution's radial geodesics are represented by a steady "flow" of water in a hole (compare with Fig. [5.11]), while the Kerr metric's coordinate grid exhibits a rotating vortex-like behavior.[29] There are many more interesting features to this solution, such as geodesics that do not stay in the same plane, like in the Schwarzschild case, a unique and rich event horizon structure, and others. It is also possible to show that, just like the Reissner–Nordström metric, spinning a Kerr black hole too fast makes its event horizons vanish, exposing a "rotating" naked singularity. And just as in the Reissner–Nordström case there are arguments that this cannot happen in reality. In other words, it also seems to fall under cosmic censorship. Spinning black holes described by the Kerr solution are thought to be a possible source of large amounts of energy. In fact, Sir Roger Penrose proposed in 1969 a process, later named after him, where this energy can be extracted from the spinning black hole by a civilization that has the necessary technology. The interested reader may consult more advanced texts.

**Exercise 8.26** Demonstrate explicitly that setting $M = 0$ in (8.169) leads to a flat spacetime metric in oblate spheroidal coordinates. You will need the results of Exercise 1.4.

---

[28]The form of the Kerr metric given in (8.169) is *not* the one originally found by Kerr, but rather a modified form written in the so-called "Boyer–Lindquist coordinates;" just because these are easier to interpret.

[29]These photographs were used in [25] for the first time to visualize the difference between the two solutions. Their original sources are the websites https://www.iliketowastemytime.com/bottomless-pit-monticello-dam-drain-hole and http://www.manfredbauer.de/?p=4756 respectively. The "Schwarzschild" photograph is of the water spillway of Lake Berryessa, a reservoir formed behind Monticello Dam in Napa County, California, the United States of America. The "Kerr" photograph is from an unknown location.

Fig. 8.7: A semi-serious comparison between the Schwarzschild black hole (left) and the Kerr black hole (right).

Finally, we present the **Kerr–Newman metric** which represents a rotating *charged* spherical mass or black hole:[30]

$$ds^2 = -\left(\frac{dr^2}{\Delta} + d\theta^2\right)\Sigma^2 + \left(cdt - a\sin^2\theta d\varphi\right)^2\frac{\Delta}{\Sigma^2} - \left[(r^2 + a^2)\,d\varphi - acdt\right]^2\frac{\sin^2\theta}{\Sigma^2},$$
(8.173)

where $a$ and $\Sigma$ are defined by (8.170) and (8.171), but $\Delta$ is now

$$\Delta = r^2 - r_s r + a^2 + r_Q^2,$$
(8.174)

and $r_Q^2$ is given by (8.163).[31] Because of the rotation, the charge of the sphere/black hole generates a current which in turn generates a magnetic field. In fact, the 4-potential vector for this solution has components

$$A^\mu = \left(\frac{rr_Q}{\Sigma^2}, 0, 0, \frac{c^2 arr_Q\sin^2\theta}{\Sigma^2 GM}\right).$$
(8.175)

Recall that the 1, 2, 3 components of $A_\mu$ are the components of the magnetic vector potential **A**, from which a magnetic field is generated (4.174).

**Exercise 8.27** Show that (8.173) reduces to the Kerr metric if $Q = 0$, to the Reissner–Nordström metric if $J = 0$, and to the Schwarzschild metric if both vanish.

**Exercise 8.28** Find the field strength tensor $F^{\mu\nu}$ that follows from the 4-potential (8.175), and write the electric and magnetic field 3-vectors **E** and **B** in Kerr–Newman coordinates.

**Exercise 8.29** Free exercise/project: Do whatever you want to the spacetime metrics in this section; show they satisfy the Einstein and/or Maxwell equations, try to plot their geodesics, study their symmetries if you haven't already done so in Exercise 6.11, or whatever you (or

---

[30]Found by the American physicist Ezra Theodore Newman (b. 1929).
[31]This metric is also given in Boyer–Lindquist coordinates.

your instructor) choose. There are so many options, both analytical and computational. Just be aware that these metrics are complex enough to engage you for quite some time. In any case, always connect the physics with the mathematics. What can you conclude from your results? How do you interpret? You may work in natural $c = 1$ or geometrized $G = c = 1$ units for simplicity.

**Exercise 8.30** The concept of mass in the general theory of relativity, as well as in generalizations like the ones we will discuss in Chapter 10, is not as clear cut as it is in flat spacetime. In fact, there are several definitions of the mass of a given specific spacetime manifold. For example, in the Schwarzschild case we chose $M$ to be the mass of the spherical object or black hole based on how it reduces to its counterpart in Newtonian mechanics, but this is not general enough to apply to any possible spacetime solution. Research this and report on why there is a problem in the first place, and discuss the various attempts to defining mass and how they differ from each other. How do these various attempts apply to the various spacetime solutions we have discussed in this book?

# 9

# Differential Forms

*Every theoretical physicist who is any good knows six or seven different theoretical represen-
tations for exactly the same physics. He knows that they are all equivalent, and that nobody is
ever going to be able to decide which one is right at that level, but he keeps them in his head,
hoping that they will give him different ideas for guessing.*

Richard Feynman

## Introduction

In today's theoretical physics literature there exists an understanding that, as Feynman's quote above portrays, there are many different ways to describe exactly the
same thing, some more economical than others. Many of these ways become intensely
mathematical very quickly. This does not necessarily mean they are "difficult to understand;" most of them are not that difficult, especially when presented efficiently.
Personally, I have always felt that being able to reduce an equation to a "simpler" (in
the sense of "shorter") form may be indicative of a deeper structure that one may or
may not be aware of. For example, consider Maxwell's electromagnetic field equations.
The way Maxwell himself wrote them in his papers was in a completely scalar form;
in other words, he wrote, for example

$$\frac{\partial B_x}{\partial x} + \frac{\partial B_y}{\partial y} + \frac{\partial B_z}{\partial z} = 0, \tag{9.1}$$

to describe what we call today the divergence of the magnetic field, or the so-called
no-monopoles law:

$$\vec{\nabla} \cdot \mathbf{B} = 0. \tag{9.2}$$

The reason he did this was that at his time the vector notation was simply not used
in physics. Well, okay, I am quite certain that the great Maxwell and his contemporaries
were aware of the vector properties of the magnetic field $\mathbf{B}$; but the point is that they
didn't call it a vector, and they most certainly did not express its divergence in the
form (9.2). Hence the Maxwell equations, the way Maxwell himself wrote them, were
*eight* equations: one for each of the two divergence equations and three for each of

*Covariant Physics: From Classical Mechanics to General Relativity and Beyond*. Moataz H. Emam,
Oxford University Press (2021). © Moataz H. Emam.
DOI: 10.1093/oso/9780198864899.003.0009

the curl equations; (which we today call the components of the curl). But the use of vectors in physics not only reduced the number of equations to four; it also revealed a deeper structure to them, that of vector spaces, and made that structure much more accessible. In a similar way, the index/tensor language reduced the Maxwell equations from four down to two, as we learned in §4.8 and §6.3, effectively revealing their covariant nature.

In the previous chapters we extensively referred to the "geometric" structure of general relativity without going into much detail. Geometry, specifically what is known as differential geometry, is a well-established branch of mathematics that was rarely used, if at all, in fundamental physics prior to the advent of relativity. Almost immediately following Einstein's discoveries of 1905 and 1915, physicists realized that differential geometry is the "natural" setting for general relativity, and can also tie in spacetime curvature with the behavior of the classical fields propagating in said spacetime. In particular, the notion of the so-called **differential forms** facilitates a deeper understanding of physics' underlying geometric structure, as well as provide some calculational convenience; once one gets used to them. The reader may think of these objects as generalizations of vectors, in a sense, and also as a special case of tensors, in a different sense. In this chapter we give what I am hoping will be an intuitive, but necessarily brief, introduction to differential forms and their methods.

## 9.1 An Easy-Going Intro

The nice thing about differential forms, as we will quickly learn, is that they can be defined in a way that makes them completely metric-independent. This property is very convenient for our purposes, as it makes them automatically covariant. But rather than complicate things, let's begin by looking at forms in a non-relativistic Euclidean setting. In this regard, one may think of the language of differential forms, or simply just "forms," as an approach to multivariable calculus different from what you have probably learned. In fact, it is possible to learn the methods of vector calculus *without* using vectors, as they are replaced, in a sense, by forms. Differential forms come with "rank," or "order," which is a number that describes their dimensionality. An "order one" differential form, or 1-form $\omega_1$, is one-dimensional, a 2-form $\omega_2$ is two-dimensional, and a $p$-form $\omega_p$ is $p$-dimensional. The simplest 1-forms one can think of are just the Cartesian differential elements $dx$, $dy$, or $dz$. Yes, the good old familiar elements of integration. The reader may think of $dx$, $dy$, and $dz$ as **basis forms**, analogous to our old friends the Cartesian unit basis vectors, $\hat{\mathbf{e}}_x$, $\hat{\mathbf{e}}_y$, and $\hat{\mathbf{e}}_z$. In that context we can define a more complex 1-form by multiplying one of the basis forms, say $dx$, with a scalar function (in itself a 0-form):

$$\omega_1 = f(x)\, dx, \tag{9.3}$$

just as multiplying the unit vector $\hat{\mathbf{e}}_x$ with a function generates a vector pointing in the $x$ direction. One major difference between forms and vectors is that a vector points in a specific direction, while a form defines a plane. For example, $\mathbf{V} = f(x)\,\hat{\mathbf{e}}_x$ points in the positive $x$, direction while $\omega_1 = f(x)\,dx$ "lives" in the plane perpendicular to the $x$ direction. More precisely, $\omega_1$ defines a *stack* of planes perpendicular to the $x$ direction, as outlined in Fig. [9.1]. Because of this relationship differential forms are sometimes said to be *dual* to vectors.

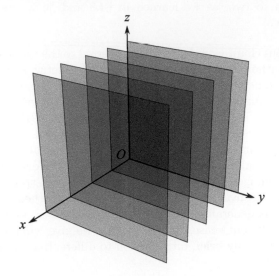

Fig. 9.1: A differential 1-form $\omega_1 = f(x)\,dx$ can be thought of as defining an infinite stack of planes orthogonal to the $x$ direction.

Having made those definitions, it is now easy to see that integrating a 1-form on a given domain is just the matter of writing

$$\int \omega_1 = \int f(x)\,dx. \tag{9.4}$$

The function $f$ does not necessarily need to be one dimensional; we could have just as easily defined $f(x,y)\,dx$ as our 1-form. In the first case one says that the 1-form is defined over an open set of $\mathbb{R}^1$, while in the latter it is $\mathbb{R}^2$ (if you insist on being mathematically precise). We can also add forms together without violating any rules of calculus; hence the most general way of writing a 1-form in three dimensions (an open set of $\mathbb{R}^3$) is

$$\omega_1 = f_x(x,y,z)\,dx + f_y(x,y,z)\,dy + f_z(x,y,z)\,dz, \tag{9.5}$$

where $f_x$, $f_y$, $f_z$ are different functions. The left-hand side of (9.4) would still look the same, while the right-hand side would have three integrals over different differential elements:

$$\int \omega_1 = \int [f_x(x,y,z)\,dx + f_y(x,y,z)\,dy + f_z(x,y,z)\,dz]. \tag{9.6}$$

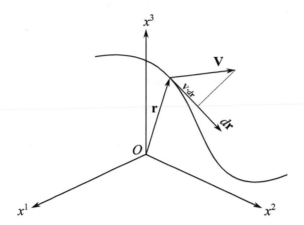

Fig. 9.2: 1-form fields as the contour projection of a vector **V**.

The reader will recognize that this is nothing more than a line or contour integral. In fact, if we think of the functions $f_i$ as the components of a vector **V**, then (9.6) is just

$$\int \omega_1 = \int \mathbf{V} \cdot d\mathbf{r} \tag{9.7}$$

and the 1-form $\omega_1$ is

$$\omega_1 = \mathbf{V} \cdot d\mathbf{r} = f_i(\mathbf{r}) \, dx^i, \tag{9.8}$$

as shown in Fig. [9.2]. It is important to emphasize that the "1" in 1-form has nothing to do with the dimension of space we are in; it just says that we have used one basis differential element per term in the definition. The fact that we can use up to three terms in the three independent basis forms $dx$, $dy$, and $dz$ is what tells us we are in three dimensions. So one can easily define 1-forms in any number of dimensions just by changing the range over which the index $i$ counts in (9.8). Also note that thinking of 1-forms as dot products of vectors immediately assigns a sign to their value, depending on whether or not the angles between the vectors are greater than or less than 90°. In a sense this can be thought of as a "directional" property of $\omega_1$.

Now pay attention; this will get very complicated very quickly. You might think that constructing a 2-form is the simple matter of sticking two 1-forms, each in a different direction, together; in other words, if you guessed that $dxdy$ is a 2-form, you would be wrong, well; maybe just *half* right. The reason is this: Consider an ordinary double integral in multivariable calculus as follows (limits assumed):

$$\int \int f(x,y) \, dxdy. \tag{9.9}$$

In this integral, the quantity $dxdy$ describes an element of area on a flat two-dimensional surface $S$. Now, just as we had a sense of direction in (9.7) and (9.8) by noting the sign of the dot product, one would also like to assign a "direction" to $dxdy$.

This is particularly useful when one thinks of such integrals as the *flux* of some vector field **V**, just as we thought of (9.6) as the line integral of a vector field. In vector calculus a direction can be set by defining a unit vector **n̂** that is perpendicular to the surface $S$ at all points, taking the dot product of **V** with **n̂** to pick out the components of **V** perpendicular to the surface, then summing up; in other words,

$$\text{Flux} = \int \int \mathbf{V} \cdot \mathbf{\hat{n}} dx dy. \tag{9.10}$$

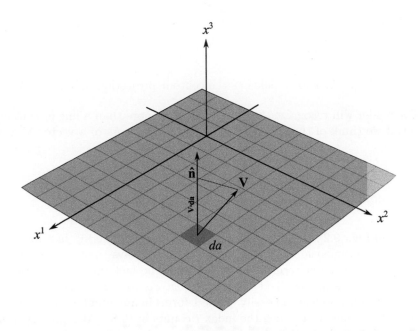

Fig. 9.3: 2-form fields as the flux element of a vector **V**.

Now, similarly to (9.7), the flux is a scalar, *not* a vector; however, it *does* have a sign that depends on whether or not the angle between **V** and **n̂** is less than or greater than 90°. To avoid confusion, let's call it the **orientation** rather than the direction. As often happens in vector calculus, one can define the element of area $da = dxdy$ itself as a vector,

$$\mathbf{da} = \mathbf{\hat{n}} dx dy, \tag{9.11}$$

such that the definition of flux becomes

$$\text{Flux} = \int \int \mathbf{V} \cdot \mathbf{da}, \tag{9.12}$$

as shown in Fig. [9.3]. In that sense, one may define a 2-form as follows:

$$\omega_2 = \mathbf{V} \cdot \mathbf{da} = \mathbf{V} \cdot \mathbf{\hat{n}} dx dy, \tag{9.13}$$

effectively assigning an orientation to the 2-form, which in a way is related to the orientation of the surface described by **da**. Hence the integral of the 2-form is simply that which we called the flux as defined in either (9.10) or (9.12); ergo

$$\int \omega_2 = \int \mathbf{V} \cdot \mathbf{da}. \tag{9.14}$$

Now, we used vectors and their dot product to introduce the concept of orientation to 2-forms in an intuitive way, but we would like to define 2-forms *without* reference to vectors, since they are, strictly speaking, not the same thing. To achieve this we introduce a new type of product known as the **exterior product**, or more popularly the **wedge product**, denoted by the symbol $\wedge$. It is just a way of splicing two forms together to generate a higher form; so, for example, one may generate a 2-form as follows:

$$\omega_2 = dx \wedge dy, \tag{9.15}$$

or more generally, if we have two 1-forms $\eta_1 = \eta_x dx + \eta_y dy$ and $\chi_1 = \chi_x dx + \chi_y dy$, then a 2-form can be constructed as follows:

$$\omega_2 = \eta_1 \wedge \chi_1 = (\eta_x dx + \eta_y dy) \wedge (\chi_x dx + \chi_y dy). \tag{9.16}$$

Note that a 0-form $f$ is just an ordinary scalar function, so we can also write

$$\omega_2 = f \wedge \kappa_2 = f\kappa_2, \tag{9.17}$$

where $\kappa_2$ is a 2-form, and so on. To pin down the orientation we assign to the wedge product the all-important property of antisymmetry:

$$dx \wedge dy = -dy \wedge dx. \tag{9.18}$$

This is where the sign of the "flux" comes in without reference to vectors: The wedge product gave an orientation to the form $\omega_2$; it is either oriented "up" or "down," in the same sense that flux is either flowing up or down with respect to $\hat{n}$. Note that this implies that the wedge product of a form with itself always vanishes, e.g.

$$dx \wedge dx = 0, \tag{9.19}$$

since (9.18) requires that $dx \wedge dx$ be equal to negative itself; which can only be true if it is zero. More generally, for any forms

$$\eta_1 \wedge \eta_1 = \omega_2 \wedge \omega_2 = 0, \tag{9.20}$$

and so on. Using (9.18) and (9.19) we can further evaluate an expression such as (9.16):

$$
\begin{aligned}
(\eta_x dx + \eta_y dy) \wedge (\chi_x dx + \chi_y dy) &= \eta_x \chi_y dx \wedge dy + \eta_y \chi_x dy \wedge dx \\
&= (\eta_x \chi_y - \eta_y \chi_x)\, dx \wedge dy.
\end{aligned} \tag{9.21}
$$

Now if we have a 2-form such as

$$\omega_2 = f(x,y)\, dx \wedge dy, \tag{9.22}$$

then its integral

$$\int \omega_2 = \int f(x,y)\, dx \wedge dy \tag{9.23}$$

has a sign that depends on the order of the differentials. Having said all of this, let's write down the most general expression for a 2-form in three dimensions:

$$\omega_2 = f_x(x,y,z)\, dx \wedge dy + f_y(x,y,z)\, dy \wedge dz + f_z(x,y,z)\, dz \wedge dx. \tag{9.24}$$

How can one write this in the index notation? Because of the antisymmetry property (9.18) one needs to be careful. In fact, the most economical way of writing (9.24) is

$$\omega_2 = \frac{1}{2}\varepsilon^i{}_{jk} f_i(x^n)\, dx^j \wedge dx^k, \tag{9.25}$$

where we have used the (also antisymmetric) Levi-Civita symbol defined in (1.94), and indices are raised and lowered by the Kronecker delta $\delta_{ij}$. Now what is the factor of one half doing here? Let's figure it out: Expand (9.25) explicitly over all possible values of the triple sum (using numbers for the indices $i$, $j$, $k$ rather than $x$, $y$, $z$):

$$\omega_2 = \frac{1}{2}\left[\varepsilon^1{}_{23}f_1 dx^2 \wedge dx^3 + \varepsilon^2{}_{31}f_2 dx^3 \wedge dx^1 + \varepsilon^3{}_{12}f_3 dx^1 \wedge dx^2 \right.$$
$$\left. + \varepsilon^1{}_{32}f_1 dx^3 \wedge dx^2 + \varepsilon^2{}_{13}f_2 dx^1 \wedge dx^3 + \varepsilon^3{}_{21}f_3 dx^2 \wedge dx^1\right], \tag{9.26}$$

then, using the explicit values of the components of the Levi-Civita symbol, as well as the antisymmetry property, we find

$$\omega_2 = \frac{1}{2}\left[f_1 dy \wedge dz + f_2 dz \wedge dx + f_3 dx \wedge dy + f_1 dy \wedge dz + f_2 dz \wedge dx + f_3 dx \wedge dy\right]$$
$$= f_1 dy \wedge dz + f_2 dz \wedge dx + f_3 dx \wedge dy, \tag{9.27}$$

which would be the same as (9.24) upon relabeling the functions $f_1 \to f_2$, $f_2 \to f_3$, and $f_3 \to f_1$. The factor of half is then used to prevent overcounting. For a 3-form one writes

$$\omega_3 = \frac{1}{3!}\varepsilon^i{}_{jkl} f_i(x^n)\, dx^j \wedge dx^k \wedge dx^l. \tag{9.28}$$

Since a 3-form is the highest order a differential form can acquire in three dimensions, this is as far as we can go here. Most generally, in three-dimensional space one can define the forms

$$\omega_1 = \omega_i dx^i$$
$$\omega_2 = \frac{1}{2}\omega_{ij} dx^i \wedge dx^j$$
$$\omega_3 = \frac{1}{3!}\omega_{ijk} dx^i \wedge dx^j \wedge dx^k, \tag{9.29}$$

where $\omega_i$, $\omega_{ij}$, and $\omega_{ijk}$ are the components of the forms. Note that $\omega_{ij}$ and $\omega_{ijk}$ are themselves necessarily totally antisymmetric; i.e. $\omega_{ij} = -\omega_{ji}$ and $\omega_{ijk} = -\omega_{jik} = -\omega_{ikj} = \omega_{kij}$, and so on.

Finally, although we have been working in Cartesian coordinates, defining forms using curvilinear coordinate basis is very straightforward. For example, in cylindrical coordinates (1.1) we can write

$$dx = \cos\varphi d\rho - \rho\sin\varphi d\varphi$$
$$dy = \sin\varphi d\rho + \rho\cos\varphi d\varphi, \tag{9.30}$$

which would lead to

$$dx \wedge dy \wedge dz = \rho d\rho \wedge d\varphi \wedge dz, \tag{9.31}$$

and so on. Note that the appearance of the *Jacobian* of cylindrical coordinates $\rho = \sqrt{|g|}$ is absolutely no surprise, since in this case the Levi-Civita symbol $\varepsilon$ in (9.25) is replaced by the Levi-Civita *tensor* following the substitution (2.147), and that automatically contains the Jacobian. We will come back to this in a moment.

**Exercise 9.1** The following are standard expressions found in textbooks on electromagnetism. For each expression identify, in Cartesian coordinates, the $p$-form hidden in the expression, its rank, and its components:

1. The potential difference $\Delta V$ in volts between two points is defined as the contour integral of the electric field: $\Delta V = -\int_1^2 \mathbf{E} \cdot d\mathbf{r}$.
2. The current intensity $I$ in Amperes is defined as the flux of the current density $\mathbf{J}$: $I = \int \mathbf{J} \cdot d\mathbf{a}$.
3. The total electric charge $Q$ in Coulombs of an object is defined as the volume integral of its charge density $\rho$: $Q = \int \rho dV$.

**Exercise 9.2** Rewrite the forms you found in Exercise 9.1 in cylindrical coordinates.

**Exercise 9.3** Rewrite the forms you found in Exercise 9.1 in the coordinates of Exercise 2.5.

## 9.2   More Generally . . .

Differential forms are a powerful tool for a variety of reasons, not the least of which is their ability to fit in any number of dimensions, as well as their metric independence, as we will see. Let's now generalize our discussion to any $n$-dimensional space *or* spacetime, whether curved or not, with the usual understanding that the use of Greek indices counts over $0, 1, 2, 3, \ldots$, while Latin indices count over $1, 2, 3, \ldots$. A differential form $\omega$ or $\omega_p$ of order $p$ on a manifold $\mathcal{M}_n$ endowed with a metric $g_{\mu\nu}$ is a totally antisymmetric tensor of rank $(0, p)$:

$$\omega = \frac{1}{p!} \omega_{\mu_1 \mu_2 \cdots \mu_p} dx^{\mu_1} \wedge \cdots \wedge dx^{\mu_p}, \tag{9.32}$$

where $\omega_{\mu_1 \mu_2 \cdots \mu_p}$ are the antisymmetric "components" of the $p$-form and it is understood that indices are raised and lowered by the given metric. The wedge product is defined such that

1. $dx^{\mu_1} \wedge \cdots \wedge dx^{\mu_p} = 0$ if some specific index $\mu_i$ appears at least twice.
2. $dx^{\mu_1} \wedge \cdots \wedge dx^{\mu_p} = -dx^{\mu_1} \wedge \cdots \wedge dx^{\mu_p}$ on the exchange of two *adjacent* indices.
3. $dx^{\mu_1} \wedge \cdots \wedge dx^{\mu_p}$ is linear in each $dx^{\mu_i}$.

The wedge products of differential forms have the following properties in *arbitrary* dimensions (note how the three-dimensional cases of the previous section fit in these):

1. $\omega_p \wedge \omega_p = 0$ if $p$ is odd.
2. $\omega_p \wedge \eta_r = (-1)^{pr} \eta_r \wedge \omega_p$.
3. $(\xi \wedge \omega) \wedge \eta = \xi \wedge (\omega \wedge \eta)$.

Note there are certain limitations to the wedge product that are imposed by the number of dimensions we are at. For example, in three dimensions an expression such as $\omega_1 \wedge \eta_3$ must vanish. The reason is that it is trying to generate a 4-form which, in this case, is one dimension larger than the space it is allowed to live in! Another way of seeing this is that if you write it out in components you find that you run into the problem of repeated basis forms, i.e.

$$\omega_1 \wedge \eta_3 = (f dx) \wedge (g dx \wedge dy \wedge dz) = fg\, dx \wedge dx \wedge dy \wedge dz, \tag{9.33}$$

which vanishes because of the repeating $dx$. This will always be true in *three* dimensions. However, if the same expression $\omega_1 \wedge \eta_3$ appears in, say, four dimensions ($x$, $y$, $z$, and $w$), then it doesn't necessarily have to vanish depending on which "direction" $\omega_1$ is at:

$$\omega_1 \wedge \eta_3 = (f dx) \wedge (g dx \wedge dy \wedge dz) = fg\, dx \wedge dx \wedge dy \wedge dz = 0$$
$$\omega_1 \wedge \eta_3 = (f dw) \wedge (g dx \wedge dy \wedge dz) = fg\, dw \wedge dx \wedge dy \wedge dz \neq 0. \tag{9.34}$$

**Exercise 9.4** In the five-dimensional Euclidean space defined by the coordinates $x$, $y$, $z$, $w$, and $v$, evaluate the following expressions. Some are given in terms of the basis forms and some in terms of arbitrary forms $\omega$, $\eta$, and $\xi$. If the answer cannot be estimated from the given information, then say so.

1. $dx \wedge dy \wedge dz \wedge dw \wedge dv$.
2. $dx \wedge dy \wedge dz \wedge dx \wedge dv$.
3. $dx \wedge dz \wedge dy \wedge dw \wedge dv$ in terms of $dx \wedge dy \wedge dz \wedge dw \wedge dv$.

4. $dx \wedge dz \wedge dw \wedge dy \wedge dv$ in terms of $dx \wedge dy \wedge dz \wedge dw \wedge dv$.
5. $\omega_2 \wedge \omega_2$.
6. $\omega_1 \wedge \omega_1$.
7. $\omega_5 \wedge \eta_3$.
8. $\omega_2 \wedge \eta_2$ in terms of $\eta_2 \wedge \omega_2$.
9. $\omega_1 \wedge \eta_3$ in terms of $\eta_3 \wedge \omega_1$.
10. $\xi \wedge (\eta \wedge \omega)$ in terms of $(\xi \wedge \omega) \wedge \eta$.

Now in order to do calculus, let's define derivatives of $p$-forms by employing the so-called **exterior derivative** $d = dx^\nu \partial_\nu$, which, in the language of mathematics, "maps" a $p$-form into a $(p+1)$-form as follows:

$$\lambda_{p+1} = d\omega_p = dx^\nu \partial_\nu \omega_p = \frac{1}{p!} \left( \partial_\nu \omega_{\mu_1 \cdots \mu_p} \right) dx^\nu \wedge dx^{\mu_1} \wedge \cdots \wedge dx^{\mu_p}. \tag{9.35}$$

For example, if $\omega_1$ is a 1-form in Euclidean three dimensions, then

$$\lambda_2 = (\partial_i \omega_j) \, dx^i \wedge dx^j = (\partial_1 \omega_2) \, dx^1 \wedge dx^2 + (\partial_2 \omega_1) \, dx^2 \wedge dx^1$$
$$= \left( \frac{\partial \omega_y}{\partial x} - \frac{\partial \omega_x}{\partial y} \right) dx \wedge dy, \tag{9.36}$$

and so on. The exterior derivative satisfies the product rule

$$d \left( \omega_p \wedge \eta_r \right) = (d\omega_p) \wedge \eta_r + (-1)^p \, \omega_p \wedge (d\eta_r), \tag{9.37}$$

as well as the very important relation

$$d^2 = 0, \tag{9.38}$$

frequently referred to as the **Poincaré lemma**. It means that applying $d$ on *any* $p$-form twice *always* vanishes. This is easy to see by taking the exterior derivative of $\lambda_{p+1}$ in (9.35), which is already an exterior derivative of $\omega_p$:

$$d\lambda_{p+1} = d^2 \omega_p = dx^\alpha \wedge dx^\nu \partial_\alpha \partial_\nu \omega_p$$
$$= \frac{1}{p!} \left( \partial_\alpha \partial_\nu \omega_{\mu_1 \cdots \mu_p} \right) dx^\alpha \wedge dx^\nu \wedge dx^{\mu_1} \wedge \cdots \wedge dx^{\mu_p}, \tag{9.39}$$

but notice that $\partial_\alpha \partial_\nu$ is symmetric, while $dx^\alpha \wedge dx^\nu$ is antisymmetric, so by the theorem discussed around eqn (1.108) their contraction must always vanish. In other words, (9.38) can be written in the more explicit form

$$dx^\mu \wedge dx^\nu \partial_\mu \partial_\nu = 0. \tag{9.40}$$

An operator satisfying (9.38) is called **nilpotent**.[1] Now differential forms that can be written as exterior derivatives of other forms, such as $\lambda = d\omega$, are called **exact**,

---

[1] If you are thinking that this sounds familiar, you are absolutely correct. We met another nilpotent operator in §7.1. There is in fact a connection between the discussion here and the one there, but explaining it will take us too far off on a tangent. We leave that to more advanced texts on geometry and topology. There is a nice introduction to these concepts in §2.9 of [10].

while forms whose exterior derivative vanishes, e.g. $d\lambda = 0$, are called **closed**. Because of (9.38), exact forms are always closed, while the converse is not necessarily true.

**Exercise 9.5** In the six-dimensional Euclidean space defined by the coordinates $x$, $y$, $z$, $w$, $v$, and $u$:

1. Expand $d\omega_1$, $d\omega_3$, and $d\omega_5$ in the components of your choice.
2. Apply the product rule on $d(\omega_1 \wedge \eta_2)$, and $d(\omega_3 \wedge \eta_3)$. Note that one of them necessarily vanishes: which one and why?

**Exercise 9.6** Find the exterior derivatives of following forms in the six-dimensional Euclidean space described by the coordinates $x$, $y$, $z$, $w$, $v$, and $u$. Are any of them closed?

1. $\omega_1 = y dx + x dy$.
2. $\omega_2 = \sqrt{z} dx \wedge dy - du \wedge dz - w^2 dv \wedge dx$.
3. $\omega_1 = -\frac{y}{x^2+y^2} dx + \frac{x}{x^2+y^2} dy$.
4. $\omega_3 = u^2 z^3 dx \wedge dy \wedge dz$.
5. $\omega_6 = \sin(uv) dx \wedge dy \wedge dz \wedge dw \wedge dv \wedge du$.

**Exercise 9.7** Consider the 1-form $\omega_1 = y dx + [z \cos(yz) + x] dy + y \cos(yz) dz$ in three-dimensional Euclidean space. First demonstrate that it is closed, then show that it is also exact by finding the 0-form $\alpha$ from which $\omega_1 = d\alpha$.

The property (9.38) is responsible for a couple of famous identities in vector calculus. It is well known that the curl of a gradient of an arbitrary scalar function $f$ is always vanishing, i.e.

$$\vec{\nabla} \times \vec{\nabla} f = 0. \tag{9.41}$$

This can be shown to be a consequence of (9.38), as follows: The scalar function $f$ is a 0-form since it carries no indices; taking its exterior derivative twice gives

$$d^2 f = dx \wedge dy \left( \frac{\partial^2 f}{\partial x \partial y} - \frac{\partial^2 f}{\partial y \partial x} \right) + dy \wedge dz \left( \frac{\partial^2 f}{\partial y \partial z} - \frac{\partial^2 f}{\partial z \partial y} \right) + dx \wedge dz \left( \frac{\partial^2 f}{\partial x \partial z} - \frac{\partial^2 f}{\partial z \partial x} \right), \tag{9.42}$$

which are exactly the components of (9.41). Similarly, it is also known that the divergence of the curl of an arbitrary vector field $\mathbf{A}$ always vanishes:

$$\vec{\nabla} \cdot \left( \vec{\nabla} \times \mathbf{A} \right) = 0. \tag{9.43}$$

Once again thinking of the components of $\mathbf{A}$ as those of a 1-form $A_1$, it is easy to show that $d^2 A_1$ leads to (9.43).

**Exercise 9.8** Show that (9.43) is a consequence of $d^2 A_1 = 0$, where $A_1 = A_{1x}(x, y, z)\, dx + A_{1y}(x, y, z)\, dy + A_{1z}(x, y, z)\, dz$ is a general 1-form.

Before we move on, let's emphasize again that even though we are working on a general manifold $\mathcal{M}_n$ with an arbitrary metric, the definitions of forms, their products, and their derivatives do not care about the metric; except when indices are raised and lowered. They are truly metric independent. For example, let's consider (9.39) for a 1-form $\omega_1$:

$$d\omega_1 = (\partial_\nu \omega_\mu)\, dx^\nu \wedge dx^\mu. \tag{9.44}$$

In a naive attempt to make it covariant, we apply the usual minimal substitution $\partial \to \nabla$:

$$(\nabla_\nu \omega_\mu)\, dx^\nu \wedge dx^\mu = (\partial_\nu \omega_\mu - \Gamma^\alpha_{\nu\mu}\omega_\alpha)\, dx^\nu \wedge dx^\mu. \tag{9.45}$$

But we find that the quantity $\Gamma^\alpha_{\nu\mu} dx^\nu \wedge dx^\mu$ vanishes because the Christoffel symbols are symmetric in their lower indices, while $dx^\nu \wedge dx^\mu$ is antisymmetric. This will happen no matter what the order of the form is; hence (9.39) as well as all definitions in this section is automatically covariant and metric independent. However, in order to be consistent, particularly when we attempt to evaluate integrals of forms, we must bring back the Levi-Civita tensor defined in §2.6:

$$\mathcal{E}_{\mu_1 \cdots \mu_n} = \sqrt{-|g|}\, \varepsilon_{\mu_1 \cdots \mu_n}$$
$$\mathcal{E}^{\mu_1 \cdots \mu_n} = \frac{1}{\sqrt{-|g|}}\, \varepsilon^{\mu_1 \cdots \mu_n}. \tag{9.46}$$

The indices of $\mathcal{E}_{\mu_1 \cdots \mu_n}$ are raised and lowered by $g_{\mu\nu}$, while those of $\varepsilon_{\mu_1 \cdots \mu_n}$ are raised and lowered by the flat metric (either Minkowski or Euclidean depending on the signature of $g_{\mu\nu}$). The ordinary volume element used in integrals can now be defined as an $n$-form, i.e. a form spanning all of space

$$\mathcal{E}_n = \sqrt{-|g|}\, dx^0 \wedge \cdots \wedge dx^{n-1}$$
$$= \frac{1}{n!}\sqrt{-|g|}\, \varepsilon_{\mu_1 \cdots \mu_n} dx^{\mu_1} \wedge \cdots \wedge dx^{\mu_n}$$
$$= \frac{1}{n!}\mathcal{E}_{\mu_1 \cdots \mu_n} dx^{\mu_1} \wedge \cdots \wedge dx^{\mu_n}, \tag{9.47}$$

which upon integration

$$\int \mathcal{E}_n = \frac{1}{n!}\int \mathcal{E}_{\mu_1 \cdots \mu_n} dx^{\mu_1} \wedge \cdots \wedge dx^{\mu_n}, \tag{9.48}$$

returns the volume (or surface or length; remember that what we call a volume can have different dimensions depending on the value of $n$) of the entire manifold $\mathcal{M}_n$, with the

Jacobian $\sqrt{-|g|}$ automatically included. In most cases the notation $\mathscr{E}_n = \sqrt{-|g|}d^n x$ is used, such as in (8.95) for $n = 4$, which is certainly more familiar, but it is generally understood that (9.47) is the correct expression. It also provides an orientation to the "volume" in question.

It is possible to avoid introducing $\mathscr{E}_{\mu_1\cdots\mu_n}$ and keep $\varepsilon_{\mu_1\cdots\mu_n}$ instead. This is done by redefining the basis forms $dx^\mu$. For instance, in the cylindrical coordinates example at the end of the previous section, one can define

$$dx^1 = d\rho, \quad dx^2 = d\phi, \quad dx^3 = dz. \tag{9.49}$$

These $dx^i$ may be referred to as the **natural basis forms**, since they are analogous to the natural basis vectors defined back in §1.6 in that they don't all have the same dimensions, for example. In this case, one *must* explicitly use $\mathscr{E}_{ijk} = \sqrt{|g|}\varepsilon_{ijk} = \rho\varepsilon_{ijk}$, leading to, for example:

$$\int \mathscr{E} = \frac{1}{3!} \int \mathscr{E}_{ijk}dx^i \wedge dx^j \wedge dx^k = \int \rho d\rho \wedge d\varphi \wedge dz. \tag{9.50}$$

The other option is to define

$$dx^1 = d\rho, \quad dx^2 = \rho d\phi, \quad dx^3 = dz, \tag{9.51}$$

which has two advantages: The first is that all three basis forms have the same dimensions of length, and as such may be referred to as the **physical basis forms**, and the second is that it is no longer necessary to introduce $\mathscr{E}_{ijk}$, and expressions such as (9.50) would immediately arise using the ordinary $\varepsilon_{ijk}$. In other words, no reference is made to the metric or its determinant, emphasizing the differential forms' independence of the metric. Either way you get the correct answer. We will use the natural basis approach in the remainder of this chapter, since it is more consistent with the choice of basis vectors we made back in §1.6.

**Exercise 9.9** For the following spacetimes define the basis forms $dx^\mu$ in both the natural and the physical choices. Confirm that in both cases you get the same results in expressions such as line elements, area elements, and volume elements.

1. The Schwarzschild solution (5.47).
2. The Robertson–Walker solution (5.102).
3. The Reissner–Nordström solution (8.165).

**Exercise 9.10** Construct a general 2-form $\omega_2$ that lives in the spacetimes of Exercise 9.9 in the natural basis, then explicitly take its exterior derivative twice to show that it will vanish.

## 9.3 Hodge Duality

Now pay close attention; we will define an operator that will not immediately make much sense, but has very important applications: It is known as the **Hodge duality** operator $\star$ (pronounced, you guessed it, "star") which maps a $p$-form $\omega_p$ to an $n - p$ form, $\kappa_{n-p}$. If a form $\omega_p$ is defined by (9.32), then its Hodge dual[2] is

$$
\kappa_{n-p} = \star \omega_p = \frac{1}{p!\,(n-p)!} \mathscr{E}_{\mu_1 \cdots \mu_p \mu_{p+1} \cdots \mu_n} \omega^{\mu_1 \cdots \mu_p} dx^{\mu_{p+1}} \wedge \cdots \wedge dx^{\mu_n}
$$
$$
= \frac{1}{(n-p)!} \kappa_{\mu_{p+1} \cdots \mu_n} dx^{\mu_{p+1}} \wedge \cdots \wedge dx^{\mu_n}, \tag{9.52}
$$

where the components of $\kappa$ are

$$
\kappa_{\mu_{p+1} \cdots \mu_n} = \frac{1}{p!} \mathscr{E}_{\mu_1 \cdots \mu_p \mu_{p+1} \cdots \mu_n} \omega^{\mu_1 \cdots \mu_p}. \tag{9.53}
$$

Note that this is dependent on the number of dimensions $n$ of the manifold, so $\omega$ spans some part of the manifold's dimensionality while $\kappa$ spans the remaining dimensions. For example, in $n = 4$ spacetime a 1-form is Hodge-dual to a 3-form, while a 2-form is Hodge-dual to another 2-form, and so on. In the simplest possible sense, this can be understood as follows: In Euclidean three-dimensional space each vector (related to a 1-form) is orthogonal to a unique plane (related to a 2-form), and the Hodge dual is a way of finding this plane. Note that the converse is also true: Any plane in three dimensions has one unique vector orthogonal to it. For this simple example, if

$$
\omega_1 = f dx, \tag{9.54}
$$

where $f$ is some arbitrary function, then

$$
\kappa_2 = \star \omega_1 = \frac{1}{1!\,(3-1)!} f \left[ \varepsilon_{23} dy \wedge dz + \varepsilon_{32} dz \wedge dy \right] = f dy \wedge dz. \tag{9.55}
$$

Now consider what happens if we take the Hodge dual of a form that already fills out all space, i.e. an $n$-form. That must result in an $n - n = 0$-form:

$$
\kappa_0 = \star \omega_n = \frac{1}{n!\,(n-n)!} \mathscr{E}_{\mu_1 \cdots \mu_n} \omega^{\mu_1 \cdots \mu_n}. \tag{9.56}
$$

Notice that the final expression in (9.56) has all of its indices contracted; in other words, it is a scalar: a 0-form. If $\omega^{\mu_1 \cdots \mu_n} = f \mathscr{E}^{\mu_1 \cdots \mu_n}$, then

$$
\kappa_0 = \star \omega_n = f, \tag{9.57}
$$

---

[2]Sir William Vallance Douglas Hodge: British mathematician (1903–1975).

since $\mathscr{E}_{\mu_1\cdots\mu_n}\mathscr{E}^{\mu_1\cdots\mu_n} = n!$, which is similar to the last of the identities (1.107). Some of these 0-forms have special significance. Consider again three-dimensional Euclidean–Cartesian space. Let's define the volume 3-form

$$\omega_3 = dx \wedge dy \wedge dz, \tag{9.58}$$

and ask what would be its Hodge dual. First write it in the index notation:

$$\omega_3 = \frac{1}{3!}\varepsilon_{ijk}dx^i \wedge dx^j \wedge dx^k. \tag{9.59}$$

In other words, the components of $\omega_3$ are just $\varepsilon_{ijk}$. Now apply (9.52)

$$\star\omega_3 = \frac{1}{3!\,(3-3)!}\varepsilon_{ijk}\varepsilon^{ijk} = \mathbf{1}, \tag{9.60}$$

where we call $\mathbf{1}$ the **identity 0-form**. Let's invert this relation by hitting both sides with another Hodge operator to find

$$\star\mathbf{1} = dx \wedge dy \wedge dz = \frac{1}{3!}\varepsilon_{ijk}dx^i \wedge dx^j \wedge dx^k, \tag{9.61}$$

in three-dimensional Euclidean space. Let's generalize this notion and say that for *any* number of dimensions on *any* manifold the volume element is always the Hodge dual of the identity 0-form:

$$\star\mathbf{1} = \frac{1}{n!}\mathscr{E}_{\mu_1\cdots\mu_n}dx^{\mu_1} \wedge \cdots \wedge dx^{\mu_n}. \tag{9.62}$$

It is common in some (fancy) research papers to use $\star\mathbf{1}$ instead of the customary $\sqrt{-|g|}d^n x$. For example, the action integral (8.117) may be rewritten as

$$S = \frac{1}{2\kappa}\int R\star\mathbf{1}. \tag{9.63}$$

**Exercise 9.11** Find the Hodge dual to the following forms in three-dimensional Euclidean space:

1. $dx$.
2. $dx \wedge dy$.
3. $dx \wedge dy \wedge dz$.

**Exercise 9.12** Find the Hodge dual to the following forms in four-dimensional Minkowski-Cartesian spacetime:

1. $dx$.
2. $cdt$.
3. $dx \wedge dy$.
4. $cdt \wedge dy \wedge dz$.

**Exercise 9.13** Find the Hodge dual to the following forms in the six-dimensional Euclidean space described by the coordinates $x$, $y$, $z$, $w$, $v$, and $u$:

1. $\omega_2 = 5dx \wedge dy - du \wedge dz + 2dv \wedge dx$.
2. $\omega_3 = u^2 z^3 dx \wedge dy \wedge dz$.
3. $\omega_6 = 5dx \wedge dy \wedge dz \wedge dw \wedge dv \wedge du$.

**Exercise 9.14** Find the Hodge dual to the forms you constructed in Exercises 9.9 and 9.10. For 9.9 use the natural basis forms.

Taking the Hodge dual twice and defining the inverse Hodge dual are generally not very straightforward. One can show that

$$\star\star\omega_p = (-1)^{p(n-p)+\varrho}\,\omega_p$$
$$\star^{-1} = (-1)^{p(n-p)+\varrho}\star, \tag{9.64}$$

where $\star^{-1}$ is the inverse Hodge dual and $\varrho$ is the number of components of the metric with a minus sign; i.e. if $\mathcal{M}_n$ is Riemannian, then $\varrho = 0$, while if it is Lorentzian, then $\varrho = 1$. In ordinary Euclidean three dimensions

$$\star\star\omega_3 = (-1)^{3(3-3)+0}\,\omega_3 = \omega_3$$
$$\star^{-1} = (-1)^{3(3-3)+0}\star = \star, \tag{9.65}$$

which we used above to find (9.61).

**Exercise 9.15** Find the inverse Hodge dual to the following forms in the given spaces/spacetimes:

1. $\omega_2 = dx \wedge dy - 3du \wedge dz + dv \wedge dx$ in the six-dimensional Euclidean space $(x, y, z, w, v, u)$.
2. $\omega_3 = t^2 dx \wedge dy \wedge dz$ in four-dimensional spacetime.
3. $\omega_4 = -8dx^1 \wedge dx^2 \wedge dx^3 \wedge dx^4$ in a six-dimensional spacetime that has *two* time dimensions.
4. The natural basis forms you found in Exercise 9.9.
5. The 2-form you constructed in Exercise 9.10.

Now the "space" where differential forms live can be thought of as a sort of vector space, in the sense that a scalar product of forms can be defined as follows:

$$(\omega_p, \eta_p) = \int_{\mathcal{M}_n} \omega_p \wedge \star\eta_p, \tag{9.66}$$

where it can be shown that

$$\omega_p \wedge \star\eta_p = \frac{1}{p!}\omega_{\mu_1\cdots\mu_p}\eta^{\mu_1\cdots\mu_p} \star 1. \tag{9.67}$$

**Exercise 9.16** Explicitly prove (9.67) using the combined action of the wedge product and the Hodge operator.

The inner product is clearly symmetrical:

$$
\begin{aligned}
&\because \omega_p \wedge \star\eta_p = \eta_p \wedge \star\omega_p \\
&\therefore (\omega_p, \eta_p) = (\eta_p, \omega_p),
\end{aligned}
\tag{9.68}
$$

as any scalar product should be. This can be used to simplify physical expressions. For example, consider the action (8.95). Using the fact that $\nabla\phi = \partial\phi$, we can write

$$
\begin{aligned}
d^4x\sqrt{-|g|}g^{\mu\nu}\left(\partial_\mu\phi\right)\left(\partial_\nu\phi\right) &= d^4x\sqrt{-|g|}\left(\partial_\mu\phi\right)\left(\partial^\mu\phi\right) \\
&= \frac{1}{4!}\mathscr{E}_{\alpha\beta\gamma\sigma}dx^\alpha \wedge dx^\beta \wedge dx^\gamma \wedge dx^\sigma\left(\partial_\mu\phi\right)\left(\partial^\mu\phi\right) \\
&= \left(\partial_\mu\phi\right)\left(\partial^\mu\phi\right) \star 1 = d\phi \wedge \star d\phi,
\end{aligned}
\tag{9.69}
$$

where we have used (9.62) as well as defined the 1-form

$$
d\phi = \left(\partial_\mu\phi\right) dx^\mu,
\tag{9.70}
$$

which in itself is just the exterior derivative of the 0-form $\phi$. Using this language the action (8.95) can be most elegantly written as

$$
S = -\frac{\alpha}{2} \int d\phi \wedge \star d\phi,
\tag{9.71}
$$

and if we add in the Einstein–Hilbert term,

$$
S = \int \left(\frac{1}{2\kappa}R \star 1 - \frac{\alpha}{2}d\phi \wedge \star d\phi\right).
\tag{9.72}
$$

Yes, yes, I know! This is all too much. Just keep in mind that, as with any new language, it takes a while to get used to things. However, one cannot deny the incredible elegance this notation enjoys, so let's keep going. One can use the Hodge dual to define the so-called **adjoint exterior derivative** operator:

$$
d^\dagger\omega_p = (-1)^{n(p+1)+\varrho} \star d \star \omega_p,
\tag{9.73}
$$

which maps a $p$-form *down* to a $(p-1)$-form as follows:

$$
\star d \star \omega_p = \star d\kappa_{n-p} = \star\tau_{n-p+1} = \varphi_{n-(n-p+1)} = \varphi_{p-1},
\tag{9.74}
$$

in contrast with the exterior derivative which does the exact opposite. Note that since $d$ is nilpotent, so is $d^\dagger$, since

$$
d^{\dagger^2} \propto \star d \star \star d\star \propto \star d^2\star = 0.
\tag{9.75}
$$

Certain useful theorems involving $d^\dagger$ can be proven. For example, one can show that

$$
(d\omega, \eta) = (\omega, d^\dagger\eta).
\tag{9.76}
$$

**Exercise 9.17** Prove (9.76).

In analogy with the exterior derivative, one says that a form $\lambda$ that can be written as $\lambda = d^\dagger \omega$ is **co-exact**, while one that satisfies $d^\dagger \lambda = 0$ is **co-closed**. Once again, co-exact forms are always co-closed, while the converse is not necessarily true.

Finally, we define the so-called **Laplace–de Rham operator**[3] on $p$-forms by[4]

$$\Delta = \left(d + d^\dagger\right)^2 = dd^\dagger + d^\dagger d. \tag{9.77}$$

A $p$-form that satisfies

$$\Delta \omega = 0 \tag{9.78}$$

is called **harmonic** and (9.78) is known as the **harmonic condition**. A form is harmonic if and only if it is both closed *and* co-closed. Harmonic forms play a fundamental role in physics. To demonstrate, consider applying the Laplace–de Rham operator to a 0-form scalar field $f$, i.e. an ordinary function, in $n = 4$ spacetime. First note that $d^\dagger f = 0$, since $f$, being a 0-form, cannot be lowered in order any further. Hence

$$
\begin{aligned}
\Delta f = d^\dagger df &= -\star d \star \left(\partial_\mu f dx^\mu\right)_{1\text{-form}} \\
&= -\star d \left[\frac{1}{1!\,(4-1)!} \mathscr{E}_{\mu\nu\rho\sigma}\left(\partial^\mu f\right) dx^\nu \wedge dx^\rho \wedge dx^\sigma\right] \\
&= -\star \frac{1}{3!} d\left[\sqrt{-|g|}\left(\partial^\mu f\right)\right] \varepsilon_{\mu\nu\rho\sigma} dx^\nu \wedge dx^\rho \wedge dx^\sigma \\
&= -\star \frac{1}{3!} \partial_\alpha\left[\sqrt{-|g|}\left(\partial^\mu f\right)\right] \varepsilon_{\mu\nu\rho\sigma} dx^\alpha \wedge dx^\nu \wedge dx^\rho \wedge dx^\sigma.
\end{aligned} \tag{9.79}
$$

Now we recognize that what's left is just $-\star \omega_4$, where $\omega_4$ is the 4-form

$$
\begin{aligned}
\omega_4 &= \frac{1}{4!} \omega_{\alpha\nu\rho\sigma} dx^\alpha \wedge dx^\nu \wedge dx^\rho \wedge dx^\sigma \\
&= \frac{1}{3!} \frac{1}{\sqrt{-|g|}} \partial_\alpha\left[\sqrt{-|g|}\left(\partial^\mu f\right)\right] \mathscr{E}_{\mu\nu\rho\sigma} dx^\alpha \wedge dx^\nu \wedge dx^\rho \wedge dx^\sigma.
\end{aligned} \tag{9.80}
$$

Hence

$$\omega_{\alpha\nu\rho\sigma} = \frac{4!}{3!} \frac{1}{\sqrt{-|g|}} \partial_\alpha\left[\sqrt{-|g|}\left(\partial^\mu f\right)\right] \mathscr{E}_{\mu\nu\rho\sigma}, \tag{9.81}$$

---

[3]Georges de Rham: Swiss mathematician (1903–1990).
[4]Sometimes referred to as simply the Laplace operator; although the usual Laplacian is a special case, as we will see in (9.85).

resulting in

$$\Delta f = -\star \omega_4 = \frac{-1}{4!\,(4-4)!}\mathscr{E}_{\alpha\beta\gamma\sigma}\omega^{\alpha\beta\gamma\sigma}$$

$$= -\frac{1}{4!}\frac{4!}{3!}\frac{1}{\sqrt{-|g|}}\partial^\alpha\left[\sqrt{-|g|}\,(\partial_\mu f)\right]\mathscr{E}_{\alpha\beta\gamma\sigma}\mathscr{E}^{\mu\beta\gamma\sigma}. \tag{9.82}$$

Now a generalization of the second identity of (1.107) can be easily checked:

$$\mathscr{E}_{\alpha\beta\gamma\sigma}\mathscr{E}^{\mu\beta\gamma\sigma} = -3!\delta^\mu_\alpha, \tag{9.83}$$

and we end up with

$$\Delta f = \frac{1}{\sqrt{-|g|}}\partial_\alpha\left[\sqrt{-|g|}\,(\partial^\alpha f)\right], \tag{9.84}$$

which is the four-dimensional generalization of (2.112), i.e. in four-dimensional space-time:

$$\Delta f = \nabla^\mu\nabla_\mu f. \tag{9.85}$$

**Exercise 9.18** Find the action of the Laplace–de Rham operator on scalar functions in:

1. Three-dimensional Euclidean space.
2. Five-dimensional spacetime.
3. 11 dimensional spacetime.[5]

## 9.4 Maxwell's Theory and Differential Forms

Before we go on, it is perhaps wise to pause and apply the language of $p$-forms to something familiar; namely Maxwell's theory of electrodynamics. We will just list the relationships and leave checking them as an exercise for the interested reader. Let's begin by defining the potential and current 1-forms in $n = 4$ spacetime:

$$A = A_\mu dx^\mu$$
$$J = J_\mu dx^\mu. \tag{9.86}$$

Indices are raised and lowered by $g_{\mu\nu}$; whatever that may be. We define the Field strength 2-form, also known as the **Faraday 2-form**, as the exterior derivative of $A$:

---

[5] In case you are thinking that choosing five and 11 spacetime dimensions in this exercise is arbitrary, let me just point out that this author had the pleasure (?) of conducting research in both of these numbers of dimensions.

$$F = dA$$
$$= (\partial_\nu A_\mu) \, dx^\nu \wedge dx^\mu = \frac{1}{2} (\partial_\nu A_\mu - \partial_\mu A_\nu) \, dx^\nu \wedge dx^\mu$$
$$= \frac{1}{2} F_{\nu\mu} dx^\nu \wedge dx^\mu, \tag{9.87}$$

which leads to (6.147). Based on this, applying the exterior derivative on $A$ one more time necessarily vanishes because of (9.38); hence

$$dF = 0. \tag{9.88}$$

In other words, $F$ is a closed exact form.[6] It's an interesting exercise to show that (9.88) is *exactly* (6.149); in other words, it leads to the Faraday and no-monopoles laws of Maxwell's equations, and this is just as a consequence of $F = dA$. The other two Maxwell equations, Gauss and Ampère–Maxwell (6.148), become

$$d^\dagger F = \mu_0 J \tag{9.89}$$

in the language of differential forms. It is important to reemphasize that the equations are all necessarily covariant and equivalent to the list in §6.3. Taking the adjoint exterior derivative of (9.89) one more time leads to

$$d^\dagger J = 0 \tag{9.90}$$

because $d^\dagger$ is nilpotent. Equation (9.90) is just the continuity eqn (6.150): the statement that the current 1-form is co-closed. The wave equation is just

$$\Delta A = \mu_0 J. \tag{9.91}$$

Finally, the Lagrangian density 4-form and the action for Maxwell's theory are

$$\mathcal{L}_M = -\frac{1}{4\mu_0} F \wedge \star F - A \wedge \star J$$
$$S_M = \int \left( -\frac{1}{4\mu_0} F \wedge \star F - A \wedge \star J \right), \tag{9.92}$$

with the understanding that this is only the matter part of the Lagrangian/action (8.119) and (8.120). The full generally covariant Einstein–Maxwell action (8.143) is hence

$$S = \int \left( \frac{1}{2\kappa} R \star 1 - \frac{1}{4\mu_0} F \wedge \star F - A \wedge \star J \right). \tag{9.93}$$

---

[6]Equation (9.88) is sometimes referred to as the **Bianchi identity**. The differential formula by the same name (the second one) in the list following eqn (7.17) has the same structure in differential form language when applied to the Riemann tensor.

**Exercise 9.19** Find the components of $\star F$ using the field strength tensor (4.184); i.e. Minkowski-Cartesian coordinates. Write them in matrix form.

**Exercise 9.20** Show that:

1. The relation (9.88) leads to (6.149).
2. The relation (9.89) leads to (6.148).
3. The relation (9.90) leads to (6.150).
4. The relation (9.91) leads to (6.151). Recall, however, that the wave equation was simplified using its gauge freedom, as alluded to earlier in footnote 32 in §4.8. As mentioned, we will not explain this as it will take us too far off track; however, for the sake of this calculation all you need to know is that you can set $d^\dagger A = 0$ as a consequence of this gauge freedom.
5. The action (9.92) leads to (8.102).

**Exercise 9.21** One cannot help but marvel at the way the *entire* theory of electrodynamics just "spills" out of the properties of differential forms; specifically, from certain forms being closed, co-closed, or both! It is worth noting that Maxwell's theory is perhaps the *simplest* such application of differential forms known. One can devise "higher" theories by simply increasing the ranks of the forms used. One such theory goes by the name "11-dimensional supergravity," which arises as a low energy limit of the sought-after "M-theory," believed to be the highest possible generalization of superstring theory and the ultimate theory of everything (*if* verified by experiment, which is yet to happen). Here is another mini-project for you. The action of supergravity theory in 11 spacetime dimensions (up to constants which we will just ignore) is:

$$ S = \int_{11} \left( R \star \mathbf{1} - \frac{1}{2} F \wedge \star F - \frac{1}{6} A \wedge F \wedge F \right). \tag{9.94} $$

Technically this is only part of the action; the remainder is made up of so-called fermionic fields, the meaning of which need not concern us. The symbol $A$ represents a potential and its field strength is $F = dA$. The third term involving $A \wedge F \wedge F$ is called the Chern–Simons term, as it does have the structure common to all Chern–Simons models, such as the one in Exercise 8.16.

1. Figure out from the action what is the rank of the potential $A$ and consequently, the rank of $F$.
2. Expand $F = dA$. Be careful, as it is not as simple as the Maxwell version.
3. Rewrite the action in the standard tensor/index/component form.
4. Vary the action with respect to $A$ to find the field equation of $F$.
5. Rewrite the equation you found in part 4 in differential form language. Compare it to (9.89). What is the quantity that acts as current in this theory?
6. Is $A$ harmonic? How would you know?
7. Vary the action with respect to the metric to find the full Einstein field equation in 11 dimensions, including $T_{\mu\nu}$.

## 9.5 Stokes' Theorem

One of the most beautiful theorems in mathematics involves the integration of $p$-forms and is known as the **generalized Stokes' theorem**. It states

$$\int_{\mathcal{M}} d\omega = \int_{\partial\mathcal{M}} \omega, \tag{9.95}$$

where $\partial\mathcal{M}$ denotes the $(n-1)$-dimensional boundary of a manifold $\mathcal{M}$ (as discussed in §7.1), unless $\mathcal{M}$ is closed, in which case $\omega$ is closed as well, i.e. $d\omega = 0$, and the right-hand side vanishes. Equation (9.95) is sometimes referred to as the **fundamental theorem of calculus on manifolds**, also known as the **Stokes–Cartan theorem**.[7] In plain English the theorem says that the integral of the exterior derivative of a $p$-form over a manifold $\mathcal{M}$ is always equal to the sum of the values (integral) of said form on the boundary of the manifold. The power of this formula is in the fact that it is a generalization of several well-known theorems in ordinary calculus; ones that were discovered independently well before (9.95). They are all special cases that can be found from (9.95) by making specific choices of manifolds and forms. The simplest case is found by letting $\omega = f(x)$ be a 0-form and letting $\mathcal{M}$ be the real number line $x$ bounded by the two points $x = a$ and $x = b$, as shown in Fig. [9.4]; hence

$$d\omega = df = \frac{df}{dx} dx. \tag{9.96}$$

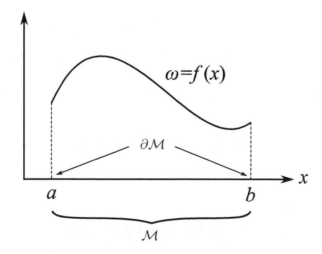

Fig. 9.4: The Stokes–Cartan theorem applied to a one-dimensional manifold: the $x$ axis.

---

[7]Élie Joseph Cartan: French mathematician (1869–1951).

The Stokes formula becomes

$$\int_a^b df = \int_a^b \frac{df}{dx} dx = [f]_a^b = f(b) - f(a), \qquad (9.97)$$

which is the familiar **fundamental theorem of one-dimensional calculus**. A generalization of this is when the domain of integration is no longer a flat line, but rather some curve in space, as in Fig. [9.5]. In this case $\omega = f(\mathbf{r})$ is still a 0-form but $d\omega$ becomes $d\omega = \partial_i f dx^i$, or in coordinate-invariant notation:

$$d\omega = \nabla_i f dx^i = \vec{\nabla} f \cdot d\mathbf{r}. \qquad (9.98)$$

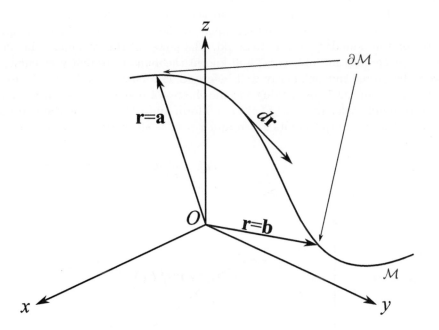

Fig. 9.5: The Stokes–Cartan theorem applied to an arbitrary one-dimensional curve.

Evaluating between the boundary points $\mathbf{r} = \mathbf{a}$ and $\mathbf{r} = \mathbf{b}$, the Stokes formula (9.95) becomes

$$\int_{\mathbf{r}=\mathbf{a}}^{\mathbf{r}=\mathbf{b}} \vec{\nabla} f \cdot d\mathbf{r} = f(\mathbf{b}) - f(\mathbf{a}), \qquad (9.99)$$

which you will find in your calculus books under the heading "the **fundamental theorem of line integrals**." We can easily see that (9.99) reduces to (9.97) in the special case when the line is flat and taken to be the $x$-axis. Moving on to two dimensions, let $\mathcal{M}$ be an arbitrary surface $S$ surrounded by some closed contour $C$, as shown in Fig.

[9.6]. Let $\omega$ be the 1-form $\omega = \mathbf{F} \cdot d\mathbf{r} = F_j dx^j$, where $\mathbf{F}$ is some arbitrary vector field. The exterior derivative of $\omega$ is hence

$$d\omega = \frac{1}{2} \left( \partial_i F_j \right) dx^i \wedge dx^j, \qquad (9.100)$$

which by direct comparison is just

$$d\omega = \left( \vec{\nabla} \times \mathbf{F} \right) \cdot \mathbf{da}. \qquad (9.101)$$

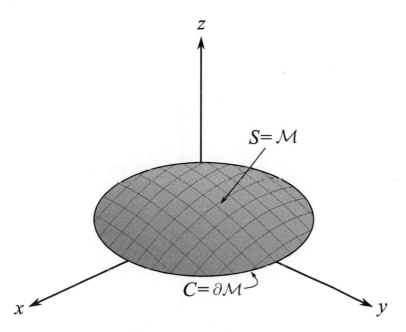

Fig. 9.6: The Stokes–Cartan theorem applied to an arbitrary two-dimensional surface $S$ bounded by the closed contour $C$.

Putting everything together, (9.95) becomes

$$\iint_S \left( \vec{\nabla} \times \mathbf{F} \right) \cdot \mathbf{da} = \oint_C \mathbf{F} \cdot d\mathbf{r}. \qquad (9.102)$$

This is known as the (ordinary) **Stokes' theorem**, or the **Kelvin–Stokes theorem**.[8] A special case of it is **Green's theorem**, when the closed region $S$ becomes flat, say in the $x$–$y$ plane:

$$\iint_S \left( \frac{\partial F_z}{\partial x} - \frac{\partial F_z}{\partial y} \right) da = \oint_C \mathbf{F} \cdot d\mathbf{r}. \qquad (9.103)$$

[8]William Thomson, First Baron Kelvin: Scots–Irish mathematical physicist (1824–1907).

**Exercise 9.22** Explicitly show that (9.100) and (9.101) are the same thing.

Finally, in three dimensions let $\omega = \mathbf{F} \cdot \mathbf{da} = \frac{1}{2}\varepsilon^i{}_{jk}F_i dx^j \wedge dx^k$ be a 2-form defined via some arbitrary vector field $\mathbf{F}$, and $\mathcal{M}$ now be some three-dimensional region of space $V$ bounded by a closed surface $\partial\mathcal{M} = S$, as shown in Fig. [9.7]. The exterior derivative of $\omega$ is then

$$d\omega = \frac{1}{2}\varepsilon^i{}_{jk}\left(\partial_l F_i\right) dx^l \wedge dx^j \wedge dx^k, \tag{9.104}$$

which can be shown to be equivalent to

$$d\omega = \vec{\nabla} \cdot \mathbf{F} dV. \tag{9.105}$$

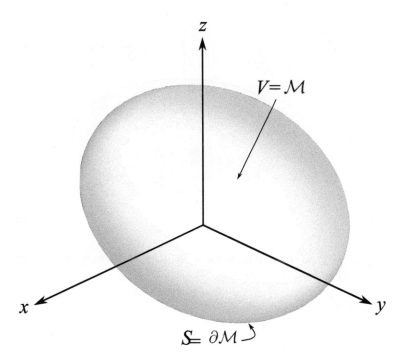

Fig. 9.7: The Stokes–Cartan theorem applied to an arbitrary three-dimensional surface $V$ bounded by the closed surface $S$.

Hence

$$\int\int\int_V \vec{\nabla} \cdot \mathbf{F} dV = \int\int_S \mathbf{F} \cdot \mathbf{da}. \tag{9.106}$$

This is the **divergence theorem**, originally due to Gauss. The degree of your familiarity with these important theorems depends on whether or not you have seen them in multivariable calculus. They are also extensively used in Maxwell's theory of electricity and magnetism; particularly in finding the differential form of Maxwell's equations from their integral form, where both the divergence theorem and the ordinary Stokes theorem are used. It is interesting that (9.95) yields all of them once one makes the specific choices of $\omega$ and $\mathcal{M}$. But (9.95) is a lot more than that, since it is obvious that it works in any arbitrary number of dimensions and on any manifold, Lorentzian manifolds included.

**Exercise 9.23** Explicitly show that (9.104) and (9.105) are the same thing.

**Exercise 9.24** As mentioned in the text, the (ordinary) Stokes (9.102) and divergence (9.106) theorems are used in electrodynamics to switch back and forth between the differential and integral forms of equations. Let's see how to do this generally.

1. Integrate both sides of the Maxwell eqn (9.88) over a manifold $\mathcal{M}$ then apply the generalized Stokes theorem to find the integral over $\partial\mathcal{M}$. This is a general result for any closed form in any number of dimensions. Explicitly show that in three-dimensional Euclidean space your result implies

$$\oint \mathbf{E} \cdot d\mathbf{r} + \int \left( \frac{\partial \mathbf{B}}{\partial t} \right) \cdot d\mathbf{a} = 0$$

$$\int \mathbf{B} \cdot d\mathbf{a} = 0, \tag{9.107}$$

which are the Faraday and the no-monopoles laws respectively.

2. Perform the same calculation you performed in part 1 on the Maxwell eqn (9.89) to find the general integral over $\partial\mathcal{M}$. To get rid of the adjoint exterior derivative, you might want to take the Hodge dual of both sides of (9.89) first, and *then* apply Stokes' theorem. Explicitly show that in three-dimensional Euclidean space your result implies

$$\oint \mathbf{E} \cdot d\mathbf{a} = \frac{1}{\varepsilon_0} \int \rho dV$$

$$\oint \mathbf{B} \cdot d\mathbf{r} - \mu_0 \varepsilon_0 \int \left( \frac{\partial \mathbf{E}}{\partial t} \right) \cdot d\mathbf{a} = \mu_0 \int \mathbf{J} \cdot d\mathbf{a}, \tag{9.108}$$

which are the Gauss and the Ampère–Maxwell laws respectively.

## 9.6 Cartan's Formalism

The language of differential forms was extensively promoted in the early twentieth century by the French mathematician Élie Cartan. He also developed what is referred to today as the **Cartan formalism, tetrad formalism,** or **bein formalism**: an alternative formulation of differential geometry that is heavily based on $p$-forms. It provides a metric-independent formulation of general relativity as well as any other theory of geometry-based gravity (Cartan's own theory of gravity included, as we will see in §10.5). In this section we will give a "lite" introduction to these methods.

Consider a general manifold $\mathcal{M}_n$, which for concreteness we take to be Lorentzian, charted by a metric tensor $g_{\mu\nu}$. As mentioned briefly in §7.1, one can imagine a flat manifold $T_x\mathcal{M}_n$ tangential to a point $x$ on $\mathcal{M}_n$, as shown in Fig. [7.2]. In physics terms $T_x\mathcal{M}_n$ is just the reference frame of some local observer at $x$, described by the flat Minkowski metric which we also take to be Cartesian for simplicity. We denote the coordinates on $\mathcal{M}_n$ by the usual Greek indices, raised and lowered by $g_{\mu\nu}$, while we denote the local coordinates in $T_x\mathcal{M}_n$ by hatted Greek indices; for example, $\hat{\alpha}$, raised and lowered by $\eta_{\hat{\alpha}\hat{\beta}}$.[9] Now let us assert that the global metric $g_{\mu\nu}$ can be expanded in terms of the local $\eta_{\hat{\alpha}\hat{\beta}}$ as follows:

$$g_{\mu\nu} = e^{\hat{\alpha}}{}_\mu e^{\hat{\beta}}{}_\nu \eta_{\hat{\alpha}\hat{\beta}}. \tag{9.109}$$

The quantities $e^{\hat{\alpha}}{}_\mu$ have many names. They are the "tetrads" or "vierbeins" if the spacetime is four-dimensional, since *tetra* is from the Greek for four and *vierbein* is German for "four legs." In higher dimensions they are referred to by the name "vielbeins" from the German for "many legs."[10] In what follows we will just refer to them generically as the **beins**. Note that if the manifold $\mathcal{M}_n$ is Riemannian, then $T_x\mathcal{M}_n$ is Euclidean and $\eta$ is replaced by $\delta$. It is easily checked that the beins satisfy orthogonality on both sets of indices:

$$e^{\hat{\alpha}}{}_\mu e_{\hat{\alpha}}{}^\nu = \delta^\nu_\mu$$
$$e^{\hat{\alpha}}{}_\mu e_{\hat{\beta}}{}^\mu = \delta^{\hat{\alpha}}_{\hat{\beta}}. \tag{9.110}$$

They can also be thought of as components of 1-forms as such:

$$e^{\hat{\alpha}} = e^{\hat{\alpha}}{}_\mu dx^\mu. \tag{9.111}$$

---

**Exercise 9.25** Invert (9.109). In other words, write $\eta_{\hat{\alpha}\hat{\beta}}$ in terms of $g_{\mu\nu}$.

---

The beins can be read off from a given metric quite easily. Starting from (9.109), contract both sides with $dx^\mu dx^\nu$ and use (9.111):

$$g_{\mu\nu}dx^\mu dx^\nu = e^{\hat{\alpha}}{}_\mu e^{\hat{\beta}}{}_\nu \eta_{\hat{\alpha}\hat{\beta}} dx^\mu dx^\nu = e^{\hat{\alpha}}e^{\hat{\beta}}\eta_{\hat{\alpha}\hat{\beta}} = ds^2. \tag{9.112}$$

In other words, the flat scalar product of the bein forms $e^{\hat{\alpha}}$ with themselves gives the full spacetime metric; alternatively, one can view $e^{\hat{\alpha}}$ as a sort of square root of

---

[9]The hatted indices are sometimes referred to as the **Lorentz indices**.

[10]In higher-dimensional geometric theories one frequently finds expressions such as "fünfbeins" in five dimensions, "sechsbeins" in six, and so on.

the metric components. As an example, consider our archetypal spacetime solution: Schwarzschild (5.52). Comparing with (9.112), we find

Schwarzschild bein forms:
$$e^{\hat{t}} = cH^{1/2}dt$$
$$e^{\hat{r}} = H^{-1/2}dr$$
$$e^{\hat{\theta}} = rd\theta$$
$$e^{\hat{\varphi}} = r\sin\theta d\varphi, \tag{9.113}$$

such that

Schwarzschild beins:
$$e^{\hat{t}}{}_{t} = cH^{1/2}$$
$$e^{\hat{r}}{}_{r} = H^{-1/2}$$
$$e^{\hat{\theta}}{}_{\theta} = r$$
$$e^{\hat{\varphi}}{}_{\varphi} = r\sin\theta. \tag{9.114}$$

**Exercise 9.26** Check that (9.114) do indeed lead back to (5.52). Explicitly check the orthogonality of the Schwarzschild beins.

**Exercise 9.27** Write down the bein forms and the beins for the spacetimes of Exercise 6.11.

The bein forms are sometimes referred to as the **non-coordinate basis**, since vectors can be written in terms of them; for example, a spacetime vector **V** can be expanded $\mathbf{V} = V^{\hat{\alpha}}e_{\hat{\alpha}}$ such that

$$V^{\hat{\alpha}} = V^{\mu}e^{\hat{\alpha}}{}_{\mu} \quad \text{and} \quad V^{\mu} = V^{\hat{\alpha}}e_{\hat{\alpha}}{}^{\mu}. \tag{9.115}$$

**Exercise 9.28** For two vectors $V^{\mu}$ and $U^{\mu}$ prove that $V^{\mu}U_{\mu} = V^{\hat{\alpha}}U_{\hat{\alpha}}$.

Based on (9.115), one can generally think of the effect of the beins as one of changing the coordinate indices $\mu$ to the flat indices $\hat{\alpha}$ and vice versa. The bein language, then, is a formulation of geometry to the coordinate/metric-based formulation we have been working with, and as such the entire structure of general covariance can be rewritten in terms of this. For example, playing the role of the Christoffel symbols in this language is the so-called **spin connection 1-form**:

$$\omega^{\hat{\alpha}\hat{\beta}} = \omega_{\mu}{}^{\hat{\alpha}\hat{\beta}}dx^{\mu}. \tag{9.116}$$

As such, the covariant derivative of, say, a $p$-form $V^{\hat{\alpha}}{}_{\hat{\beta}}$ in this language is

$$\nabla_\mu V^{\hat{\alpha}}{}_{\hat{\beta}} = dV^{\hat{\alpha}}{}_{\hat{\beta}} + \omega^{\hat{\alpha}}{}_{\hat{\gamma}} \wedge V^{\hat{\gamma}}{}_{\hat{\beta}} - (-1)^p\, V^{\hat{\alpha}}{}_{\hat{\gamma}} \wedge \omega^{\hat{\gamma}}{}_{\hat{\beta}} \tag{9.117}$$

instead of (2.83). We will see how to calculate the spin connections in a moment. Now, we recall that based on Riemannian geometry any given manifold can be completely characterized by two quantities: its curvature, which we are familiar with, and its **torsion**—which we have only very briefly discussed. In simple terms, torsion is what happens when the Christoffel connection fails to satisfy

$$\Gamma^\rho_{\mu\nu} = \Gamma^\rho_{\nu\mu}, \tag{9.118}$$

which we called the torsion-free condition back around eqn (2.91), as well as emphasized as the the fifth property of derivatives in §7.2. Based on this failure, we can define the so-called **torsion tensor** as follows:

$$\mathcal{T}^\rho{}_{\mu\nu} = \frac{1}{2}\left(\Gamma^\rho_{\mu\nu} - \Gamma^\rho_{\nu\mu}\right), \tag{9.119}$$

which vanishes in general relativity, or equivalently,

$$\mathcal{T}^\rho{}_{\mu\nu} \nabla_\rho f = -\left(\nabla_\mu \nabla_\nu f - \nabla_\nu \nabla_\mu f\right). \tag{9.120}$$

**Exercise 9.29** Although made up of non-tensorial parts, the torsion tensor is in fact a tensor; in other words, it transforms following (2.36). Prove this.

**Exercise 9.30** Show that (9.120) is a consequence of (9.119).

The torsion tensor is antisymmetric in its lower indices—i.e. $\mathcal{T}^\rho{}_{\mu\nu} = -\mathcal{T}^\rho{}_{\nu\mu}$—as can be easily checked. Without the torsion-free condition, the Christoffel symbols are no longer defined via the usual (4.33). In fact, one can show that the correct formula is now

$$\Gamma^\rho_{\mu\nu} = \frac{1}{2}g^{\rho\sigma}\left(\partial_\mu g_{\nu\sigma} + \partial_\nu g_{\sigma\mu} - \partial_\sigma g_{\mu\nu} - \mathcal{T}_{\mu\nu\sigma} - \mathcal{T}_{\nu\mu\sigma} - \mathcal{T}_{\sigma\mu\nu}\right), \tag{9.121}$$

where the antisymmetric parts of the connection are carried by the components of $\mathcal{T}_{\mu\nu\sigma}$.

**Exercise 9.31** Derive (9.121).

To understand what torsion means, consider parallel transporting a vector $B^\sigma$ along a given curve using the equation of parallel transport (7.4). A straightforward, but tedious, calculation shows that if the torsion is non-vanishing, then the vector will *necessarily* rotate, or precess, about the curve itself; as shown in Fig. [9.8]. If the vector

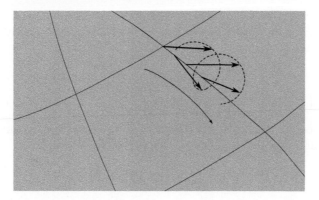

Fig. 9.8: A vector parallel transported along a curve in a space with torsion will precess about the curve.

happens to be part of a specific volume of spacetime, this has the effect of *twisting* the volume about an axis tangential to the curve; which justifies the use of the term "torsion" to describe this property.

If torsion is non-vanishing, everything else we discussed will effectively change: The Riemann tensor will contain components of the torsion tensor and the Ricci tensor in particular will no longer necessarily be symmetric. It could still be found by taking the trace of the Riemann curvature (7.21) but will now satisfy

$$R_{\mu\nu} = R_{\nu\mu} - 3\left(\nabla_\mu \mathcal{T}^\rho{}_{\rho\nu} - \nabla_\nu \mathcal{T}^\rho{}_{\mu\rho} - \nabla_\rho \mathcal{T}^\rho{}_{\nu\mu}\right) + \mathcal{T}^\rho{}_{\rho\sigma} \mathcal{T}^\sigma{}_{\mu\nu}. \tag{9.122}$$

Based on these two major geometric properties, curvature and torsion, Cartan introduced the **torsion 2-form**,

$$\mathcal{T}^{\hat\alpha} = \frac{1}{2}\mathcal{T}^{\hat\alpha}{}_{\mu\nu} dx^\mu \wedge dx^\nu, \tag{9.123}$$

and the **curvature 2-form**,

$$\mathcal{R}^{\hat\alpha\hat\beta} = \frac{1}{2}\mathcal{R}^{\hat\alpha\hat\beta}{}_{\mu\nu} dx^\mu \wedge dx^\nu, \tag{9.124}$$

both of which are defined by what is today known as the **Cartan structure equations**:

$$\mathcal{T}^{\hat\alpha} = de^{\hat\alpha} + \omega^{\hat\alpha}{}_{\hat\beta} \wedge e^{\hat\beta} \tag{9.125}$$

$$\mathcal{R}^{\hat\alpha\hat\beta} = d\omega^{\hat\alpha\hat\beta} + \omega^{\hat\alpha}{}_{\hat\gamma} \wedge \omega^{\hat\gamma\hat\beta}. \tag{9.126}$$

Taking the exterior derivative of (9.125) and (9.126) leads to the Bianchi identities,

$$d\mathcal{T}^{\hat\alpha} + \omega^{\hat\alpha}{}_{\hat\beta} \wedge de^{\hat\beta} - d\omega^{\hat\alpha}{}_{\hat\beta} \wedge e^{\hat\beta} = 0$$

$$dR^{\hat\alpha\hat\beta} + \omega^{\hat\alpha}{}_{\hat\gamma} \wedge d\omega^{\hat\gamma\hat\beta} - d\omega^{\hat\alpha}{}_{\hat\gamma} \wedge \omega^{\hat\gamma\hat\beta} = 0, \tag{9.127}$$

the second of which we have already seen in component form in §7.3. The components of the torsion and curvature 2-forms are related to their respective tensors as follows:

$$\mathcal{T}^{\hat{\alpha}}_{\ \mu\nu} = e^{\hat{\alpha}}_{\ \rho} T^{\rho}_{\ \mu\nu}$$

$$\mathcal{R}^{\hat{\alpha}\hat{\beta}}_{\ \ \mu\nu} = e^{\hat{\alpha}}_{\ \rho} e^{\ \hat{\beta}\kappa} R^{\rho}_{\ \kappa\mu\nu}. \tag{9.128}$$

We can now say that general relativity is described by a geometry where the first structure equation vanishes; thus the beins satisfy the condition

$$de^{\hat{\alpha}} = -\omega^{\hat{\alpha}}_{\ \hat{\beta}} \wedge e^{\hat{\beta}}. \tag{9.129}$$

In this case the components of the spin connection are found by

$$\omega_{\mu}^{\ \hat{\alpha}\hat{\beta}} = e^{\hat{\alpha}\nu} \nabla_{\mu} e^{\hat{\beta}}_{\ \nu} = e^{\hat{\alpha}\nu} \left( \partial_{\mu} e^{\hat{\beta}}_{\ \nu} - \Gamma^{\rho}_{\mu\nu} e^{\hat{\beta}}_{\ \rho} \right), \tag{9.130}$$

where the Christoffel symbols are defined in the usual way (4.33). Also, a consequence of the metric compatibility condition (7.7) is

$$\omega^{\hat{\alpha}\hat{\beta}} = -\omega^{\hat{\beta}\hat{\alpha}}. \tag{9.131}$$

**Exercise 9.32** Calculate the spin connections for the Schwarzschild metric (5.47) and the metrics of Exercise 6.11.

**Exercise 9.33** Explicitly show that the spin connections you found in Exercise 9.32 satisfy the torsion-free condition (9.129).

**Exercise 9.34** Find the components of the Riemann 2-form (9.124) for the Schwarzschild spacetime. You may use (7.18). Confirm that the second Bianchi identity (9.127) is satisfied.

In this language one can show that the action describing physics in a curved background with vanishing torsion (i.e. general relativity) is[11]

$$S = \int \frac{1}{2\kappa} \varepsilon_{\hat{\alpha}\hat{\beta}\hat{\gamma}\hat{\sigma}} \left( e^{\hat{\alpha}} \wedge e^{\hat{\beta}} \wedge R^{\hat{\gamma}\hat{\sigma}} + \Lambda e^{\hat{\alpha}} \wedge e^{\hat{\beta}} \wedge e^{\hat{\gamma}} \wedge e^{\hat{\sigma}} \right) + S_M, \tag{9.132}$$

where $S_M$ is the action of the matter/field content, $\kappa$ is the Einstein constant (7.88), and $\Lambda$ is the cosmological constant. Varying this action with respect to the beins, i.e. applying

$$\frac{\delta S}{\delta e^{\hat{\alpha}}_{\ \mu}} = 0, \tag{9.133}$$

---

[11]Following the notation of [26]. Another good review of general relativity in the bein language is [27].

yields *two* equations: The first is the torsion-free condition (9.129) and the second is the Einstein field eqn (7.100) in its bein form

$$\varepsilon_{\hat{\alpha}\hat{\beta}\hat{\gamma}\hat{\sigma}}\left(e^{\hat{\alpha}} \wedge R^{\hat{\beta}\hat{\gamma}} + \Lambda e^{\hat{\alpha}} \wedge e^{\hat{\beta}} \wedge e^{\hat{\gamma}}\right) = \kappa T_{\hat{\sigma}}, \tag{9.134}$$

where the **energy-momentum 3-form** is

$$T_{\hat{\sigma}} = \frac{1}{3!}T_{\hat{\sigma}}{}^{\mu}\varepsilon_{\mu\nu\rho\kappa}dx^{\nu} \wedge dx^{\rho} \wedge dx^{\kappa}, \tag{9.135}$$

found by

$$T_{\hat{\sigma}}{}^{\mu} = \frac{\delta S_M}{\delta e^{\hat{\alpha}}{}_{\mu}}. \tag{9.136}$$

Clearly this alternative formulation, being coordinate independent, is a lot more powerful than the ordinary tensor language. But just like any other, it takes a while to get used to. Some modern theories of gravity, such as loop quantum gravity, heavily use Cartan's formalism, as it seems to be the best suited for their purposes.

**Exercise 9.35** The exercises in this chapter kept getting longer and longer, to the point that most exercises in this subsection have become mini-projects of their own. Here are two more:

1. Explicitly show that (9.134) is equivalent to (7.100).
2. Show that varying the first two terms of (9.132) leads to (9.129) and the left-hand side of (9.134).

# 10

# Generalizing General Relativity

*I do not know what I may appear to the world; but to myself, I seem to have been only like a boy playing on the seashore, and diverting myself now and then in finding a smoother pebble or prettier shell than ordinary, while the great ocean of truth lay all undiscovered before me.*

Sir Isaac Newton

## Introduction

The general theory of relativity, following the concept of general covariance, has been shown time and time again to be our best understanding of how spacetime "works." Experimental and observational evidence have repeatedly supported the theory ever since that fateful day in 1919 when Sir Arthur Eddington and his team took pictures of the solar eclipse that showed an amount of gravitational lensing that is as close to the general relativistic prediction as one can get within allowed errors.[1] Since then, the repetition of this particular test, as well as many other observations and experiments, has provided even more support for the theory. General relativity is now well established as a pillar of humanity's intellectual achievements, as well as one of two foundations upon which we place our most fundamental understanding of the universe; the other is of course quantum mechanics.

Having said this, it is understood that general relativity cannot be the end of the story; in fact, one may wonder if there will ever be an end to our quest for understanding the laws of nature. Right now there seem to be two different ways with which one can extend one's understanding of spacetime. The first is to successfully join our two most fundamental theories into one; in other words, find and confirm the elusive theory of quantum gravity. We will not be discussing this, as it requires too much knowledge of quantum mechanics—well above the assumed background of the reader of this book. The interested student may consult any review article on the topic.[2] The second path would be to generalize general relativity itself by asking this question: Is the general theory of relativity a special case of some more complex theory, in the same sense as

---

[1] As told briefly in §6.2.9.
[2] For example, [28].

*Covariant Physics: From Classical Mechanics to General Relativity and Beyond*. Moataz H. Emam, Oxford University Press (2021). © Moataz H. Emam.
DOI: 10.1093/oso/9780198864899.003.0010

Newton's theory was a special case of it? The motivation for going down this road is two-fold: There is just the usual curiosity to see how to make things more general or more precise. There are also practical motivations such as the one we briefly discussed in §5.8 and §7.7: Some of these modified theories attempt to explain the accelerated expansion of the universe not by assuming some cosmological constant or an unknown type of energy, but rather by modifying the geometric properties of spacetime itself. The assumption is that a modified theory acts gravitationally like general relativity on the scales of galaxies and less, while producing a "repulsive" effect on larger scales. Many of these theories are now of only historical interest; others are still popular as possible contenders. The ones we will review are the most popular ones of the latter category. For the first time in this book we will work in geometrized units (see §4.7), where $c = 1$ and $G = 1$. This means that the Einstein constant becomes just $\kappa = 8\pi$.

## 10.1 Brans–Dicke Theory

Developed in the 1960s, the Brans–Dicke theory of gravitation[3] is an example of a **scalar–tensor theory** in that it contains a scalar field $\phi$ in addition to the usual metric tensor.[4] The scalar field is not of the same type generated by, for example, (8.95), but is rather a field that contributes to the curvature of spacetime itself. Being dynamic, $\phi$ has the effect of allowing the "strength" of the gravitational field, i.e. the curvature of spacetime, to vary depending on the value of $\phi(x^\mu)$. In other words, the Einstein constant $\kappa$ is replaced, or more accurately supplemented, by a varying field. The predictions of Brans–Dicke seem to be in agreement with all observations thus far. So until new evidence that favors the theory arise, most theoretical physicists prefer to work with the simpler general relativity. A complete description of Brans–Dicke theory is given by the action

$$S_{BD} = \frac{1}{2\kappa} \int d^4x \sqrt{-|g|} \left[ \phi R - \frac{\omega}{\phi} (\partial_\mu \phi)(\partial^\mu \phi) - V(\phi) \right] + S_M, \tag{10.1}$$

where $\omega$ is the dimensionless **Dicke coupling constant**, $V$ is some arbitrary potential for $\phi$, and the action $S_M$ contains within it the description of all the *other* matter/energy content. Varying the action with respect to $\phi$ gives the scalar field equation

$$\nabla_\mu \nabla^\mu \phi = \frac{1}{3 + 2\omega} \left( \kappa T + \phi \frac{dV}{d\phi} - 2V \right), \tag{10.2}$$

where $T = g^{\mu\nu} T_{\mu\nu}$ is the trace of the energy momentum tensor found in the usual way (8.112) from $S_M$. Varying with respect to the metric gives the **Brans–Dicke gravitational field equations**

---

[3]Carl Henry Brans: American mathematical physicist (b. 1935) and Robert Henry Dicke: American physicist (1916–1997).

[4]Yes I know: this is an abuse of notational conventions since scalars are *also* tensors of zero rank.

$$G_{\mu\nu} = \frac{\kappa}{\phi}T_{\mu\nu} + \frac{\omega}{\phi^2}\left[(\partial_\mu\phi)(\partial_\nu\phi) - \frac{1}{2}g_{\mu\nu}(\partial_\alpha\phi)(\partial^\alpha\phi)\right]$$

$$+ \frac{1}{\phi}[\nabla_\mu\nabla_\nu\phi - g_{\mu\nu}\nabla_\alpha\nabla^\alpha\phi] - \frac{V}{2\phi}g_{\mu\nu}. \tag{10.3}$$

**Exercise 10.1** Vary the action appropriately to derive (10.2) and (10.3).

Note that the appearance of the scalar field $\phi$ in the first term of the right-hand side of (10.3) is what we meant by replacing the Einstein *constant* with a variable quantity. In other words, we can write $\tilde{\kappa}(\varphi) = \kappa/\phi$ and thus have a new gravitational coupling/strength that is dependent on the scalar field $\phi$. The gravitational field, more precisely the curvature of spacetime, is generated by the matter/energy content described by $T_{\mu\nu}$, in addition to that described by $\phi$ (the remaining terms of the right-hand side). Let's also point out that (10.2), being essentially a Poisson equation, implies that the source of the scalar field $\phi$ is also the matter/energy content described by $T$ plus whatever interactions are allowed by the potential $V$. Recall from our discussion in §8.2 that a scalar field acts as a potential and it is its field strength $F = d\phi$ (8.77) that can presumably be measured experimentally. As such, solutions of (10.2) and (10.3) with a constant $\phi$ should automatically reduce to their respective general relativistic solutions. In fact this can be readily seen by putting any constant value for $\phi$ in the above equations and taking $\omega \to \infty$ in (10.2). Yes, the appearance of $\phi$ without derivatives in the action and in the gravitational field equations will not go away, but one can just absorb it in the definition of $\kappa$; in other words, simply set $\phi = 1$.

In his paper [29] Brans himself found *four* different classes of static spherically symmetric solutions to the vacuum ($T_{\mu\nu} = 0$) field eqns (10.2) and (10.3), with vanishing $V$ potential. We will focus on one of them: the so-called class I solution,

$$ds^2 = -H^{m+1}dt^2 + H^{n-1}dr^2 + r^2H^n d\Omega^2, \quad \text{and} \quad H(r) = 1 - 2\frac{r_0}{r}$$

$$\phi(r) = \phi_0 H^{-\frac{1}{2}(m+n)}, \tag{10.4}$$

where $m$, $n$, $\phi_0$, and $r_0$ are arbitrary constants and Dicke's coupling is constrained by

$$\omega = -2\frac{m^2 + n^2 + nm + m - n}{(m+n)^2}. \tag{10.5}$$

**Exercise 10.2** Derive the condition (10.5) by substituting the class I solution into the Brans–Dicke field equations.

This solution was studied extensively in [30] and it was shown to have different interpretations depending on the values of the constants. Firstly, one can check that both the metric and the scalar field satisfy asymptotic flatness for all values of $m$ and $n$. The reader will recall that this is a desirable physical property. In the case of the metric it becomes flat at radial infinity, while in the case of the scalar potential its field strength $F = d\phi$ vanishes. Another important property is that for $m = n = 0$ and the identification $2r_0 \to r_s$ it is easy to see that $\phi$ becomes a constant and the metric immediately reduces to the ordinary Schwarzschild solution. The behavior of (10.4) for $r = 2r_0$ is just as anomalous as in the Schwarzschild case; however, by checking the components of the Riemann tensor it can be shown that this is a Schwarzschild-like coordinate singularity *only* for $n \le -1$. For other values of $n$ it becomes a true naked singularity. As the reader may recall, we briefly discussed the issue of naked singularities back in §8.6 and referred to the general "belief" among physicists and cosmologists that they cannot exist. This so-called "cosmic censorship" is just a postulate, though, and there is no rigorous mathematical reasoning behind it. Even the not-too-rigorous evidence for it that we have[5] all falls within the context of general relativity. The naked singularity solution here cannot then be ruled out as a possibly physical object. If found, that would be strong evidence in support of the whole theory. Interestingly, by studying the geodesics of (10.4) the authors of [30] were able to show that *if* and only *if* $m - n + 1 > 0$ will the sphere $r = 2r_0$ become an actual event horizon for all values of $m$ and $n$, in the sense of becoming a "point of no return." Because of this, the solution is usually referred to today as the **Brans–Dicke black hole** or just the **Brans black hole**, even if actually being a black hole is just a subclass of what is essentially an entire family of solutions.

It was pointed out in various sources, including [30], that some effects, such as X-ray emission from an accretion disk, would have signatures *different* from their counterparts in general relativity. These can then be used by astronomers to detect and differentiate Brans–Dicke objects from the ordinary general relativistic black holes. So far this has not happened, and Brans–Dicke theory is still unconfirmed as an actual generalization of general relativity. There are several other known exact solutions to the theory, including cosmological ones that lead to upgraded cosmological evolution equations replacing (5.112). This turns out to be important because it has long been known that a scalar field could generate the negative pressure required to account for the accelerated expansion of the universe.[6] We refer the interested reader to the literature.

---

**Exercise 10.3** Verify that for $m = n = 0$ and $2r_0 = rs$ the scalar potential $\phi$ vanishes and the Brans–Dicke class I metric reduces to the ordinary Schwarzschild solution.

**Exercise 10.4** Find all of the non-vanishing components of the Brans–Dicke class I Riemann tensor and check that it has a coordinate singularity only for the case $n \le -1$.

---

[5] Such as Wald's calculation referred to in the same section.

[6] As well as the much earlier phase of inflation: in itself an extreme case of acceleration. Review point 3 in the list toward the end of §5.8.

## 10.2  $f(R)$ **Theory**

This is one of the most popular modifications of general relativity. It is also general enough to be considered not just a theory, but rather a class of theories. The modification is simple: just replace the Ricci scalar $R$ in the Einstein–Hilbert action (8.117) with an arbitrary function in $R$; thus

$$S_{f(R)} = \frac{1}{2\kappa} \int d^4x \sqrt{-|g|} f(R) + S_M, \tag{10.6}$$

where $S_M$ is the usual matter/energy action. One can think of the $f(R)$ function as a power series,

$$f(R) = \sum_{n=-\infty}^{n=+\infty} c_n R^n = \cdots + \frac{c_{-2}}{R^2} + \frac{c_{-1}}{R} + c_0 + c_1 R + c_2 R^2 + \cdots, \tag{10.7}$$

where all $c_n$ are arbitrary constants. General relativity itself would be the special case $c_1 = 1/2\kappa$, with all other constants vanishing, or if one wishes to include the cosmological constant (8.130), then one also takes $c_0 = -2\Lambda$. Of course $f(R)$ does not necessarily need to be a power series; it can be any function of the Ricci scalar one wishes. Applying the least action principle (8.121) yields the $f(R)$ field equations

$$f'(R) R_{\mu\nu} - \frac{1}{2} f(R) g_{\mu\nu} - (\nabla_\mu \nabla_\nu - g_{\mu\nu} \nabla_\alpha \nabla^\alpha) f'(R) = \kappa T_{\mu\nu},$$

$$\text{where} \quad f'(R) = \frac{df}{dR}. \tag{10.8}$$

---

**Exercise 10.5**  Vary the $f(R)$ action to derive the field eqns (10.8). As pointed out in footnote 7 a boundary term will appear. You may ignore it.

---

This can easily be seen to give the Einstein field equations if $f(R) = R$ and consequently $f'(R) = 1$.[7] It is possible to rewrite (10.8) in the more familiar form

$$G_{\mu\nu} = R_{\mu\nu} - \frac{1}{2} g_{\mu\nu} R = \frac{\kappa}{f'(R)} \left( T_{\mu\nu} + T_{\mu\nu}^{eff} \right), \tag{10.9}$$

where the "effective" stress tensor is defined by

$$T_{\mu\nu}^{eff} = \frac{1}{\kappa} \left[ \frac{1}{2} (f - Rf') g_{\mu\nu} + (\nabla_\mu \nabla_\nu - g_{\mu\nu} \nabla_\alpha \nabla^\alpha) f' \right], \tag{10.10}$$

as can be easily checked. The theory has gathered a lot of interest in recent years for various reasons; chief among them is that it is possible to rewrite the action in a form

---

[7]In deriving (10.8), which the reader is urged to attempt, an unwanted boundary term arises, just as it did in the pure Einstein–Hilbert case. This term in this case is even more persistent and problematic than the general relativistic one, and there seem to be several ways of dealing with it in the physics literature. See, for example, [31].

that is dependent on a scalar field $\phi$ with appropriate redefinitions of variables. This field, just like in Brans–Dicke theory, can be a possible generator of the universe's accelerated expansion. To show how this field arises, first define the action

$$S_\chi = \frac{1}{2\kappa} \int d^4x \sqrt{-|g|}\, [f(\chi) + f'(\chi)(R - \chi)] + S_M, \qquad (10.11)$$

parameterized by the scalar field $\chi$. It is straightforward to show that this is equivalent to (10.6). Varying (10.11) with respect to $\chi$, i.e.

$$\frac{\delta S_\chi}{\delta \chi} = 0, \qquad (10.12)$$

gives the condition

$$f''(\chi)(R - \chi) = 0, \qquad (10.13)$$

from which we see that $\chi = R$ if $f''(\chi) \neq 0$, which reduces (10.11) to the $f(R)$ action (10.6). Now redefine further:

$$\phi = f'(\chi) \qquad (10.14)$$

and introduce a "potential,"

$$V(\phi) = \chi(\phi)\phi - f(\chi), \qquad (10.15)$$

to end up with the action

$$S_\phi = \frac{1}{2\kappa} \int d^4x \sqrt{-|g|}\, [\phi R - V(\phi)] + S_M. \qquad (10.16)$$

Comparing this action, which is totally equivalent to (10.6), with (10.1), we see that it is a subclass of Brans–Dicke theories with vanishing Dicke coupling $\omega$. We can take it even further by more redefinitions:

$$d\tilde{\phi} = \sqrt{\frac{3}{2\kappa}} \frac{d\phi}{\phi} \qquad (10.17)$$

and

$$\tilde{g}_{\mu\nu} = \phi g_{\mu\nu}. \qquad (10.18)$$

This last is known in mathematical circles as a **conformal transformation**, which simply means scaling the object in question, in this case the metric tensor, by a variable quantity, in this case $\phi$. These redefinitions lead to the action

$$\tilde{S} = \int d^4x \sqrt{-|\tilde{g}|} \left[ \frac{1}{2\kappa} \tilde{R} - \frac{1}{2} \left( \partial^\mu \tilde{\phi} \right) \left( \partial_\mu \tilde{\phi} \right) - \tilde{V}\left( \tilde{\phi} \right) \right] + S_M. \qquad (10.19)$$

**Exercise 10.6** Using the conformal reparameterizations (10.17) and (10.18), derive (10.19) from (10.16).

This is equivalent to ordinary general relativity with a dynamic scalar field $\tilde{\phi}$, kinetic term and all, as can be seen by comparing with (8.95) and (8.120). In summary, all of the actions seen here, derived by certain redefinitions of the fields, are exactly equivalent to the original $f(R)$ action (10.6). Using any one of them is just a matter of calculational convenience, but the important thing to note here is that this equivalence is indicative of the fact that $f(R)$ theory can yield the negative pressure required to account for the accelerated expansion of the universe, since it is known that a scalar field can in fact produce such an acceleration. Note, however, that $f(R)$ theory in its original form puts all of the dynamics on the geometry of spacetime, but in its reduced form the dynamics is partially represented by a matter content; namely the scalar field. This is just a matter of interpretation, though: the mathematics doesn't care how we view it.

Going back to the notation of the original action (10.6) and field eqns (10.8), let's consider as an example the flat Robertson–Walker metric (5.103) with the energy momentum tensor of an ideal fluid (7.67) to find the cosmological evolution equations according to $f(R)$ theory; in other words, "upgrading" (5.112). In terms of the Hubble parameter (5.114), these are

$$H^2 = \frac{\kappa}{3f'} \left[ \rho + \frac{1}{2}(Rf' - f) - 3H\dot{R}f'' \right]$$

$$2\dot{H} + 3H^2 = -\frac{\kappa}{f'} \left[ p + \dot{R}^2 f''' + 2H\dot{R}f'' + \ddot{R}f'' + \frac{1}{2}(f - Rf') \right]. \qquad (10.20)$$

**Exercise 10.7** Derive the Friedmann-like eqns (10.20).

In order for these to have an accelerated expansion solution, the function $f(R)$ must be chosen carefully. There are several contenders. One popular choice is known as the Starobinsky model:

$$f(R) = R + \frac{1}{36\alpha}R^2, \qquad (10.21)$$

where $\alpha$ is some arbitrary constant. Analytical analysis as well as computational solutions of eqns (10.20) using (10.21) and others has been performed[8] and shows rates of change of the scale factor $a$ that conform with the requirements of an accelerated expansion of the universe, whether in the inflationary epoch or in the current accelerated phase.

**Exercise 10.8** Substitute (10.21) in the Friedmann-like eqns (10.20) and interpret: Can one get a positive acceleration from the resulting equations?

---

[8]See [32] and the references within.

## 10.3   Gauss–Bonnet Gravity

In this model, the Einstein–Hilbert action is modified with the addition of the so-called **Gauss–Bonnet term**, constructed from the "squares" of the Ricci scalar, the Ricci tensor, and the Riemann tensor as follows:[9]

$$\mathscr{G} = R^2 - 4R^{\mu\nu}R_{\mu\nu} + R^{\mu\nu\rho\sigma}R_{\mu\nu\rho\sigma}. \tag{10.22}$$

The action (8.131) becomes

$$S_{GB} = \frac{1}{2\kappa} \int d^n x \sqrt{-|g|}\,(R - 2\Lambda + \alpha\mathscr{G}) + S_M, \tag{10.23}$$

where $\alpha$ is an arbitrary coupling constant and we have written the action for an arbitrary number of spacetime dimensions $n$. The significance of the Gauss–Bonnet term $\mathscr{G}$ comes from a theorem by the same name which relates the *topology* of a manifold with its geometry. Topology is that branch of mathematics that deals with the possibility of "deforming" geometric shapes into each other. For example, a sphere cannot be continuously deformed into a torus without "punching" a hole into it; hence a sphere and a torus are topologically distinct. On the other hand, a sphere can be deformed into, say, a cube without damage; hence topologically speaking a cube and a sphere are the same object. Now consider any closed Lorentzian manifold $\mathcal{M}_n$; i.e. one without a boundary. Its *geometry* is described by the usual tensors $R_{\mu\nu\rho\sigma}$, $R_{\mu\nu}$, and $R$. We also define its **total curvature** as the integral of $\mathscr{G}$ over the entire manifold. The manifold's topology, on the other hand, is described by the so-called **topological invariants**, the most important of which is the **Euler characteristic** $\chi(\mathcal{M})$. It is an invariant in the sense that it doesn't change if the manifold is deformed without changing its topology. In admittedly oversimplified terms, $\chi(\mathcal{M})$ counts the number of "holes" in the manifold and gives different values for spheres and tori but the same values for spheres and cubes, demonstrating that they are essentially the same thing, *topologically*. The **Gauss–Bonnet theorem** connects $\chi(\mathcal{M})$ with the *total* curvature of a given manifold; in other words, it connects topology with geometry. What this means is that if we deform $\mathcal{M}_n$ in any way *without* changing its topology, then its total curvature does not change. The mathematical statement is just

$$\chi(M) \propto \int_{\mathcal{M}} d^n x \sqrt{-|g|}\mathscr{G}. \tag{10.24}$$

If we focus on our usual case of interest, $n = 4$, what are the field equations that we get by varying (10.23)? Surprisingly, these are *exactly* the ordinary Einstein field eqns (7.100)! The reason is that in four dimensions (or less) $\mathscr{G}$ is just a surface term that vanishes at infinity. However, for more dimensions than four, the field equations are no

---

[9]Pierre Ossian Bonnet: French mathematician (1819–1892).

longer trivial. Hence this theory is only useful in studying the structure of spacetime if we are interested in higher-dimensional gravity theories. Gauss–Bonnet gravity is in fact a subclass of the more general **Lovelock theory of gravity**, which is also non-trivial *only* in higher dimensions.[10] These theories are useful in that they may have some connections to theories that are naturally set in dimensions higher than four, particularly superstring theory.

As an example of a non-trivial Gauss–Bonnet result, let's look at static "spherically" symmetric solutions of (10.23) in five dimensions, $n = 4 + 1$; i.e. four spatial dimensions plus the usual one of time. In other words, $x^\mu = (t, r, \psi, \theta, \varphi)$, where $r$ is the usual radial coordinate and $\psi$, $\theta$, $\varphi$ are angles, effectively generalizing spherical coordinates in $3 + 1$ spacetime. Spherical symmetry here refers to 3-spheres in four-dimensional space, i.e. rotations about the three angles. These are of course impossible to visualize; however, analogy with ordinary three-dimensional space should convince the reader that static spherically symmetric solutions in this spacetime should have the general form

$$ds^2 = -f^2\left(r\right) dt^2 + \frac{dr^2}{g^2\left(r\right)} + r^2 d\Sigma_3^2, \tag{10.25}$$

where here $d\Sigma_3^2 = g_{ij} dx^i dx^j$ $(i, j, = \psi, \theta, \varphi)$ plays the role that $d\Omega^2$ played in, say, (5.47). The reader will recall that $d\Omega^2$ was the metric of the unit sphere, as required by spherical symmetry. In higher dimensions, however, $d\Sigma_3^2$ can be *any* metric that satisfies the condition $R_{ij} \propto g_{ij}$. A manifold that satisfies this condition is called an **Einstein manifold**. Now the simplest case of $d\Sigma_3^2$ occurs when $\alpha = 0$, i.e. the Gauss–Bonnet term goes away and we are left with the ordinary general relativistic action. In this case, $d\Sigma_3^2$ becomes a direct generalization of $d\Omega^2 = d\theta^2 + \sin^2\theta d\varphi^2$: in other words, a 3-sphere with metric $d\Sigma_3^2 = d\psi^2 + \sin^2\psi \left(d\theta^2 + \sin^2\theta d\varphi^2\right)$. Furthermore, if in addition to this the cosmological constant vanishes, then (10.25) becomes a Schwarzschild-like solution in $n = 5$ with $f^2 = g^2 = 1 - C/r^2$, with $C = $ constant. Note that Schwarzschild in $n = 5$ has an inverse $r$-*squared* dependence, as opposed to the inverse $r$ dependence of the ordinary $n = 4$ Schwarzschild. In fact, in an arbitrary $n$-dimensional spacetime, it can be shown that a Schwarzschild-like metric will have inverse $r^{n-3}$ dependence.[11]

Varying (10.23) with $S_M = 0$ leads to the vacuum **Einstein–Gauss–Bonnet field equations**:

$$G_{\mu\nu} + \Lambda g_{\mu\nu} + 4\kappa \left[ R R_{\mu\nu} - 2R_{\mu\rho\nu\sigma} R^{\rho\sigma} + R_{\mu\rho\sigma\kappa} R_\nu{}^{\rho\sigma\kappa} - 2R_{\mu\rho} R_\nu{}^\rho - \frac{1}{4} g_{\mu\nu} \mathscr{G} \right] = 0. \tag{10.26}$$

---

[10]David Lovelock: British theoretical physicist and mathematician (b. 1938).

[11]This is a consequence of spherical symmetry in higher dimensions. In fact, it can be further shown that Newton's gravitational potential formula, as well as Coulomb's electrostatic potential formula, all follow the same pattern in arbitrary dimensions.

**Exercise 10.9** If you really *really* want to, derive (10.26).

In [33], several cases of the ansatz (10.25) that satisfy (10.26) were found. For example,

$$f^2 = g^2 = \frac{\Lambda}{3}r^2 - \mu, \tag{10.27}$$

where $\mu$ is an arbitrary integration constant, represents a black hole in $n = 5$ if $\Lambda > 0$. The event horizon is clearly located at $r = \sqrt{\frac{3\mu}{\Lambda}}$. The solution

$$f^2 = \left( \sqrt{\frac{\Lambda}{3}}r + a\sqrt{\frac{\Lambda}{3}r^2 + \mathscr{R}_\Sigma} \right)^2$$

$$g^2 = \frac{\Lambda}{3}r^2 + \mathscr{R}_\Sigma \tag{10.28}$$

was also found, where $a$ is an arbitrary integration constant and $\mathscr{R}_\Sigma$ is the Ricci scalar of $d\Sigma_3^2$. In this case interpretation varies depending on the values of the constants. For example, if $\Lambda > 0$, $\mathscr{R}_\Sigma = -1$, and $a < 1$, the resulting metric can be shown to represent a "tunnel" connecting two different points in spacetime, usually referred to as a **wormhole**. The case $\mathscr{R}_\Sigma = 0$ was also analyzed and has an even stranger interpretation. The authors of [33] showed that if $\Lambda > 0$ and $a > 0$, the metric looks like it is describing a "horn"-like manifold; hence they called it the "horn spacetime solution."

Gauss–Bonnet, as well as other higher-dimensional theories, is somewhat "exotic" and has solutions of theoretical interest in the context of higher-dimensional physics only. As alluded to earlier, they can have considerable overlap with theories that *require* the existence of higher dimensions, such as superstring theory. Since at the time of this writing we still have no observational or experimental evidence that spacetime dimensions higher than four exist, these models are not very useful in describing our observed universe and will continue to be of only theoretical interest until otherwise proven. However, some higher-dimensional modified gravity theories have, at least historically, provided some hope for a better generalization of general relativity. We discuss one such model in the next section.

## 10.4 Kaluza–Klein Theory

We know that there are four fundamental forces in nature: gravity, electromagnetism, the strong nuclear force, and the weak nuclear force. Finding a unification between the four, in the sense of showing that they are all essentially the same thing, is the current "Holy Grail" of theoretical physics. As briefly mentioned earlier, we have only one candidate to such a theory of everything, superstring theory—although unfortunately there doesn't seem to be any experimental evidence that supports it (yet). The theory starts in 10-dimensional spacetime, i.e. with six extra *unobserved* spatial dimensions.[12] Then it is assumed that the extra dimensions are too "small" for direct experience, in the sense that they are wrapped over extremely tiny six-dimensional manifolds. This wrapping process induces what appear to be fields in four-dimensional spacetime. The hope is that these fields would account for the four forces that we observe. This is as yet unrealized; however, it is a popular concept and has many merits. The idea that wrapped higher dimensions could account for lower-dimensional fields was first proposed by Theodor Kaluza[13] in 1919, and further improved by Oskar Klein[14] in around 1926. At the time, the discovery of the two nuclear forces was still in the future, and as such their model was an attempt to unify gravity with the electromagnetic force only. While the Kaluza–Klein theory ultimately failed to describe the desired unification, it is still considered an important foundation to the ideas of higher dimensions and a precursor to the methods of superstring theory. So much so, in fact, that the technique of dimensional reduction is still referred to in the physics literature as the "Kaluza–Klein procedure."

The idea is simple as well as revolutionary: Start with pure general relativity in *five* spacetime dimensions—one of time and four of space—then "dimensionally reduce" the theory back to four by "wrapping" the extra dimension on a circle. In so doing, we retrieve ordinary four-dimensional relativity *plus* new fields that arise as a consequence of the wrapping. To begin, define an arbitrary five-dimensional spacetime metric as follows:

$$ds_5^2 = \tilde{g}_{MN} dx^M dx^N, \tag{10.29}$$

where the upper case Latin indices count over $(0, 1, 2, 3, 4)$. Explicitly,

$$[\tilde{g}_{MN}] = \begin{pmatrix} \tilde{g}_{00} & \tilde{g}_{01} & \tilde{g}_{02} & \tilde{g}_{03} & \tilde{g}_{04} \\ \tilde{g}_{10} & \tilde{g}_{11} & \tilde{g}_{12} & \tilde{g}_{13} & \tilde{g}_{14} \\ \tilde{g}_{20} & \tilde{g}_{21} & \tilde{g}_{22} & \tilde{g}_{23} & \tilde{g}_{24} \\ \tilde{g}_{30} & \tilde{g}_{31} & \tilde{g}_{32} & \tilde{g}_{33} & \tilde{g}_{34} \\ \tilde{g}_{40} & \tilde{g}_{41} & \tilde{g}_{42} & \tilde{g}_{43} & \tilde{g}_{44} \end{pmatrix}. \tag{10.30}$$

Next, recognize that this can be written in the form

$$[\tilde{g}_{MN}] = \begin{pmatrix} g_{\mu\nu} + \phi^2 A_\mu A_\nu & \phi^2 A_\mu \\ \phi^2 A_\nu & \phi^2 \end{pmatrix}, \tag{10.31}$$

where the Greek indices count over the usual $(0, 1, 2, 3)$, $\phi$ is a scalar field, and $A_\mu$ is what appears to be a 4-vector field. Hence

---

[12] A more general version of superstring theory, known as M-theory, is set in 11 spacetime dimensions.

[13] Theodor Franz Eduard Kaluza: German mathematician and physicist (1885–1954).

[14] Oskar Benjamin Klein: Swedish theoretical physicist (1894–1977).

$$\tilde{g}_{\mu\nu} = g_{\mu\nu} + \phi^2 A_\mu A_\nu$$
$$\tilde{g}_{4\nu} = \tilde{g}_{\nu 4} = \phi^2 A_\nu$$
$$\tilde{g}_{44} = \phi^2. \tag{10.32}$$

What is a purely "gravitational" spacetime metric in five dimensions would appear to us in four dimensions as the ordinary metric plus a vector "potential" $A_\mu$ and a scalar field $\phi$. All fields, including gravity, are assumed to be only dependent on the unwrapped four spacetime dimensions; in other words, they satisfy

$$\partial_4 \tilde{g}_{MN} = 0, \tag{10.33}$$

which is known as the **cylinder condition**. This is important because the fifth dimension is unobserved, and four-dimensional fields cannot be allowed to depend on it; otherwise we would have detected the fifth dimension a long time ago. If we can show that this 4-vector $A_\mu$ is the same as the electrodynamic vector potential, then we have essentially achieved unification between gravity and electrodynamics. The separation between $g_{\mu\nu}$ and the extra fields becomes apparent when we write the line element:

$$
\begin{aligned}
ds_5^2 &= g_{\mu\nu} dx^\mu dx^\nu + \phi^2 A_\mu A_\nu dx^\mu dx^\nu + \phi^2 A_\mu dx^\mu dx^4 + \phi^2 A_\nu dx^\nu dx^4 + \phi^2 \left(dx^4\right)^2 \\
&= ds_4^2 + \phi^2 \left(A_\mu dx^\mu + dx^4\right)\left(A_\nu dx^\nu + dx^4\right) \\
&= ds_4^2 + \phi^2 \left(A_\mu dx^\mu + dx^4\right)^2. \tag{10.34}
\end{aligned}
$$

From a four-dimensional perspective, one can imagine that at every four-dimensional spacetime point there exists a *very* small circle representing the wrapped fifth dimension, as shown in Fig. [10.1]. The circle *has* to be extremely small, otherwise we would have felt its presence in our everyday life or in experiments. Its only effect on us is the appearance of the fields $A_\mu$ and $\phi$.

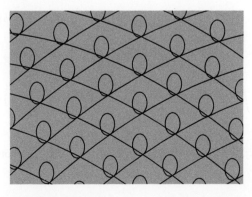

Fig. 10.1: The fifth dimension wrapped over a circle, i.e. in every four-dimensional spacetime point there exists a small circle.

Now the purely gravitational five-dimensional Einstein–Hilbert action is

$$S_5 = \frac{1}{2\tilde{\kappa}} \int d^5x \sqrt{-|\tilde{g}|}\, \tilde{R}, \tag{10.35}$$

where $\tilde{\kappa}$ is the five-dimensional version of the Einstein constant and $\tilde{R}$ is the five-dimensional Ricci scalar found from the five-dimensional Ricci tensor in the usual way:

$$\tilde{R} = \tilde{g}^{MN} \tilde{R}_{MN}. \tag{10.36}$$

The tensor $\tilde{R}_{MN}$ is calculated using $g_{MN}$ via the five-dimensional analogue of (7.22):

$$\tilde{R}_{MN} = \left(\partial_N \tilde{\Gamma}^K_{MK}\right) - \left(\partial_K \tilde{\Gamma}^K_{MN}\right) + \tilde{\Gamma}^A_{MK} \tilde{\Gamma}^K_{AN} - \tilde{\Gamma}^A_{MN} \tilde{\Gamma}^K_{AK}, \tag{10.37}$$

where of course

$$\tilde{\Gamma}^K_{MN} = \frac{1}{2} \tilde{g}^{KS} \left(\partial_M \tilde{g}_{NS} + \partial_N \tilde{g}_{SM} - \partial_S \tilde{g}_{MN}\right) \tag{10.38}$$

and the inverse metric is

$$[\tilde{g}^{MN}] = \begin{pmatrix} g^{\mu\nu} & -A^\mu \\ -A^\nu & A^\beta A_\beta + \frac{1}{\phi^2} \end{pmatrix}, \tag{10.39}$$

such that orthogonality is satisfied. The action (10.35) represents empty five-dimensional spacetime; in other words, varying it leads to the (five-dimensional) Einstein field equations $\tilde{R}_{MN} = 0$, and nothing else. Now the detailed calculation of $\tilde{R}$ shows the splitting of its components into three pieces; the first involves *only* the four-dimensional derivatives of the four-dimensional Christoffel symbols, while the remaining two pieces involve derivatives of $A_\mu$ and $\phi$. Explicitly:

$$\tilde{R} = R - \frac{1}{4} \phi^2 F_{\mu\nu} F^{\mu\nu} - \frac{1}{\phi^2} \left(\partial_\mu \phi\right) \left(\partial^\mu \phi\right), \tag{10.40}$$

where $F^{\mu\nu} = \partial^\mu A^\nu - \partial^\nu A^\mu$, as one would hope. Substituting in the action (10.35):

$$S_5 = \frac{1}{2\tilde{\kappa}} \int d^5x \sqrt{-|\tilde{g}|} \left[R - \frac{1}{4} \phi^2 F_{\mu\nu} F^{\mu\nu} - \frac{1}{\phi^2} \left(\partial_\mu \phi\right) \left(\partial^\mu \phi\right)\right]. \tag{10.41}$$

**Exercise 10.10** Explicitly verify that (10.39) satisfies the orthogonality of the metric (10.31).

**Exercise 10.11** Derive (10.40).

It is straightforward to show from (10.31) that

$$|\tilde{g}|_{5\text{D}} = \phi^2 \, |g|_{4\text{D}} \, ; \tag{10.42}$$

hence (10.41) becomes

$$S_5 = \frac{1}{2\tilde{\kappa}} \int dx^4 \int d^4x \sqrt{-|g|} \, \phi \left[ R - \frac{1}{4}\phi^2 F_{\mu\nu} F^{\mu\nu} - \frac{1}{\phi^2} \left( \partial_\mu \phi \right) \left( \partial^\mu \phi \right) \right], \tag{10.43}$$

where $dx^4$ is the extra fifth dimension and $d^4x = dx^0 dx^1 dx^2 dx^3$ is the four-dimensional element (sorry about the notational confusion; $dx^4$ is not the same thing as $d^4x$). Now the integral over $dx^4$ is just the circumference of the circle wrapping the extra dimension (assumed a constant). If we now define

$$\frac{1}{2\tilde{\kappa}} \int d^4x = \frac{1}{2\kappa}, \tag{10.44}$$

where $\kappa$ is the usual four-dimensional Einstein constant, then the action becomes

$$S_4 = \int d^4x \sqrt{-|g|} \, \phi \left[ \frac{1}{2\kappa} R - \frac{1}{4\mu_0} \phi^2 F_{\mu\nu} F^{\mu\nu} - \frac{1}{\phi^2} \left( \partial_\mu \phi \right) \left( \partial^\mu \phi \right) \right], \tag{10.45}$$

where $\mu_0$ is the usual magnetic permeability, introduced by appropriately scaling the field $\phi$. Now, aside from the scalar field, is this a clear unification between gravity and electromagnetism? Unfortunately not quite; keep reading to find out why. The field equations that arise from varying (10.45) are

$$G_{\mu\nu} = \kappa \phi^2 T_{\mu\nu} - \frac{1}{\phi} \left( \nabla_\mu \nabla_\nu \phi - g_{\mu\nu} \nabla_\alpha \nabla^\alpha \phi \right),$$

$$\nabla_\mu F^{\mu\nu} = -\frac{1}{\phi} \left( \partial_\mu \phi \right) F^{\mu\nu},$$

$$\nabla_\alpha \nabla^\alpha \phi = \frac{1}{4} \phi^3 F_{\mu\nu} F^{\mu\nu}, \tag{10.46}$$

where the tensor $T_{\mu\nu}$ is *exactly* the electromagnetic stress tensor (7.70). Note that setting a constant 4-vector $A^\mu$ leads to vanishing $F^{\mu\nu}$ and the action reduces to

$$S_4 = \int d^4x \sqrt{-|g|} \left[ \frac{1}{2\kappa} \phi R - \frac{1}{\phi} \left( \partial_\mu \phi \right) \left( \partial^\mu \phi \right) \right], \tag{10.47}$$

which is a subclass of the Brans–Dicke action (10.1). From the perspective of keeping the electromagnetic field, the appearance of $\phi$ in (10.45) is obviously unwelcome, since this seems to be a field completely unobserved in nature. The easiest way to get rid of it is to simply assume that it is a constant; specifically unity $\phi = 1$. This immediately reduces the first two equations of (10.46) to the vacuum Einstein–Maxwell equations[15]

$$G_{\mu\nu} = 0 \rightarrow R_{\mu\nu} = 0, \qquad \nabla_\mu F^{\mu\nu} = 0. \tag{10.48}$$

This is promising; however, it is inconsistent, since a constant $\phi$ necessarily leads to $F_{\mu\nu} F^{\mu\nu} = 0$ in the third equation of (10.46). In other words, getting rid of the scalar

---

[15]Compare with (6.148) and (7.85) with vanishing $J^\mu$ and $T_{\mu\nu}$ respectively.

field completely ruins the theory. As such, one is led to assume that the scalar field $\phi$ is not constant, and consequently its field strength $\partial_\mu \phi$ is non-vanishing and hence becomes potentially detectable. The fact that such a field was never observed is a clear failure of the theory, despite its obvious elegance. In addition, the discovery of the two nuclear forces made it clear that the theory is no longer a theory of "all" forces. For these reasons the Kaluza–Klein model fell out of favor in the few decades following its discovery. It was only revived in the much more complex, although essentially similar, situation of the dimensional reduction of superstring theory, as mentioned earlier.

**Exercise 10.12** Vary (10.45) to derive the Kaluza–Klein field eqns (10.46). You do not need to reinvent the wheel; in other words, whatever parts of this calculation you have done before, such as varying the $R$ term, you do not need to redo.

## 10.5   Einstein–Cartan Theory

We have reviewed examples of gravitational theories where the Einstein–Hilbert action is modified, and theories whose natural settings are in higher dimensions. It is time to look at a theory that lives in the ordinary four spacetime dimensions, and *has* the exact action as general relativity; yet is *so* fundamentally different. This modified gravity theory was proposed by Élie Cartan in 1922, less than seven years after Einstein published the general theory of relativity. In §9.6 we learned that Riemannian manifolds have two properties that characterize them: their curvature and their torsion. We also asserted that the geometry described by general relativity requires that torsion vanishes. It is normal, then, to ask: What theory of gravity would arise if we relaxed this condition and allowed the torsion tensor to have non-trivial values? This is exactly what Cartan did. In this theory, eqns (9.119), (9.120), (9.121), and (9.122) replace their usual torsion-free counterparts. The theory is highly complex, as you might imagine, as well as poorly understood. We will only give a very brief outline of it.

With these conditions in mind, varying the action (8.120) with respect to the metric still gives the Einstein field equations

$$G_{\mu\nu} = R_{\mu\nu} - \frac{1}{2}g_{\mu\nu} = \kappa P_{\mu\nu}, \tag{10.49}$$

except that the second-rank tensor $P_{\mu\nu}$ is not the usual symmetric stress tensor, but rather the so-called **canonical stress energy tensor**. In contrast to $T_{\mu\nu}$, it is not symmetric: $P_{\mu\nu} \neq P_{\nu\mu}$. The asymmetric components of $P_{\mu\nu}$ can be shown to be related to the so-called **spin tensor** $S^{\mu\nu\rho}$, defined by

$$\partial_\mu S^{\mu\nu\rho} = \frac{1}{2}\left(P^{\nu\rho} - P^{\rho\nu}\right), \tag{10.50}$$

meaning that in ordinary general relativity $S^{\mu\nu\rho}$ is just an ineffective constant. The spin tensor is known by that name since it can be shown to be the density per unit volume of the angular momentum of spinning objects in spacetime; i.e. objects spinning as they follow the twisted spacetime itself, as described briefly in §9.6.[16] Now, it is useful to think of the torsion tensor as a separate tensorial field by itself, effectively independent of the (still symmetric) metric tensor, in which case we can vary the action with respect to it, giving the **Cartan equations**:

$$\mathcal{T}^\rho{}_{\mu\nu} + \delta^\rho_\mu \mathcal{T}^\alpha{}_{\nu\alpha} - \delta^\rho_\nu \mathcal{T}^\alpha{}_{\mu\alpha} = \kappa S^\rho{}_{\mu\nu}. \tag{10.51}$$

These equations relate the torsion of spacetime to the spin of objects or fields in pretty much the same way the Einstein equations relate the curvature of spacetime to the matter/energy content. It is not, however, a differential equation like Einstein's, but rather an algebraic condition on the components of $\mathcal{T}$.

To motivate the possible use of the Einstein–Cartan theory, perhaps it is sufficient to note that it deals with spacetime singularities in a completely different way. For example, using the theory to formulate a cosmological model, it can be shown that a singularity at $t \to 0$ no longer exists. It is replaced with a bouncing universe; i.e. a universe that was initially contracting reaches a minimum volume (*not* zero as general relativity assumes), and then bounces back into an expansion. In pretty much the same vein, a mass collapsing into a black hole never reaches zero volume in Einstein–Cartan theory, but also bounces into another part of our spacetime manifold, or even another universe entirely, effectively forming a wormhole. This is all we will say about this very interesting model. The interested reader may consult the literature for more.

**Exercise 10.13** Cartan's theory of gravity is much too complicated to force our readers to do any calculations in it. I think I have already pushed the limits of your mathematical abilities as far as they can possibly go in these last few chapters. However, one thing that can perhaps be done with relative painlessness is this: Write a report on the history of Cartan's theory of gravity, how it was used on and off over the years, and the major theoretical achievements that may keep it around as a viable physical possibility. On the flip side, what are its cons? Why is it that it is not as popular as, say, $f(R)$ gravity with the modified gravity crowd?

---

[16]Spin here is in a classical sense, not a quantum one, in case you were wondering.

# Afterthoughts

If you arrived at this page by reading the entire book (and didn't skip), then you most certainly deserve a very hearty:

Congratulations!

Oh well, let's admit it, even if you cheated a bit and skipped some chapters, you still deserve a (perhaps not so hearty) cheer!

I do hope, dear reader, that you have enjoyed this book. I write, so I am told, the way I speak; as such, I hope you feel, like I do, that you and I have just had a very long conversation. And yes I do consider it a conversation, not just me lecturing and you listening. I consider the very turnings of the gears inside your head as you learned this material, as well as your contributions to the problems and derivations, to be *your* side of the conversation. And what a fruitful conversation this was. But, like all fun things, it must come to an end. Let me then bring this discussion to a close by addressing *you* personally:

## If you are an undergraduate physics student:

You are the main audience of this book! I wrote this for you. I am hoping that you will take this material and use it in future studies or research, or just to gain a general understanding of one of the most exciting of human endeavors: theoretical physics. We covered so many different topics, from classical mechanics to higher dimensions. Was it exhaustive? Not even close. All of this was only just skimming the surface; the tip of the proverbial iceberg. I am really hoping that you have your diving gear ready and are eager to explore the rest of it. I know *I* am still exploring, and it's unlikely that I will learn all I want to in only one lifetime. But hey! It's the trip that counts. So much fun. My best wishes to you.

## If you are a graduate physics student:

Good for you starting easy! Yes, this book is, in a sense, an easy version of others that you should be starting right about now. Having said that, however, let me claim that some of the concepts in this book will not be found in even more advanced texts. Well, not explained in so many words, anyway. As I said in the preface, the concept of covariance is rarely discussed point blank at any level. I hope that you, having finished this book, have a better understanding now of what covariance means, and perhaps some of the physics that you have learned over many years is starting to finally make sense. In fact, it is indeed right about now in your education that the light bulbs of understanding are finally beginning to glow! You have been through many years of math and physics. Most of it may have felt random or disconnected. And only now, in your graduate studies, do the separate strings start to resonate together in harmony. I hope this book contributed just a bit to this overall understanding. Good luck!

**If you are an engineer or a student of engineering:**

Thank you for joining us on this wonderful journey. True, most of the material in this book would not be of practical use to you (and I am speaking from personal experience, having a degree in engineering myself); however, being a mathematically inclined professional, you may have found it interesting and feel it has contributed to your overall knowledge. Some of it, however, *may* be of practical use to you. The first three chapters, for instance, could have provided you with some insight into topics that most engineers deal with on a daily basis, while being unaware of their overall organization. Like physicists (and in some cases even more than physicists), you too work with coordinate systems and tensors. Some of the techniques developed in these early chapters may be of use to you in your next calculation. Or, if not useful, at least more insightful.

**If you are an expert on the material:**

Thank you very much, my dear colleague, for taking the time to read this. And although I am hoping that you liked it, I am afraid that you are probably somewhat confused over the "care-free" way in which I've presented the material. I did warn you in the preface: It is not as rigorous as a math book, nor is it as full with experimental examples as a physics book. It is somewhere in between, designed with a specific gap that needs to be filled. I sincerely hope that you agree that it has been filled, or at least partially so.

*And finally,*

**If you are none of the above:**

Wow! I am officially impressed! You are teaching yourself? Kudos to you, my friend. No one has claimed that the material in a book such as this is easy. And if you have previously taught yourself enough to approach this book, then you are my hero indeed. Well done! What's next now? More advanced topics of course. On to geometry, topology, quantum mechanics, superstrings, and more. The world is your oyster. Enjoy it!

# References

[1] Serway, R. *et al.* (2014). *Physics for Scientists and Engineers.* Boston: Brooks/Cole.

[2] Ludyk, G. (2013). *Einstein in Matrix Form: Exact Derivation of the Theory of Special and General Relativity without Tensors.* BerlinHeidelberg: Springer-Verlag.

[3] Beezer, R. (2009). *A First Course in Linear Algebra.* Gainsville, FL: University Press of Florida.

[4] Corben, H. *et al.* (1977) *Classical Mechanics.* New York: Dover Press.

[5] White, F. (2011). *Fluid Mechanics.* New York: McGraw Hill.

[6] Weinberg, S. (1972). *Gravitation and Cosmology: Principles and Applications of the General Theory of Relativity.* New York: Wiley.

[7] Einstein, A. (1921). *Relativity: The Special and General Theory.* New York: Henry Holt & Company.

[8] Schwabl, F. (1965). *Statistical Mechanics.* BerlinHeidelberg: Springer-Verlag.

[9] Jackson, J. (1999). *Classical Electrodynamics,* 3rd ed. New York: Wiley.

[10] Ryder, L. (1985). *Quantum Field Theory.* Cambridge: Cambridge University Press.

[11] Mukhanov, V. (2005). *Physical Foundations of Cosmology.* Cambridge: Cambridge University Press.

[12] Goldstein, H. (2014). *Classical Mechanics,* 3rd ed. New York: Pearson.

[13] Carroll, S. (2004). *Spacetime and Geometry. An Introduction to General Relativity.* Boston: Addison-Wesley.

[14] Crelinsten, J. (2006). *Einstein's Jury. The Race to Test Relativity.* Princeton, NJ: Princeton University Press.

[15] Janssen, M. and Renn, J. "Arch and scaffold: How Einstein found his field equations." *Physics Today* **68**, 11, 30 (2005): https://doi.org/10.1063/PT.3.2979.

[16] Walters, S. "How Einstein got his field equations." arXiv:1608.05752.

[17] Wald, R. (1984). *General Relativity.* Chicago: University of Chicago Press.

[18] Gershenfeld, N. (1999). *The Nature of Mathematical Modeling.* Cambridge: Cambridge University Press.

[19] Thorne, K. (2014). *The Science of Interstellar.* New York: W. W. Norton & Company.

[20] Gamow, G. (1970). *My World Line: An Informal Autobiography.* New York, NY: Viking Press.

[21] Riess, A. *et al.* (1998). "Observational evidence from supernovae for an accelerating universe and a cosmological constant." *Astron. J.* **116**, 1009–38.

[22] Feynman, R. (1970). *The Feynman Lectures on Physics.* New York: Pearson.

[23] Barut, A. (1964). *Electrodynamics and Classical Theory of Fields and Particles*. New York: Dover Press.

[24] Wald, R. (February 1974). "Gedanken experiments to destroy a black hole." *Annals of Physics*, **82**, 2: 54856.

[25] Heinicke, C. *et al.* (2014). "Schwarzchild and Kerr solutions of Einstein's field equation: An introduction." *Int. J. Mod. Phys.* **D24**, 02, 1530006.

[26] Rovelli, C. (2004). *Quantum Gravity*. Cambridge: Cambridge University Press.

[27] Yepez, J. (2011). "Einstein's vierbein field theory of curved space." arXiv: 1106.2037.

[28] Emam, M. (2011). *Are We There Yet? The Search for a Theory of Everything*. Bentham ebooks.

[29] Brans, C. (1962). "Mach's principle and a relativistic theory of gravitation. II." *Phys. Rev.* **125**, 2194.

[30] Campanelli, M. *et al* (1993). "Are black holes in Brans-Dicke theory precisely the same as in general relativity?" *Int. J. Mod. Phys.* **D2**, 451–462.

[31] Sotiriou, T. *et al.* (2010). "$f(R)$ Theories of Gravity." *Rev. Mod. Phys.* **82**, 451.

[32] Jaime, L. *et al.* "$f(R)$ Cosmology revisited." arXiv:1206.1642.

[33] Dotti, G. *et al.* (2007). "Exact solutions for the EinsteinGaussBonnet theory in five dimensions: Black holes, wormholes and spacetime horns." *Phys. Rev.* **D76**, 064038.

[34] Trendafilova, C. *et al.* (2011). "Static solutions of Einstein's equations with cylindrical symmetry." *Eur. J. Phys.* **32**, 1663–1677.

# Index